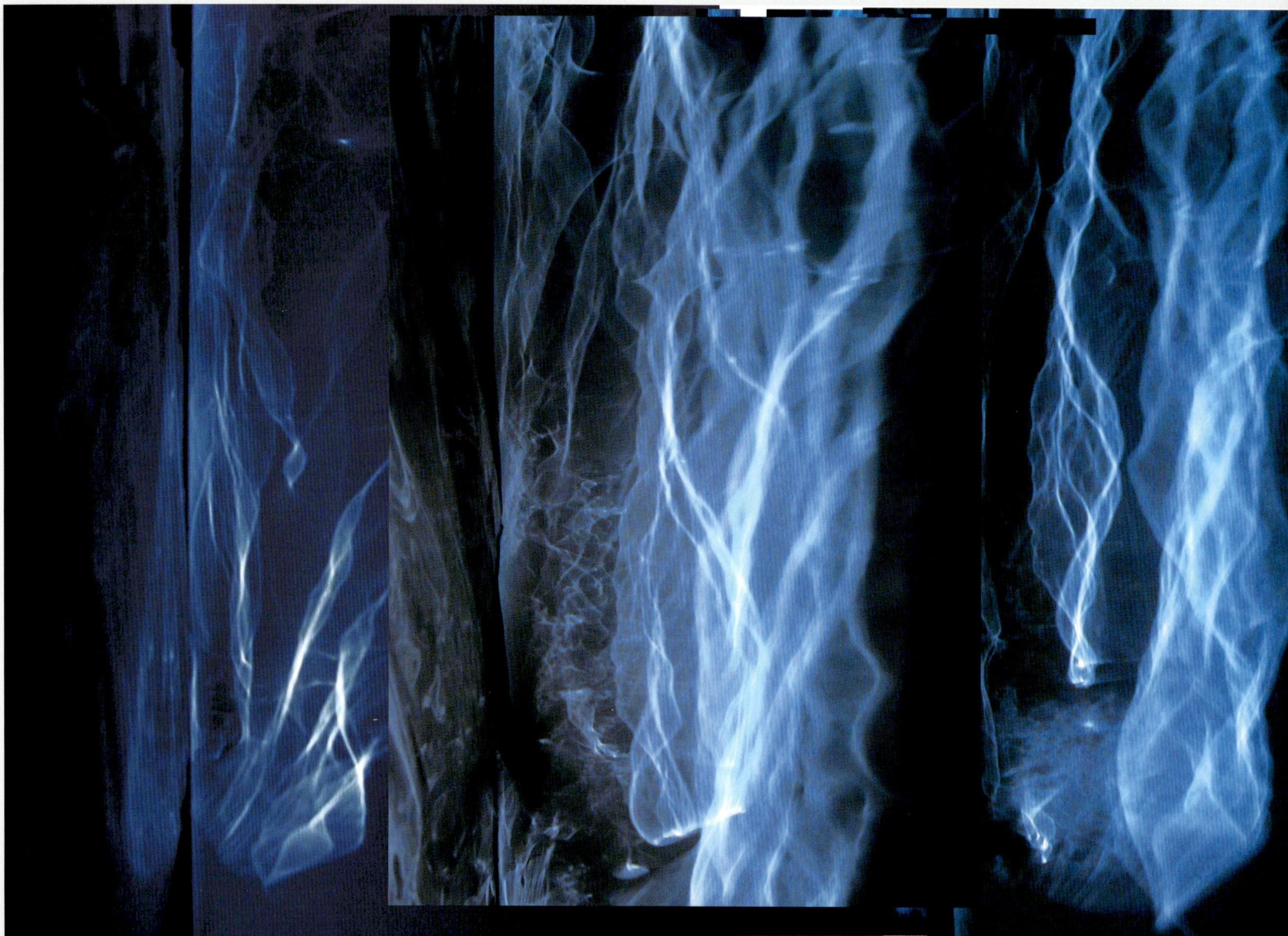

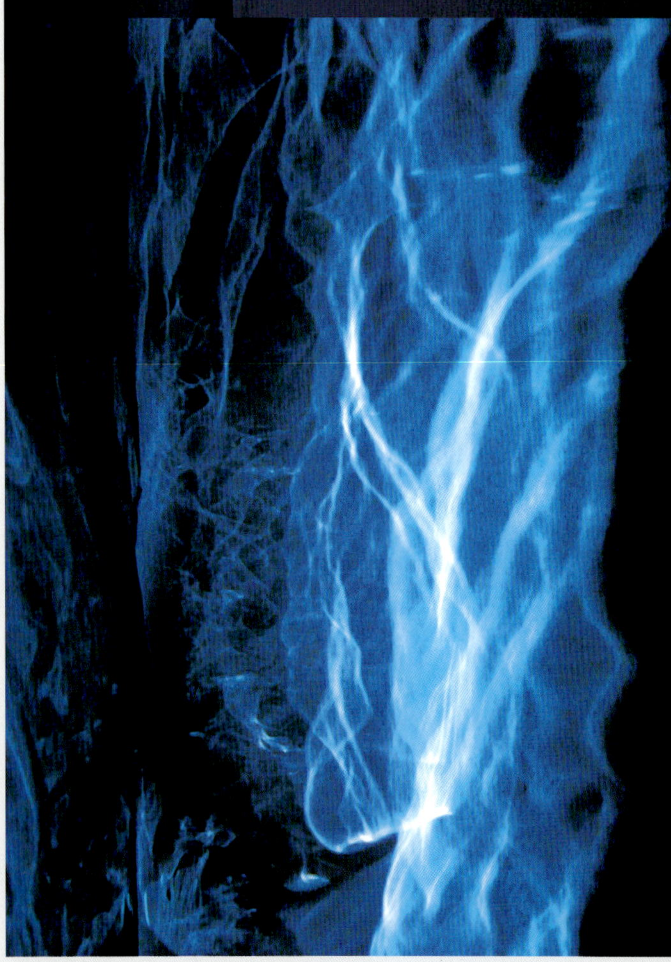

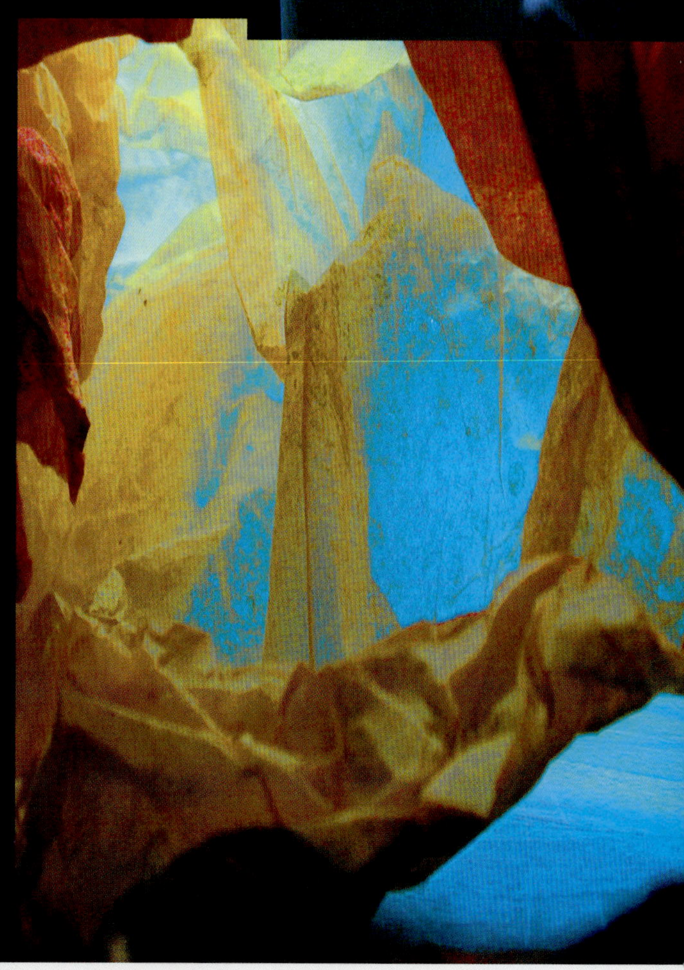

simulated cavitation

the process of forming micro-bubbles in a liquid by generating intense ultrasound waves.

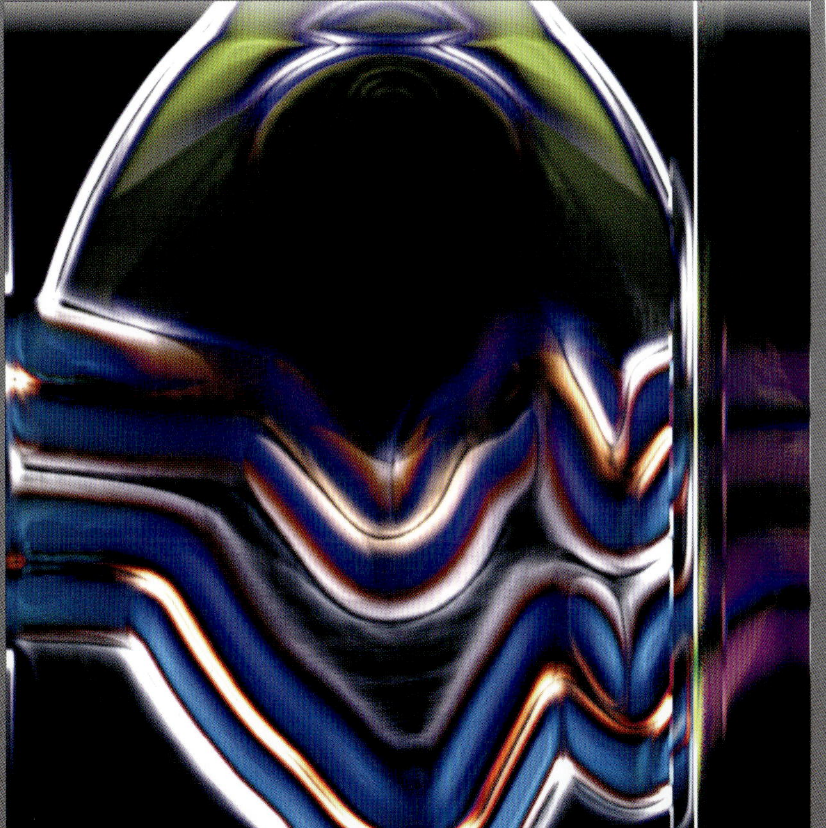

DEVICES BASED UPON ABSTRACTS BY MR. JACK VEACH:
Furthered the instrumentation software to isolate more expansive ultrasonic-radiation pressure

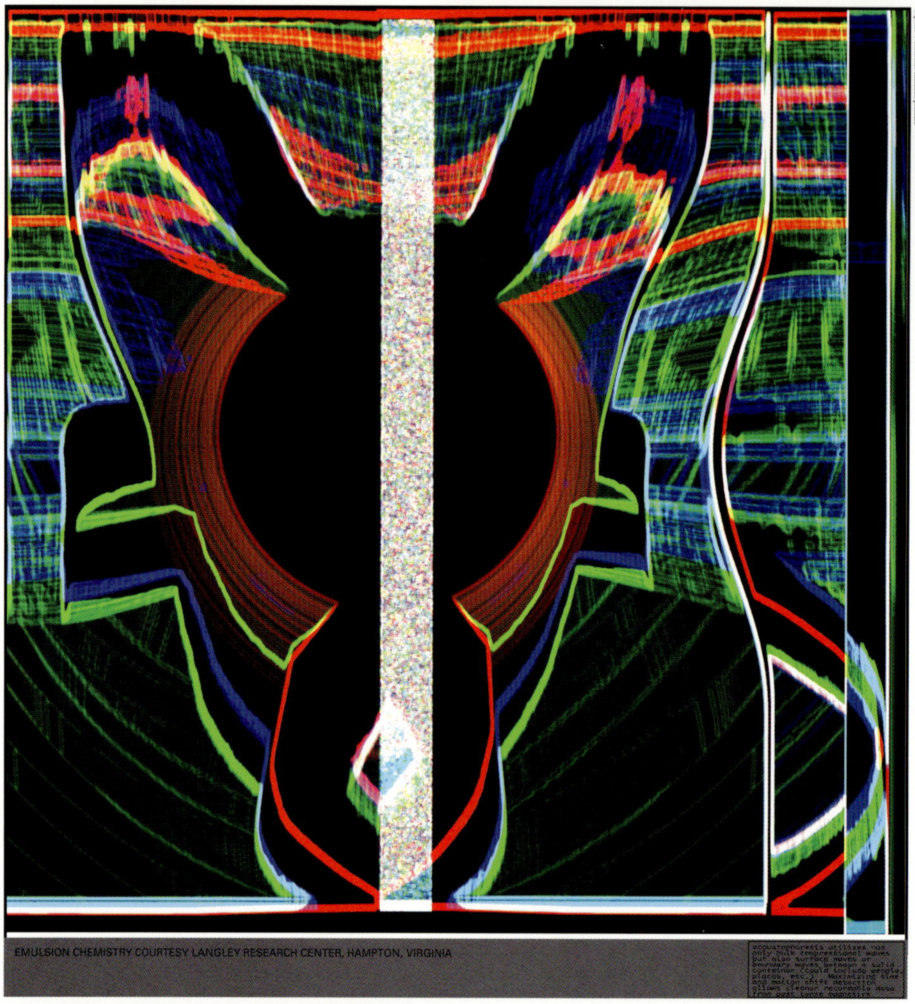

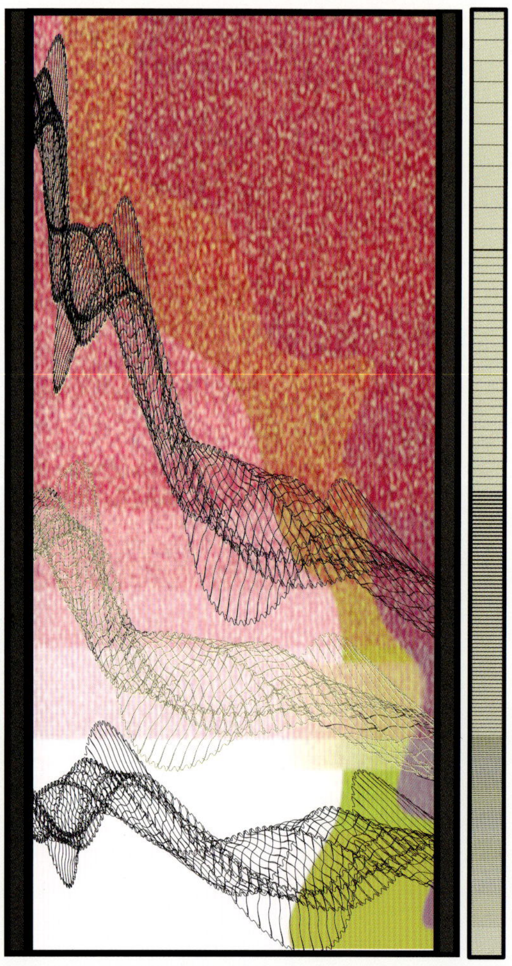

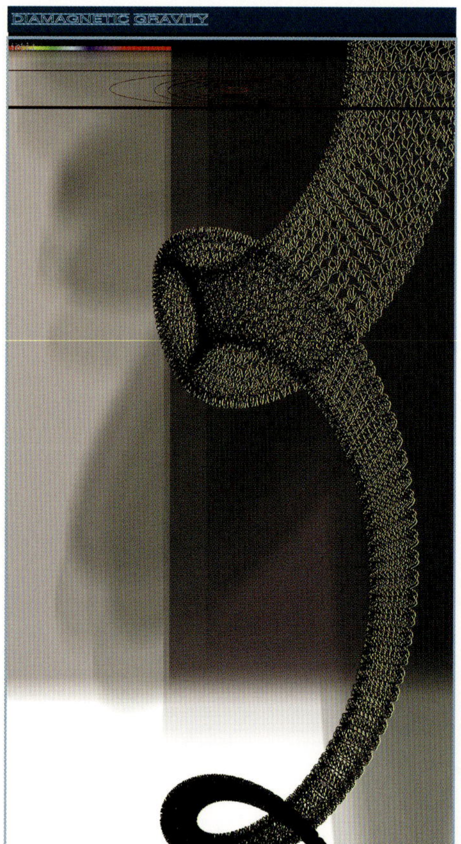

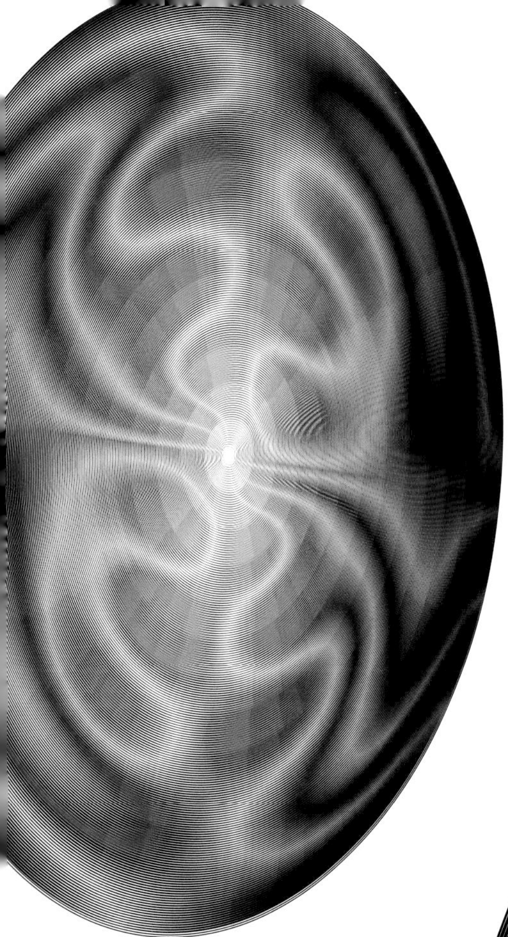

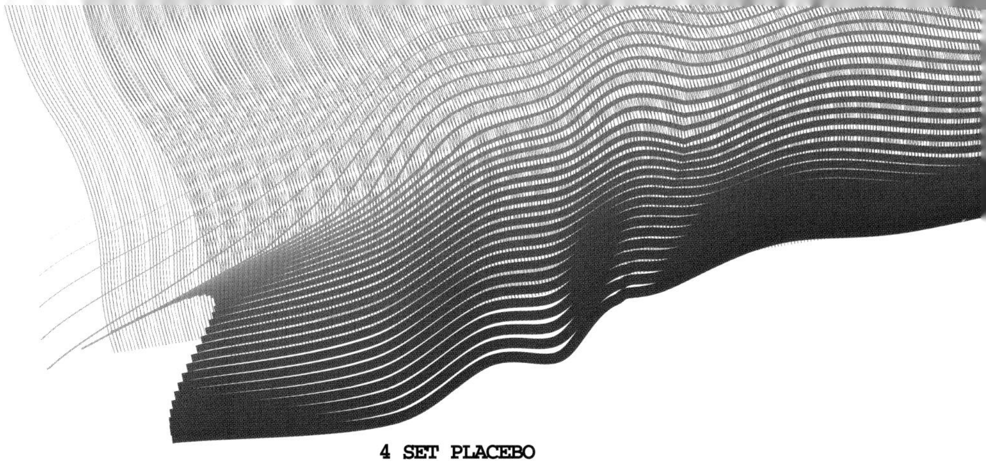

4 SET PLACEBO

note: spontaneity of moire
10/30a interrupts compliance
with perception

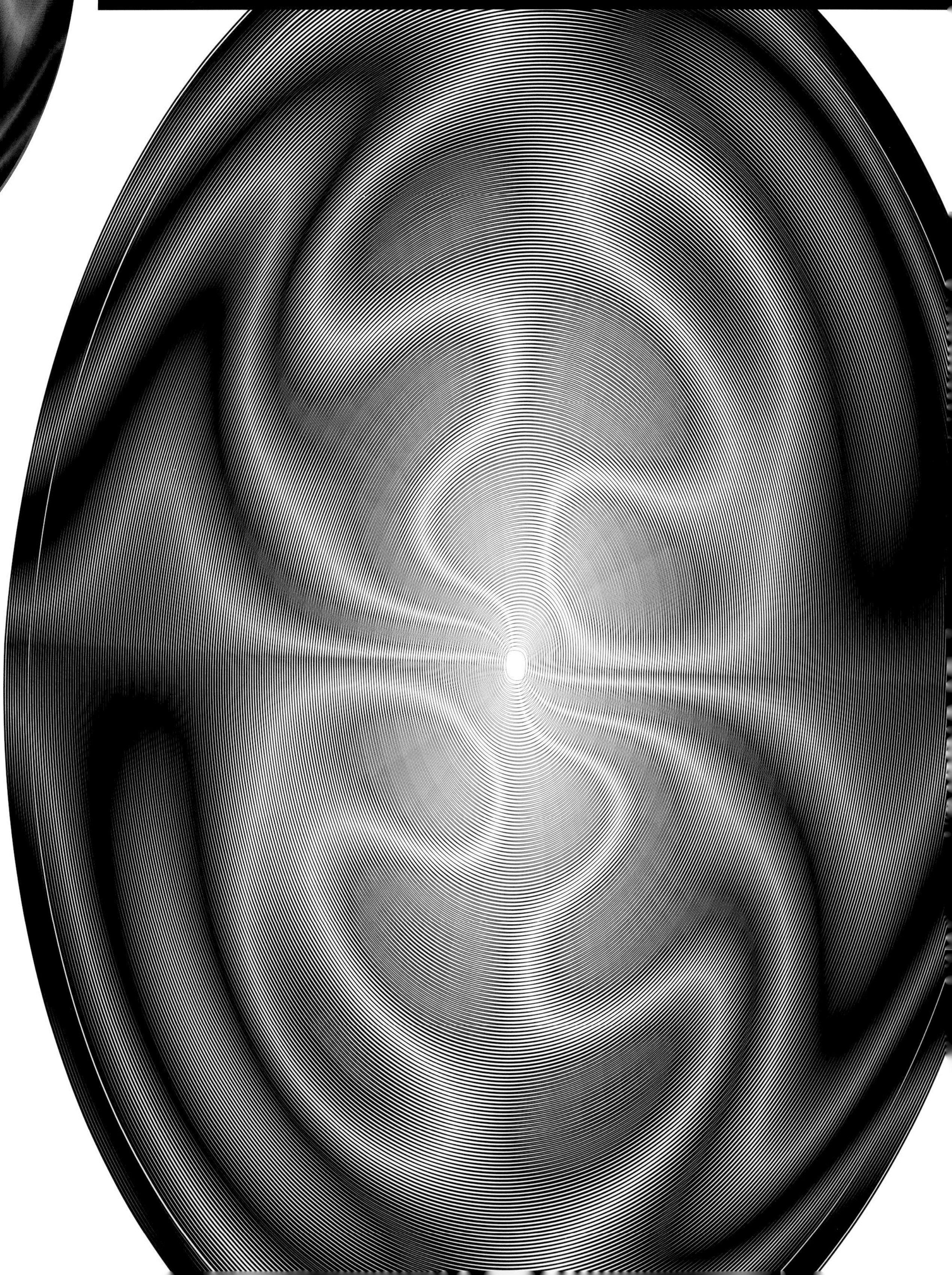

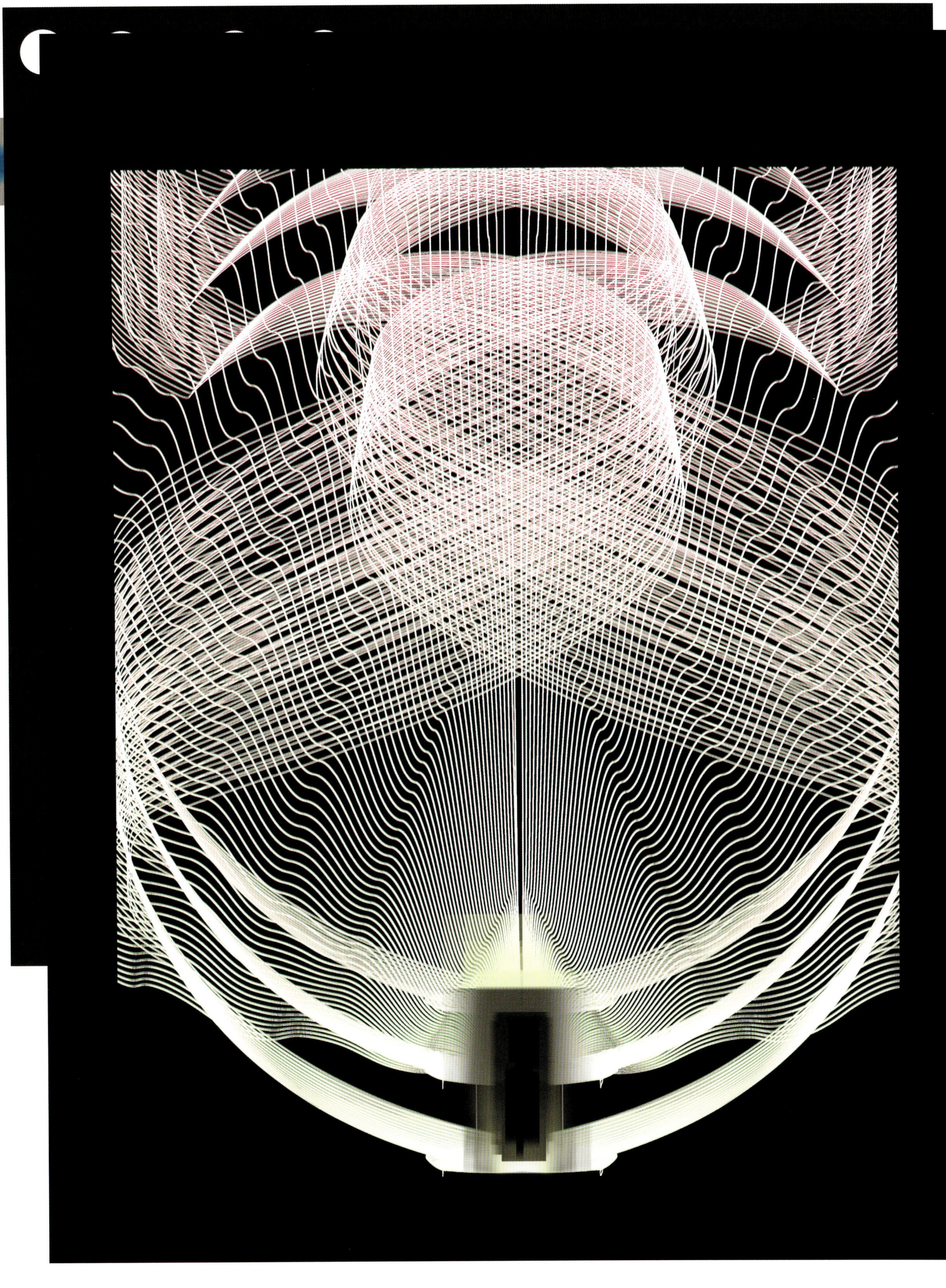

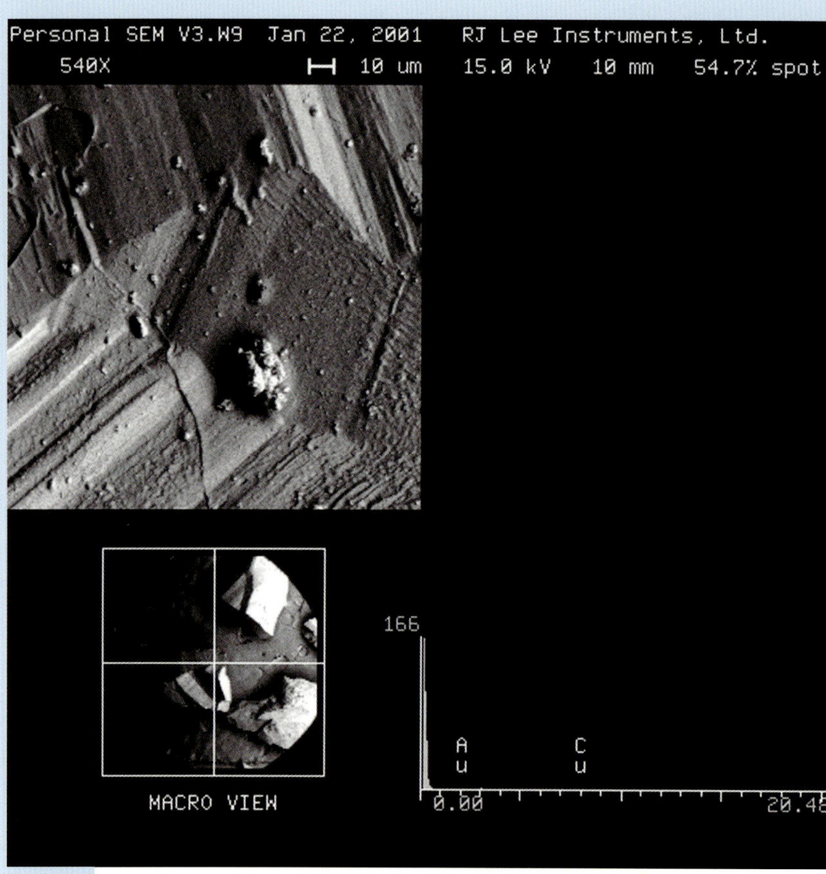
Personal SEM V3.W9 Jan 22, 2001
RJ Lee Instruments, Ltd.
540X
10 um
15.0 kV
10 mm
54.7% spot
MACRO VIEW
166
Au
Cu
0.00
20.48

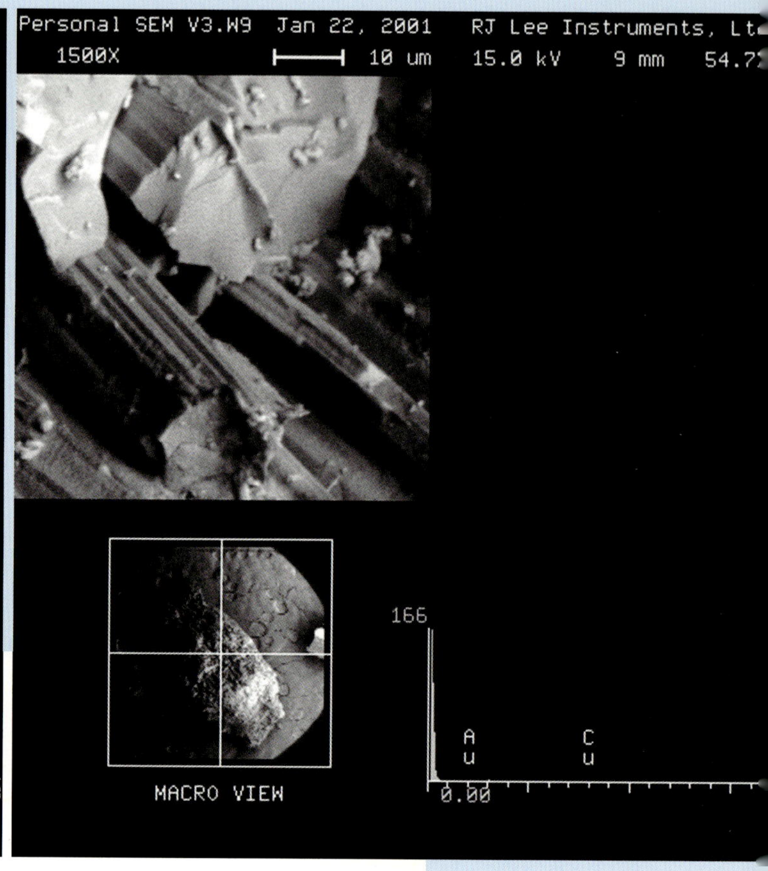
Personal SEM V3.W9 Jan 22, 2001
RJ Lee Instruments, Lt
1500X
10 um
15.0 kV
9 mm
54.7
MACRO VIEW
166
Au
Cu
0.00

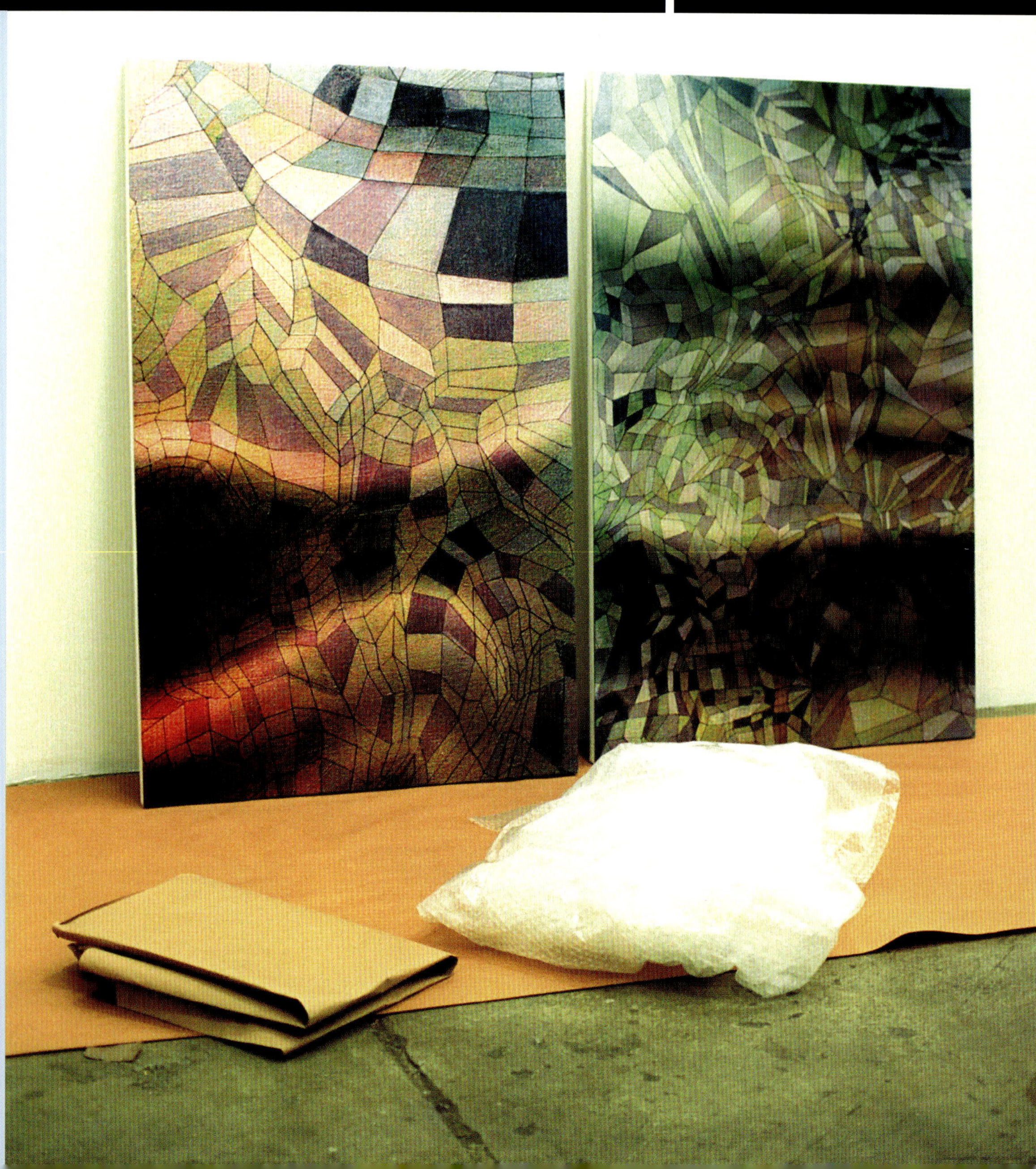

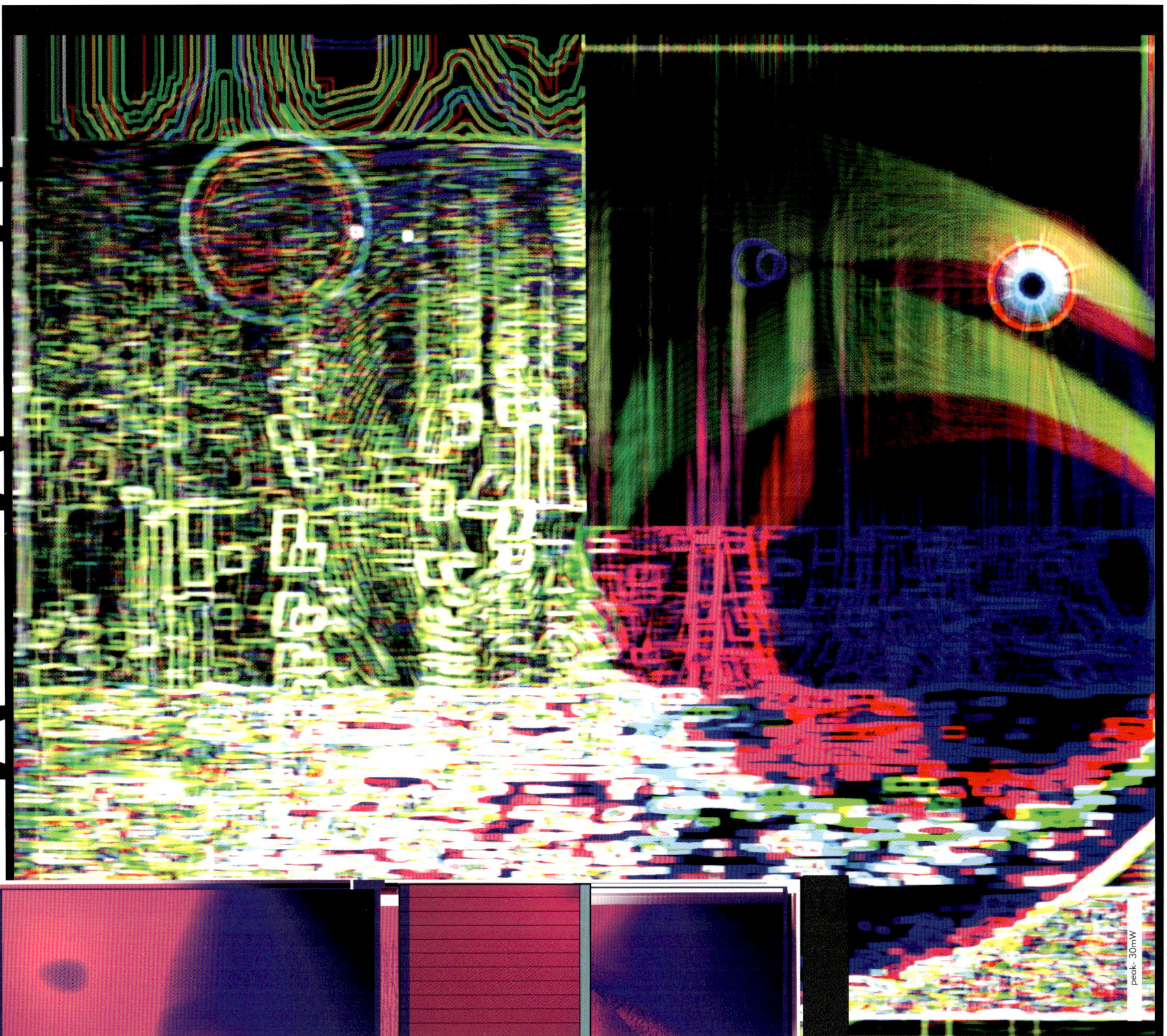


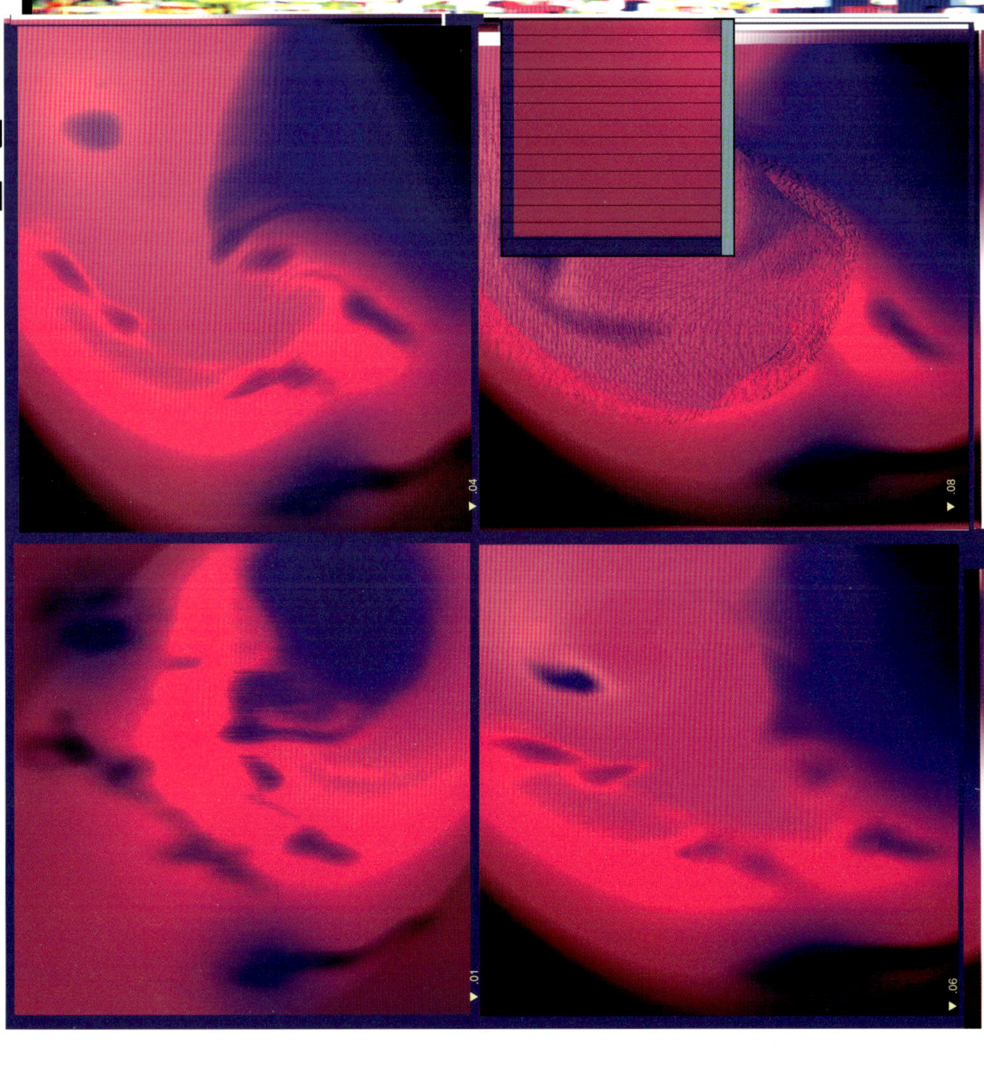


SPECTRAL STRESS CAUSED BY ELEKTRA

DELAWARR

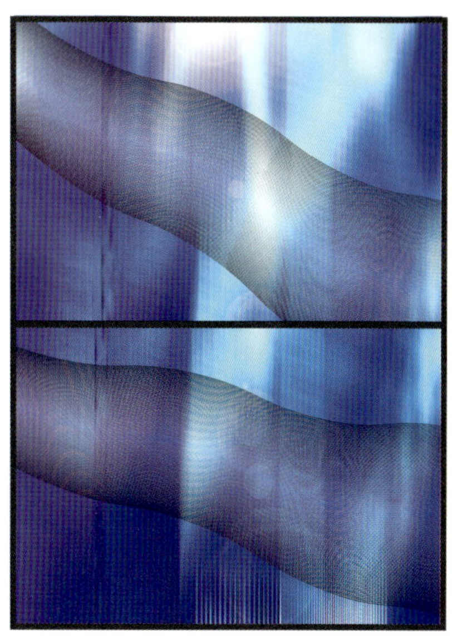

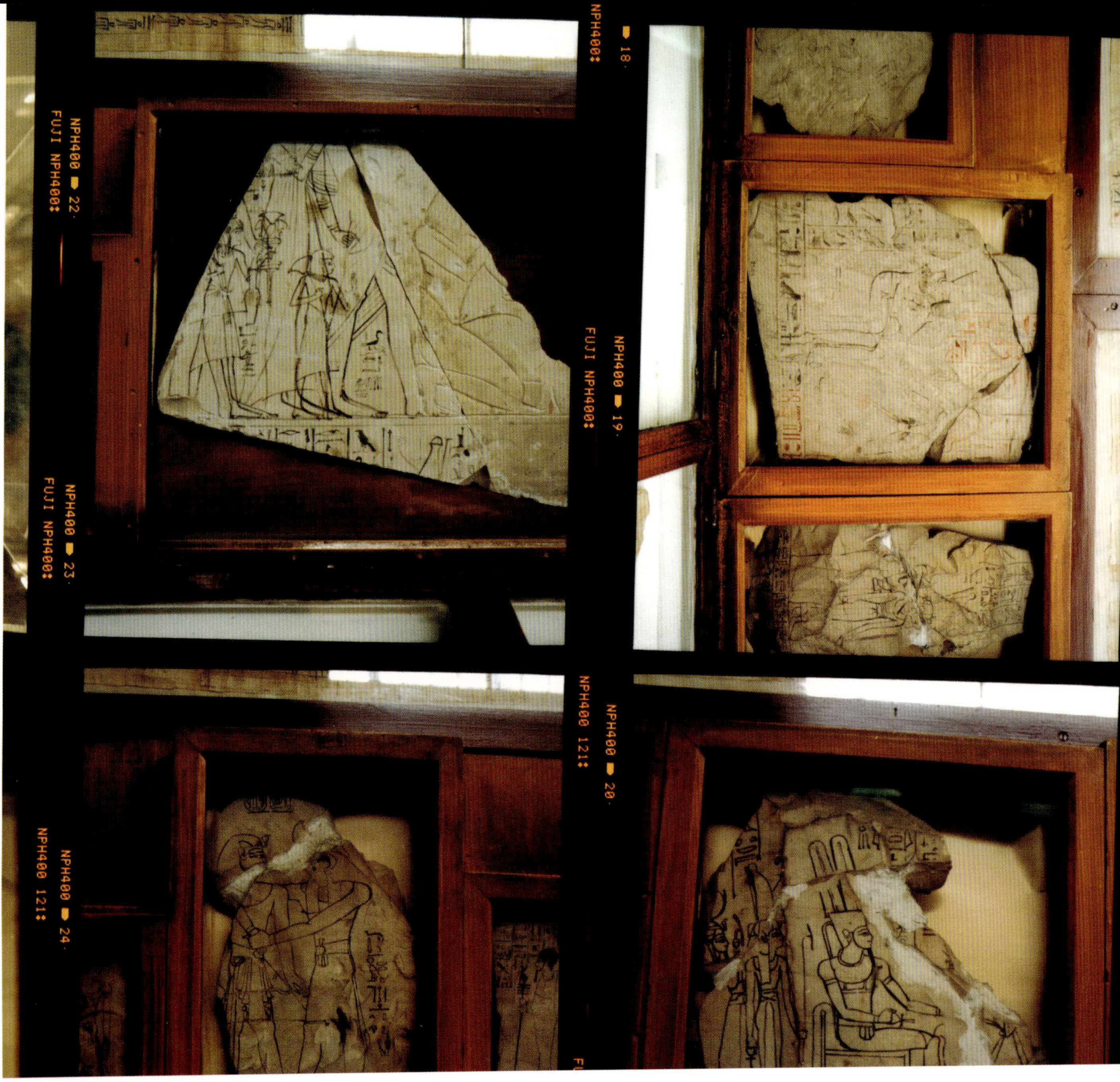
NPH400 ▶ 18
NPH400
NPH400 ▶ 22
FUJI NPH400
NPH400 ▶ 19
FUJI NPH400
NPH400 ▶ 23
FUJI NPH400
NPH400 ▶ 20
NPH400 121
NPH400 ▶ 24
NPH400 121

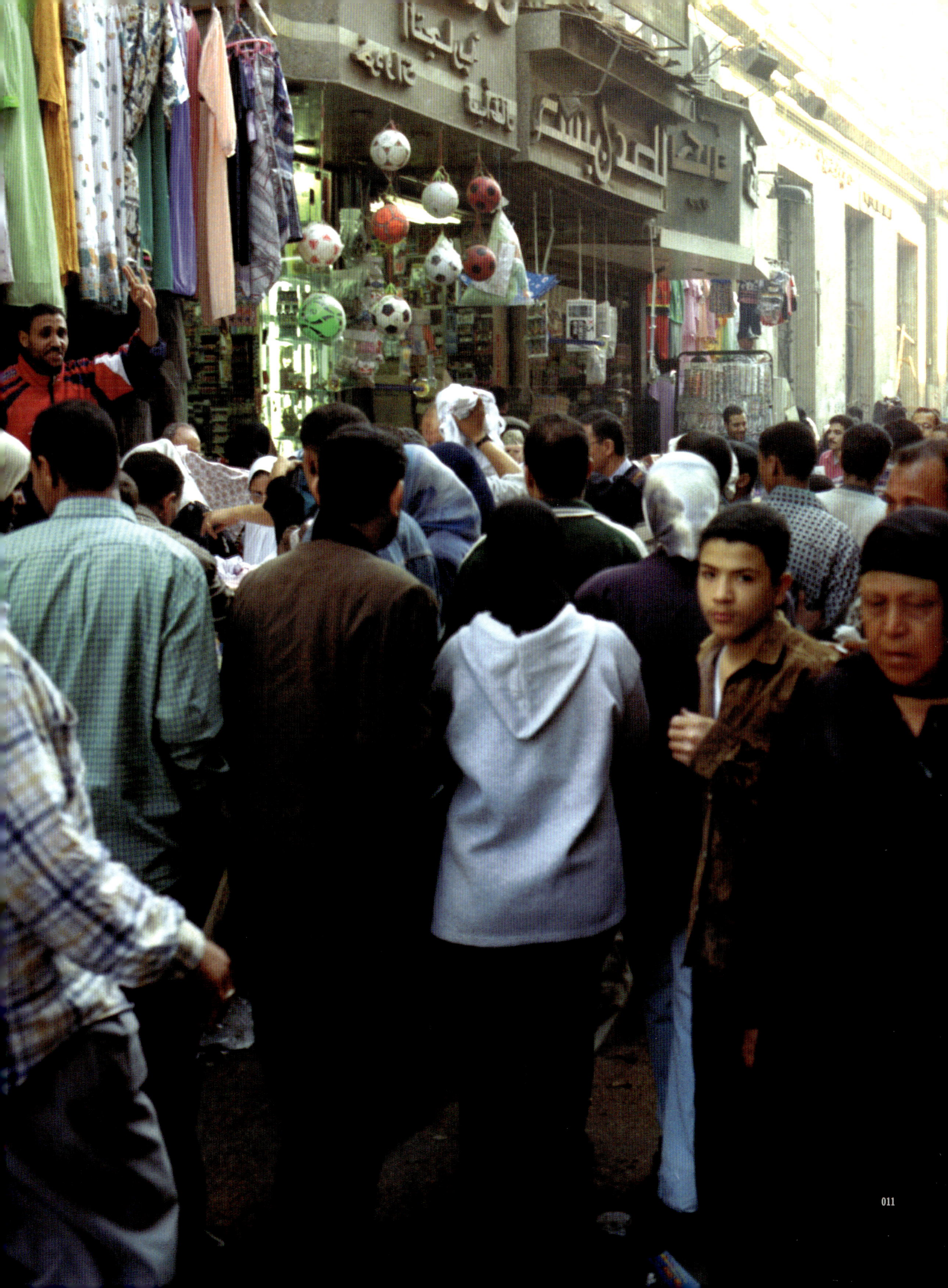

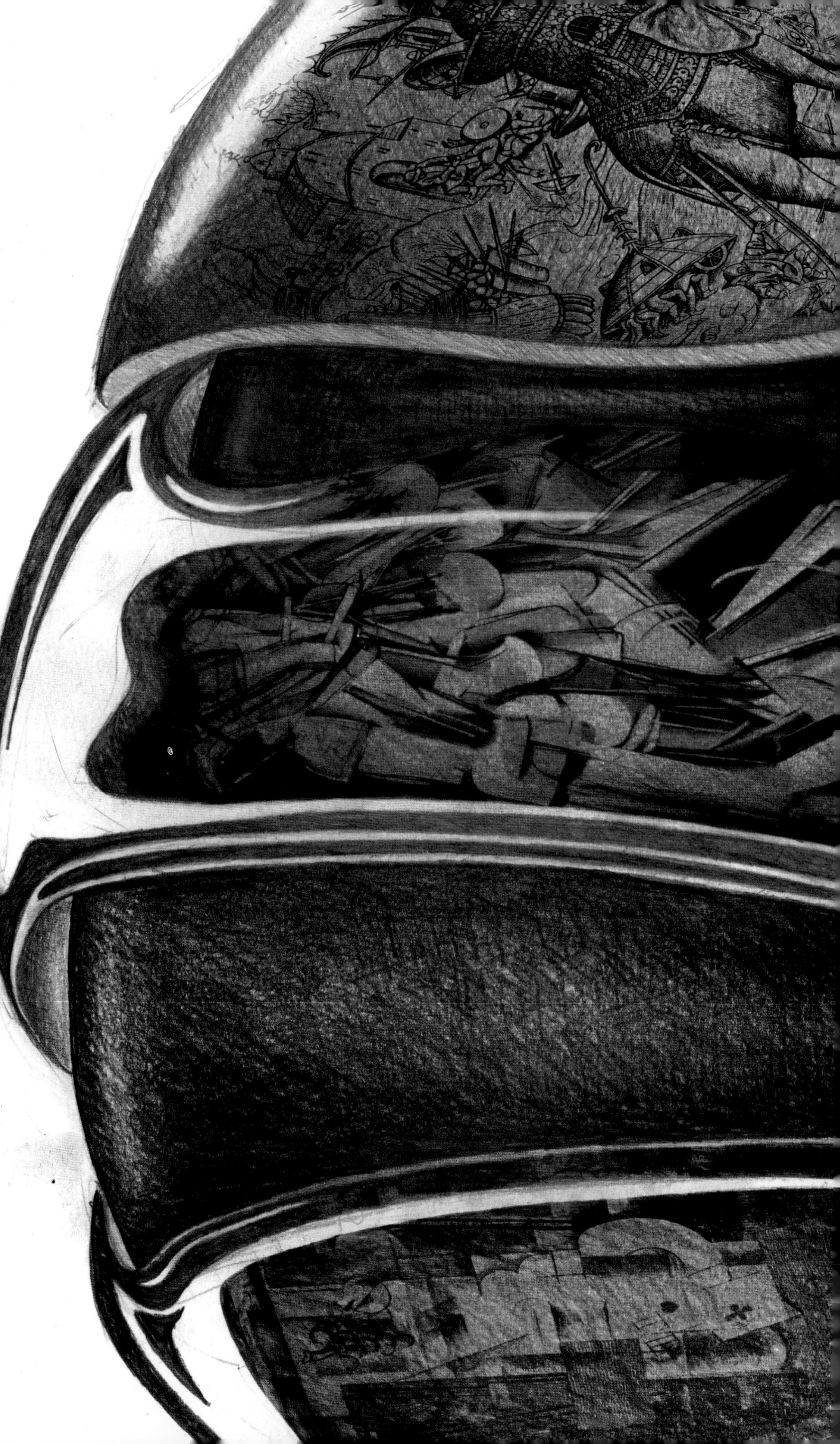

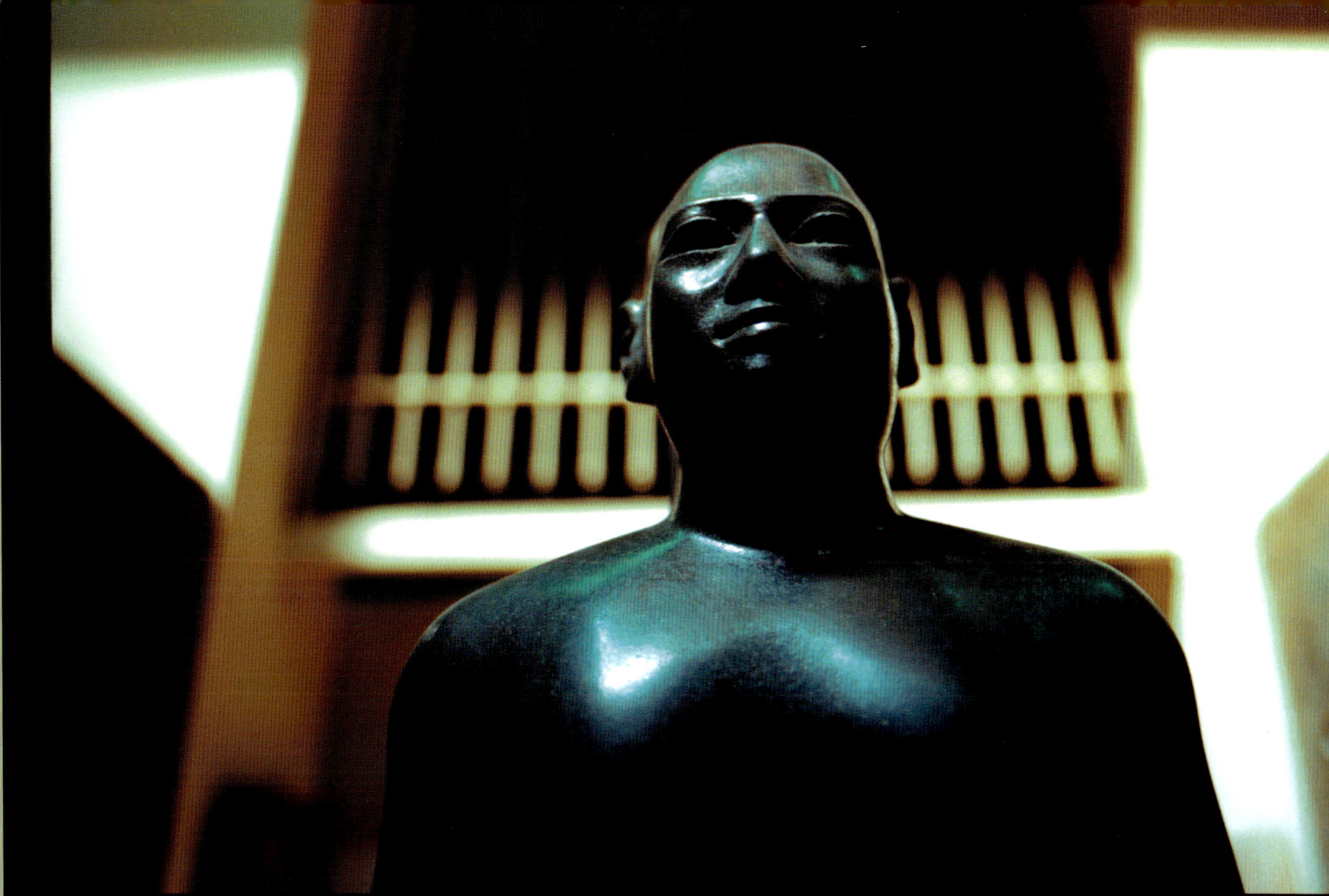

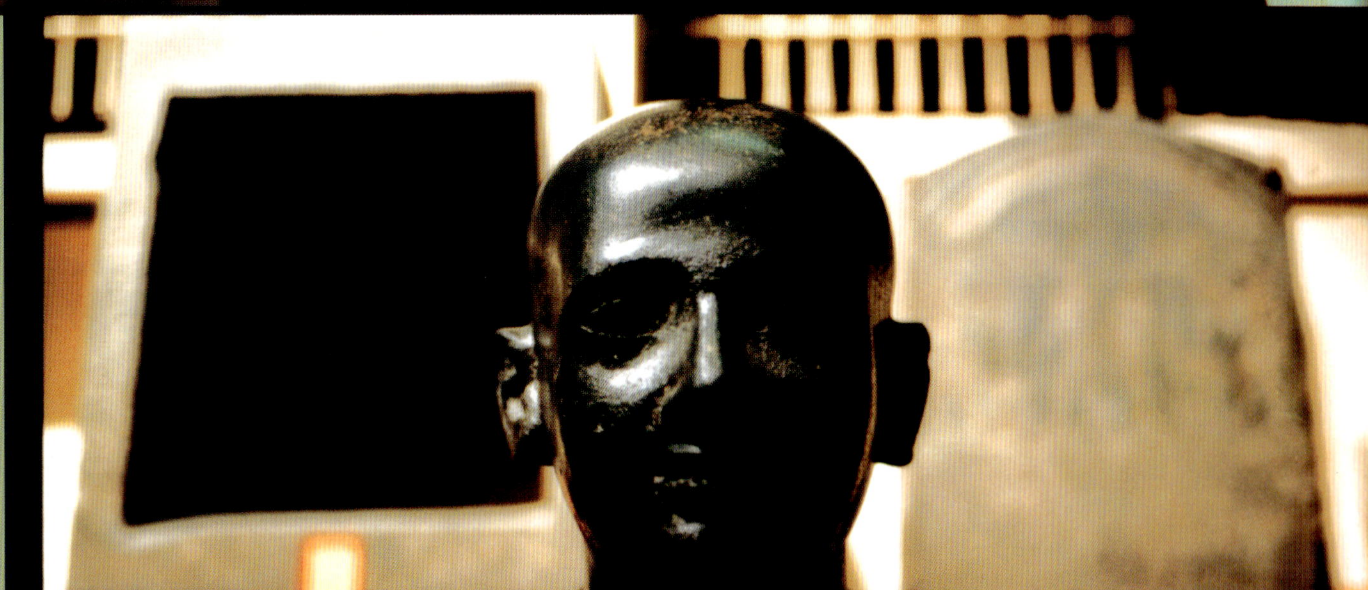
NPH400
4
NPH400

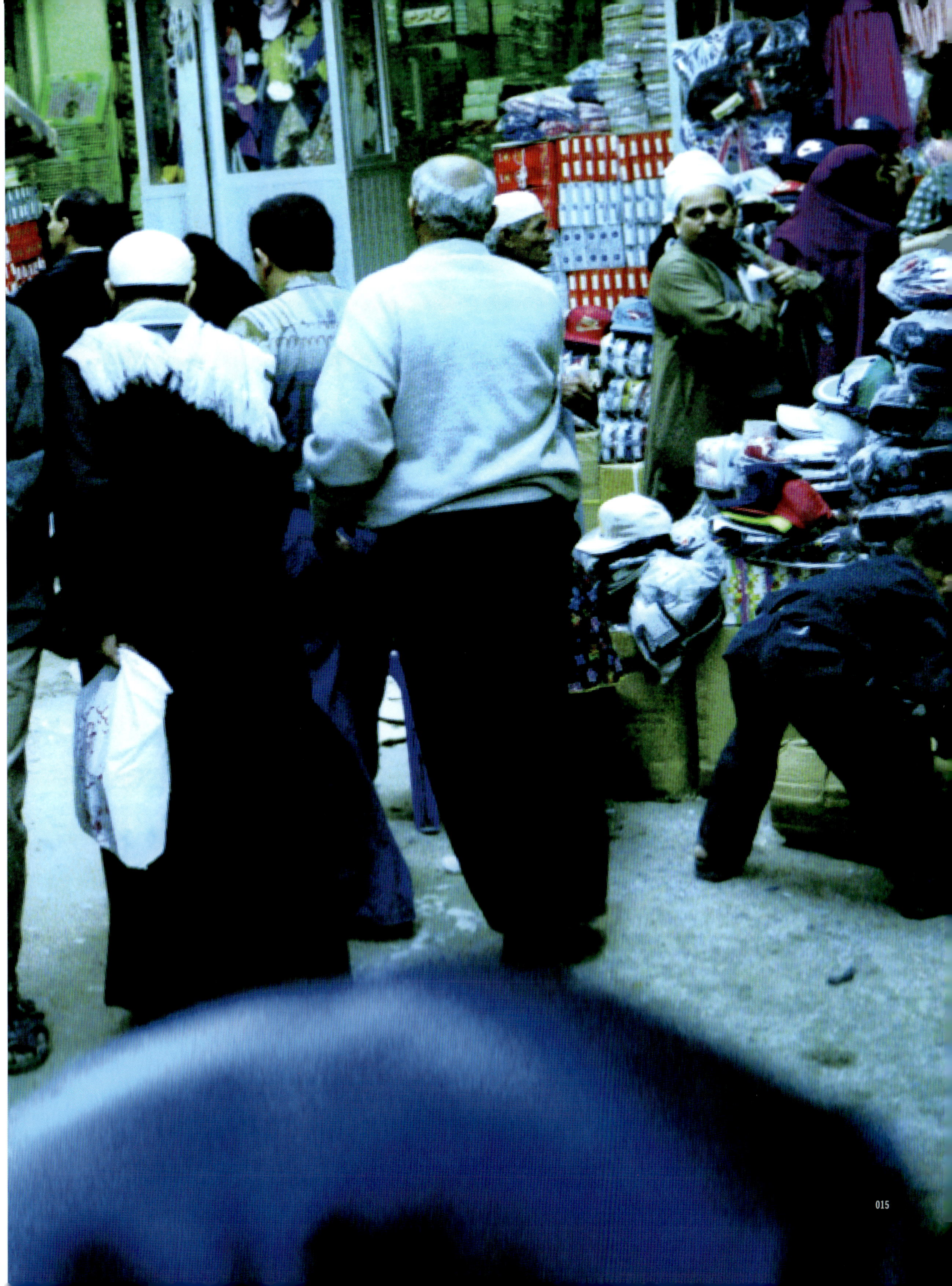

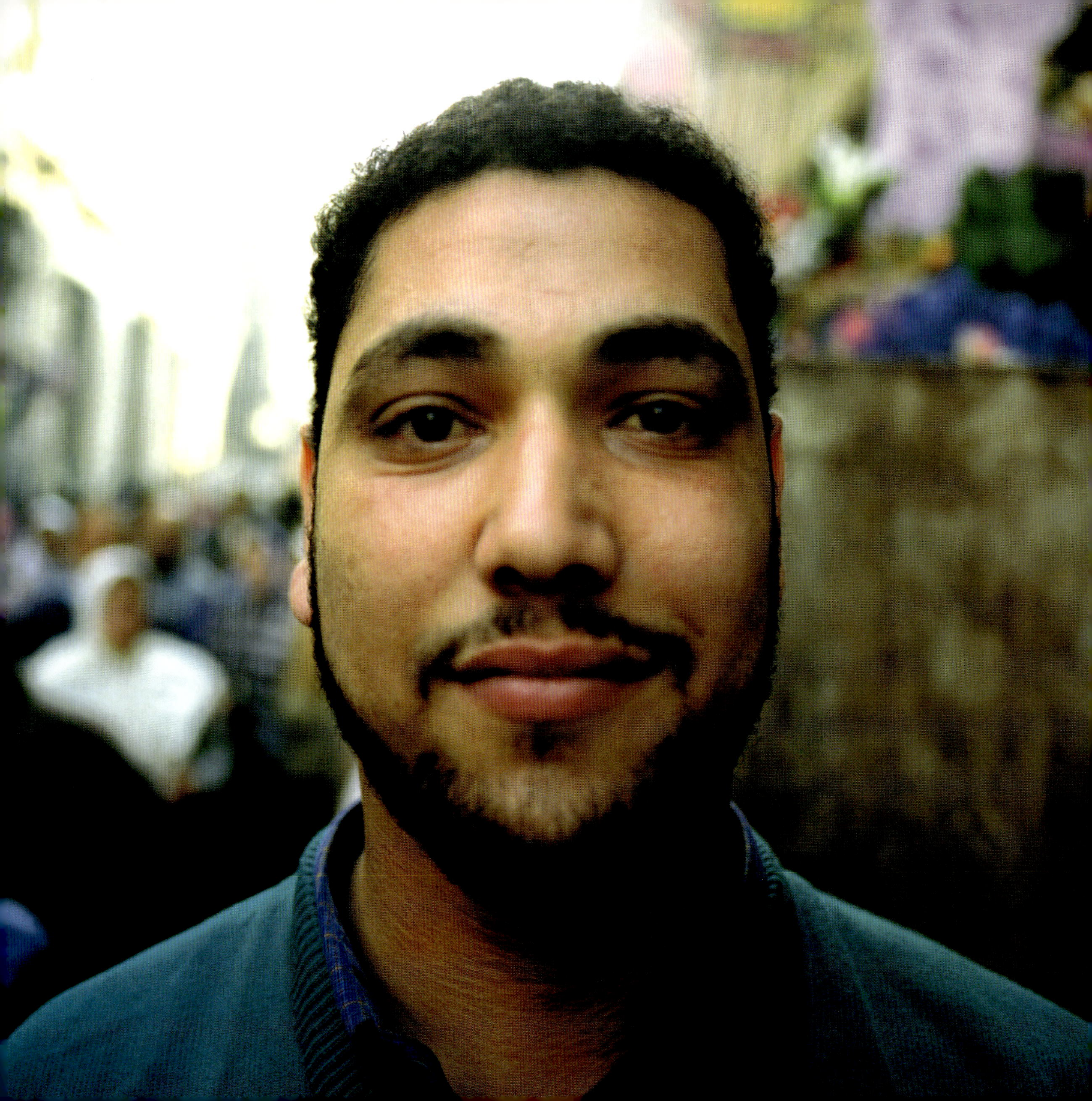

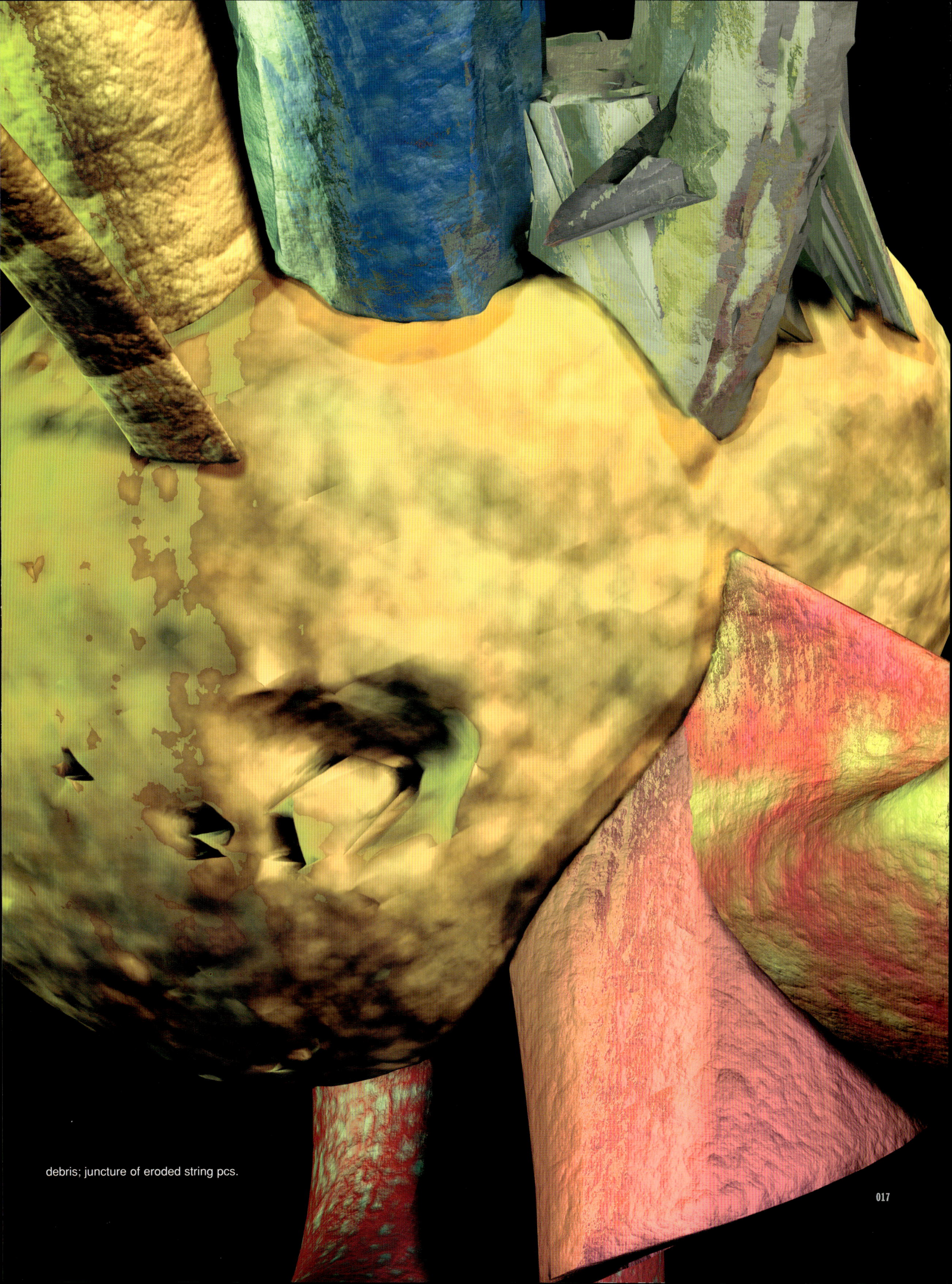

debris; juncture of eroded string pcs.

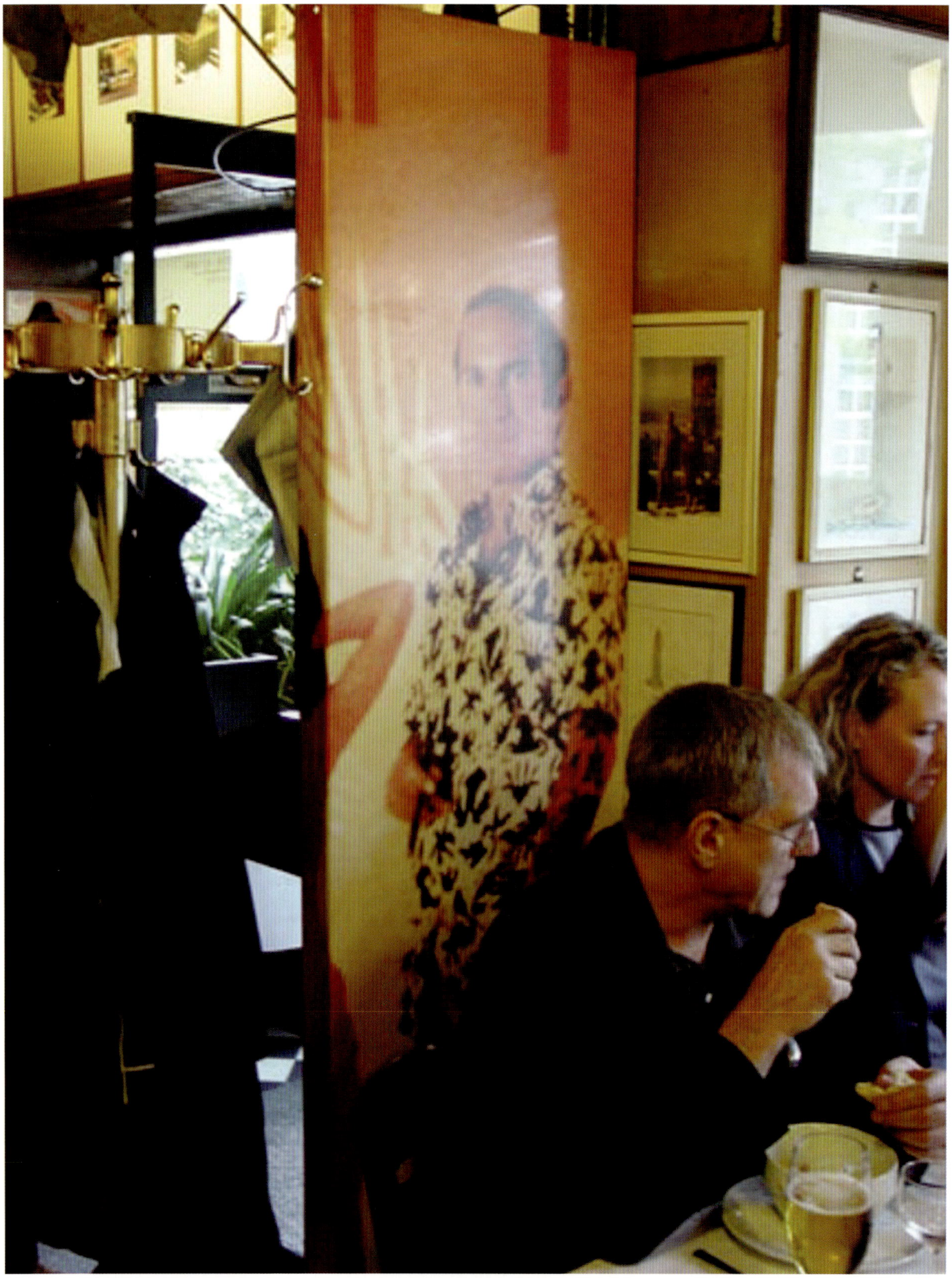

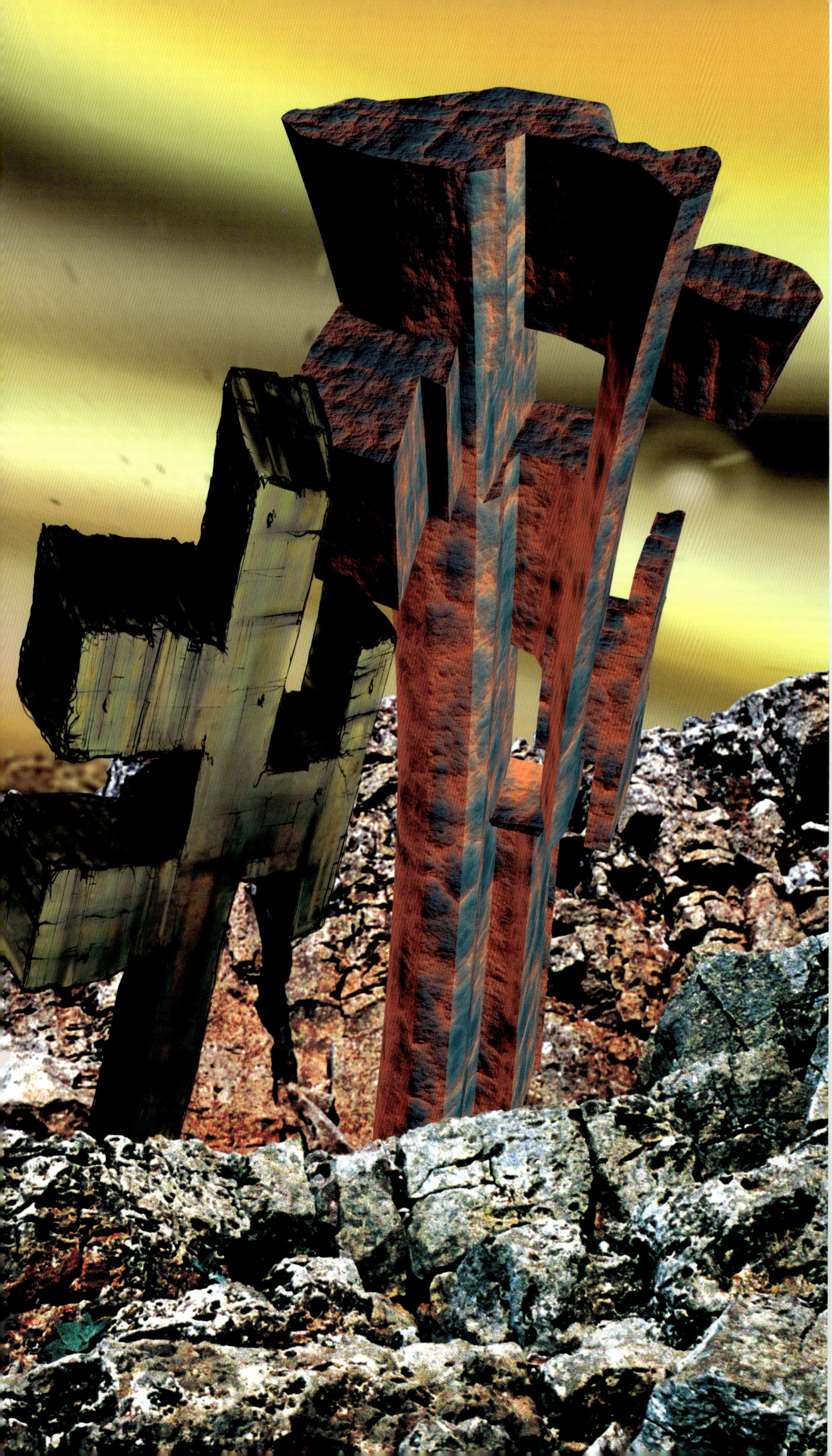

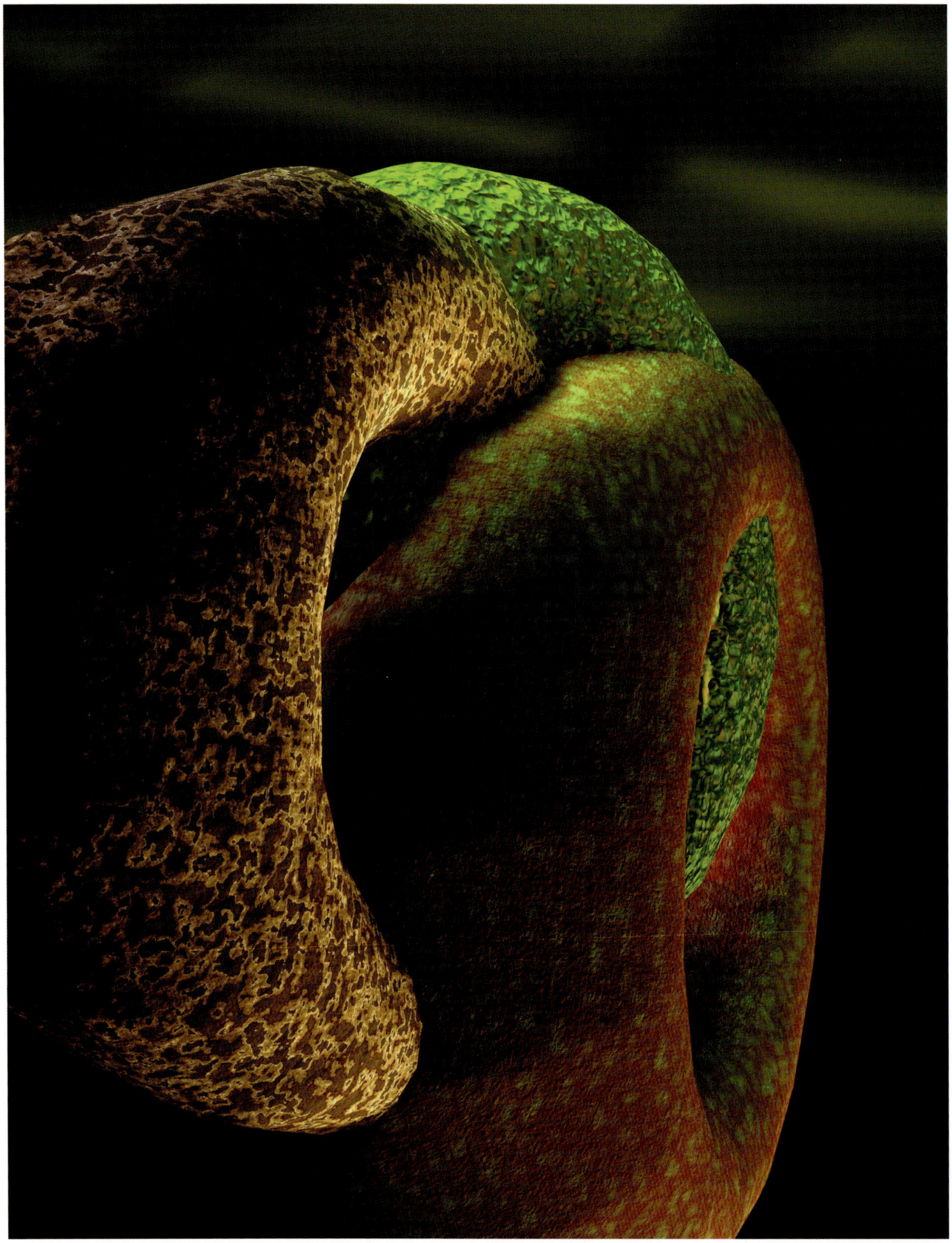

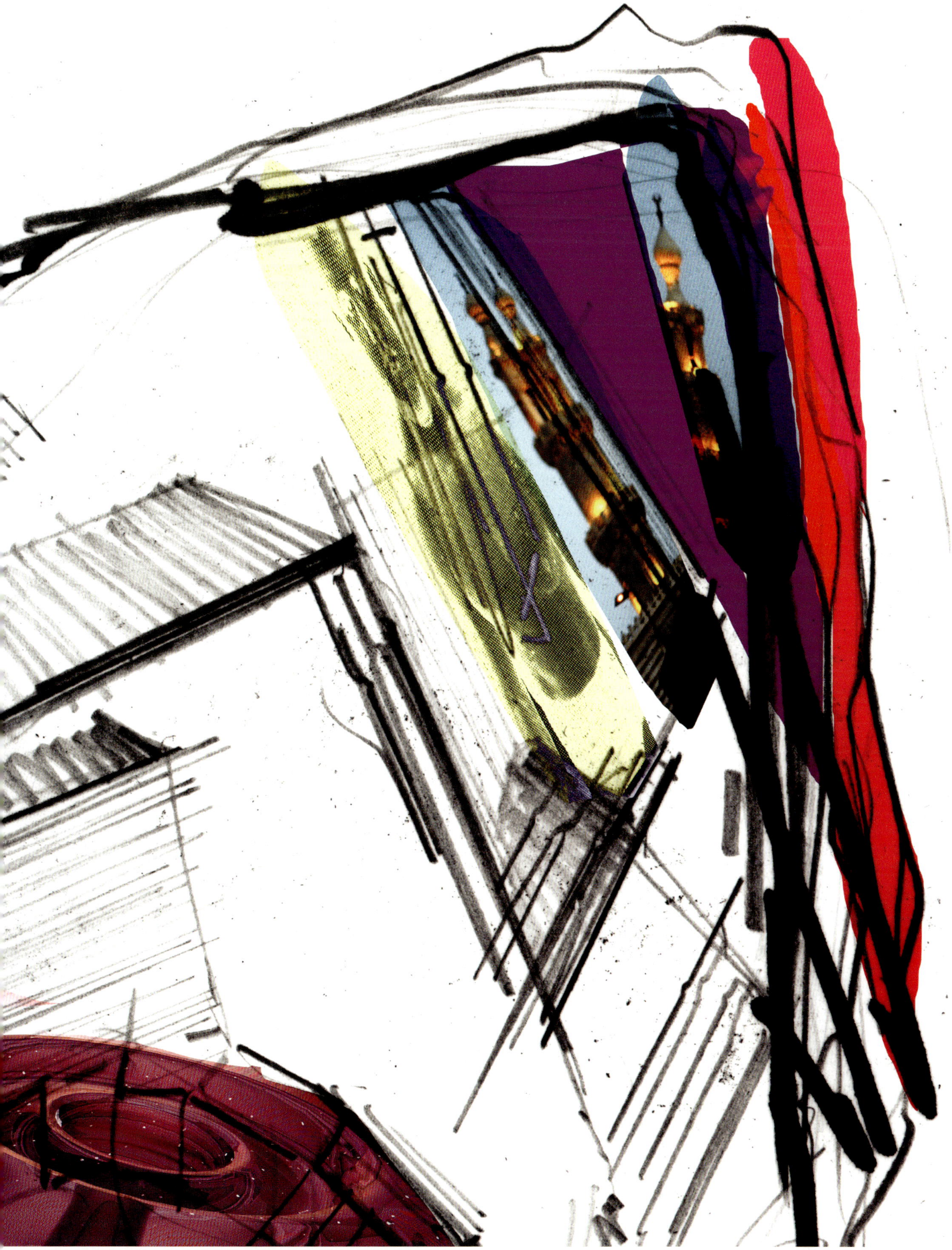

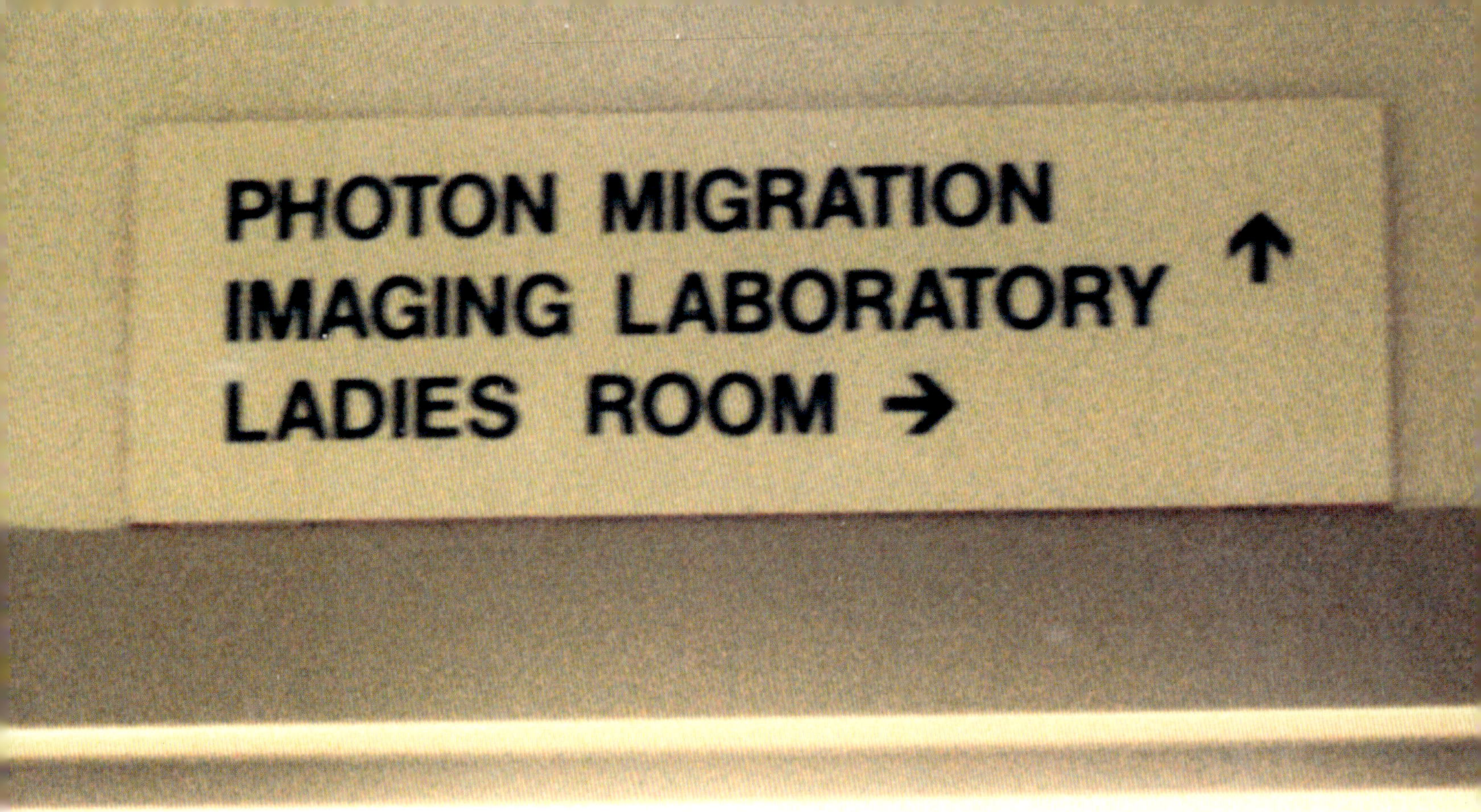
PHOTON MIGRATION
IMAGING LABORATORY
LADIES ROOM

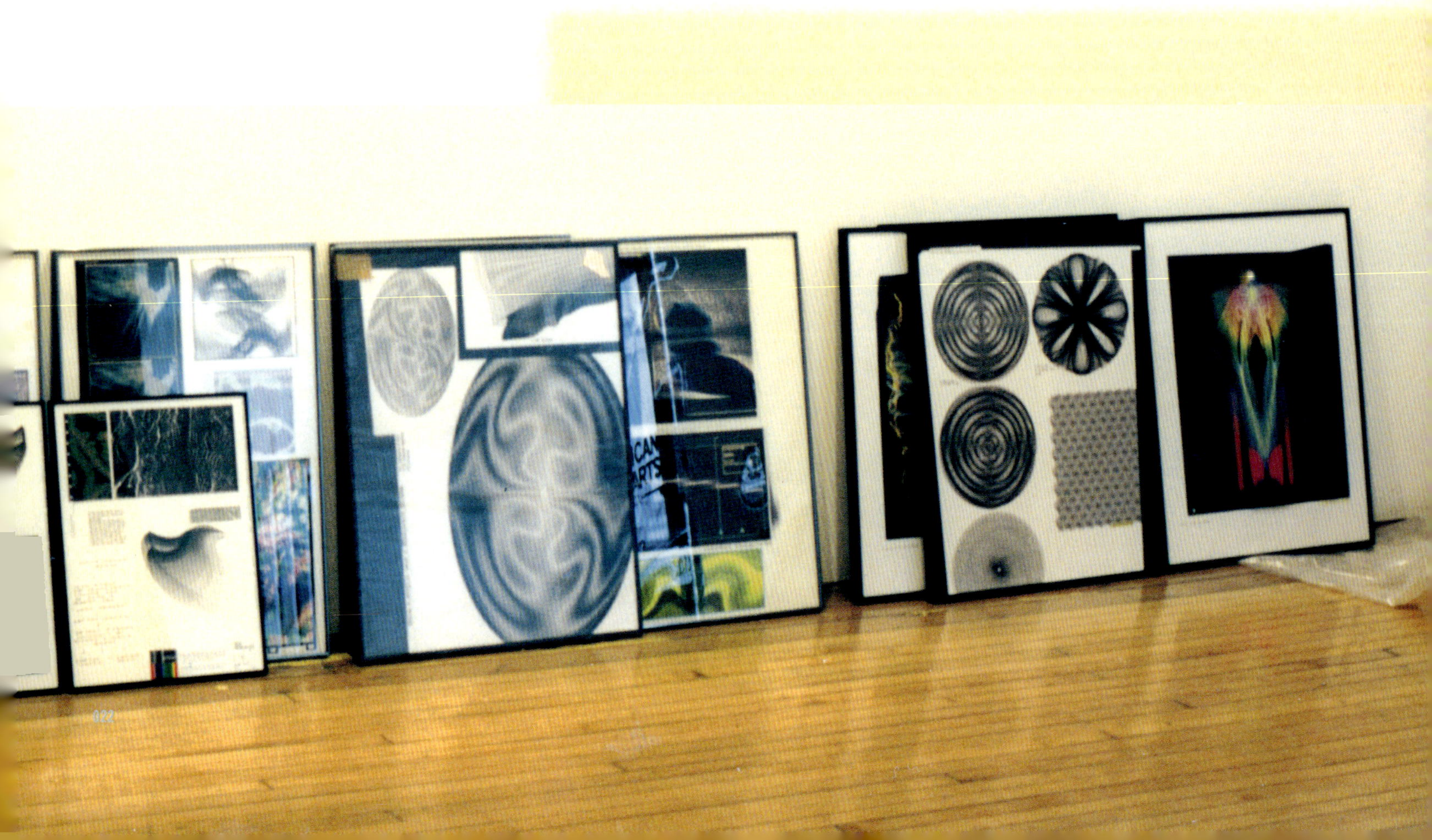

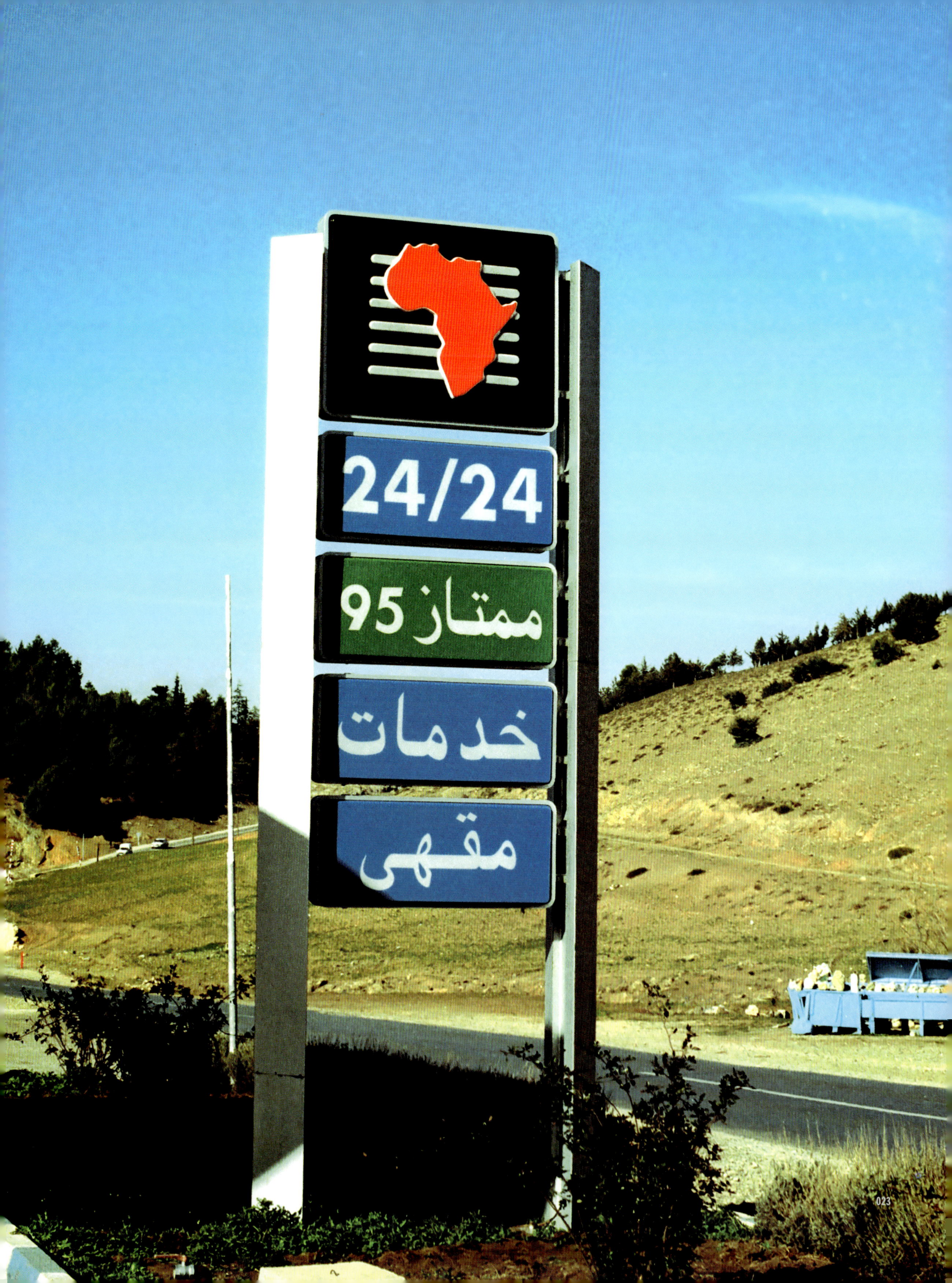
24/24
ممتاز 95
خدمات
مقهى

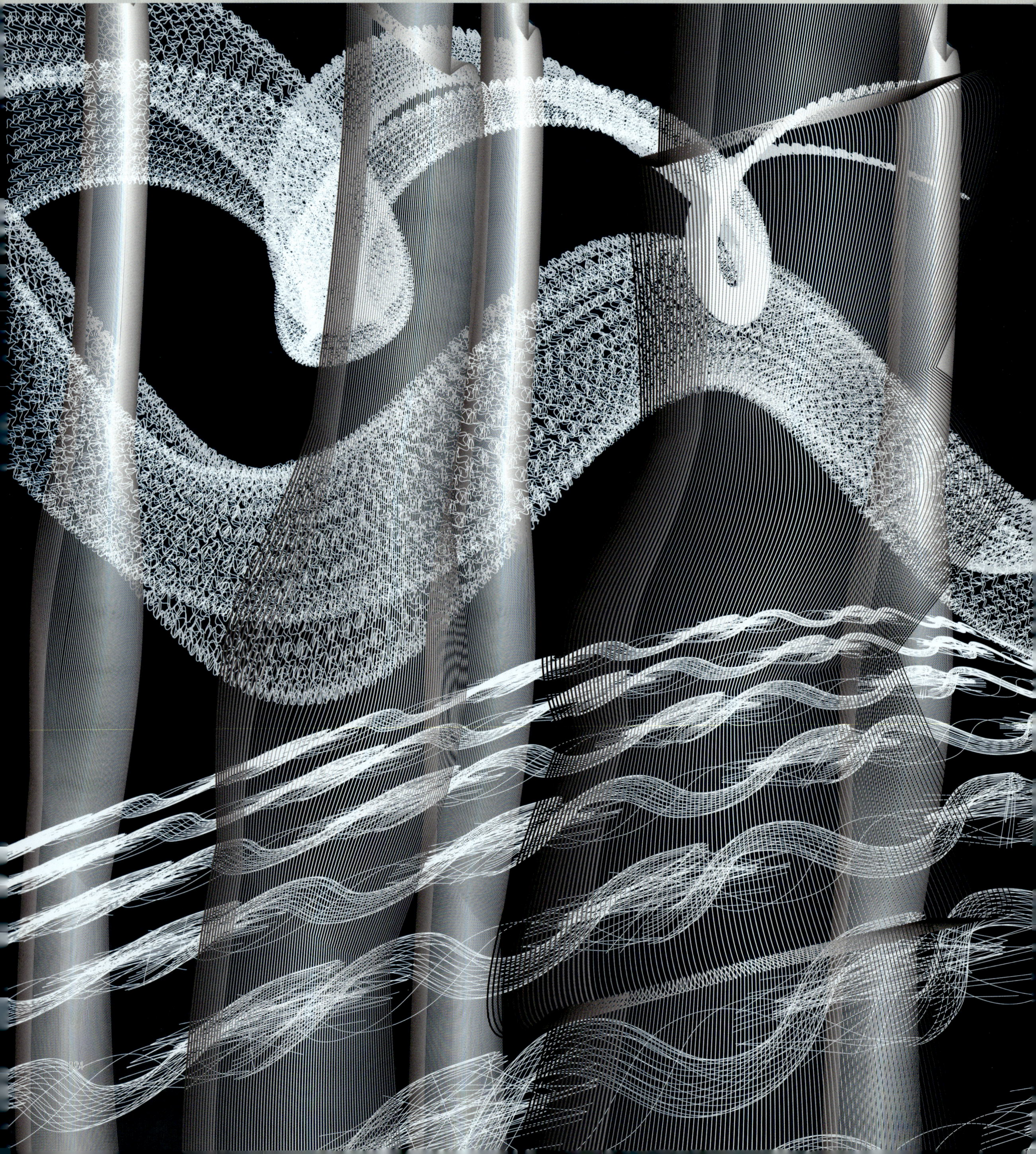

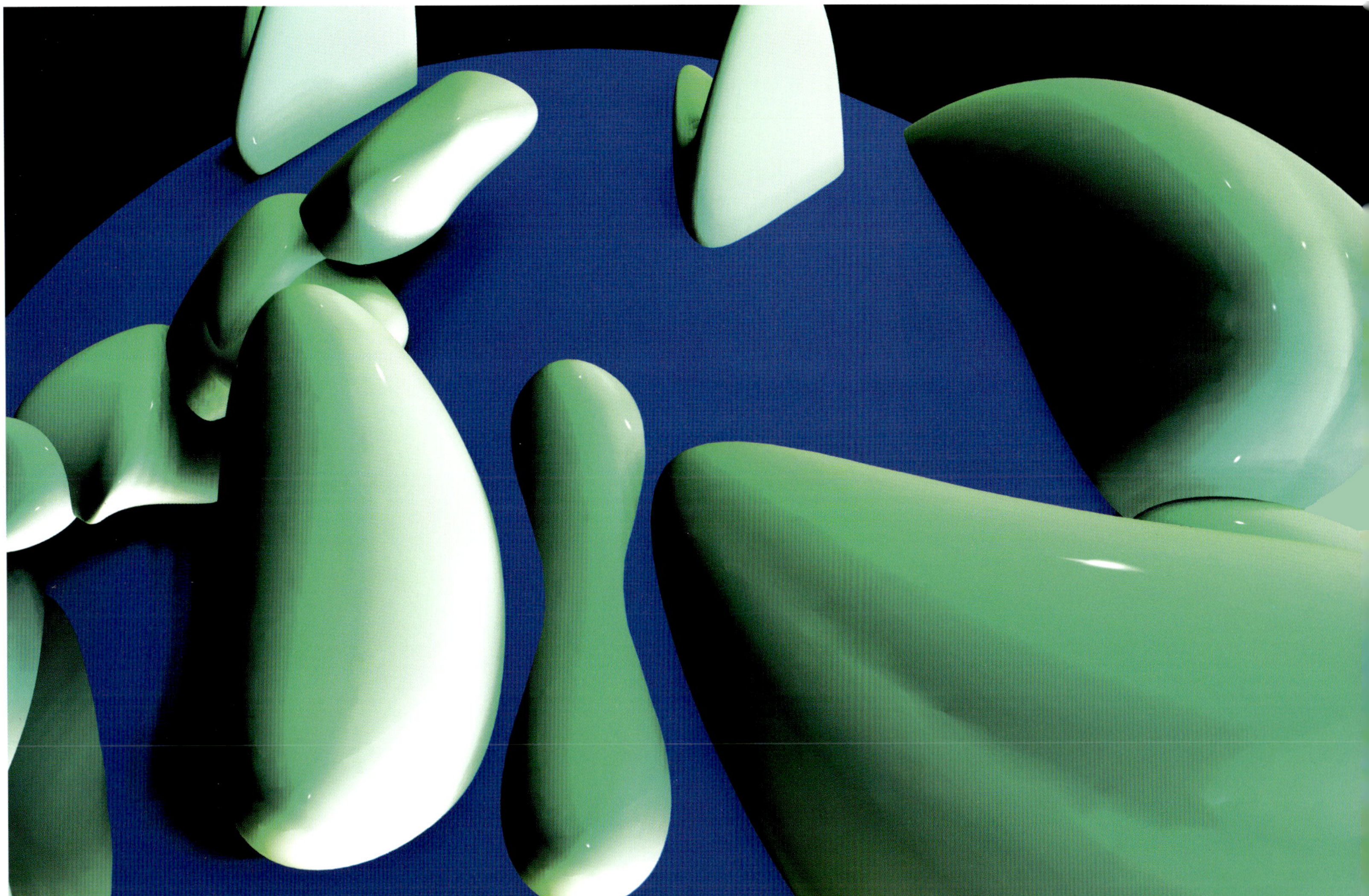

Label: Culture planet weather-schema; 2nd person perspective

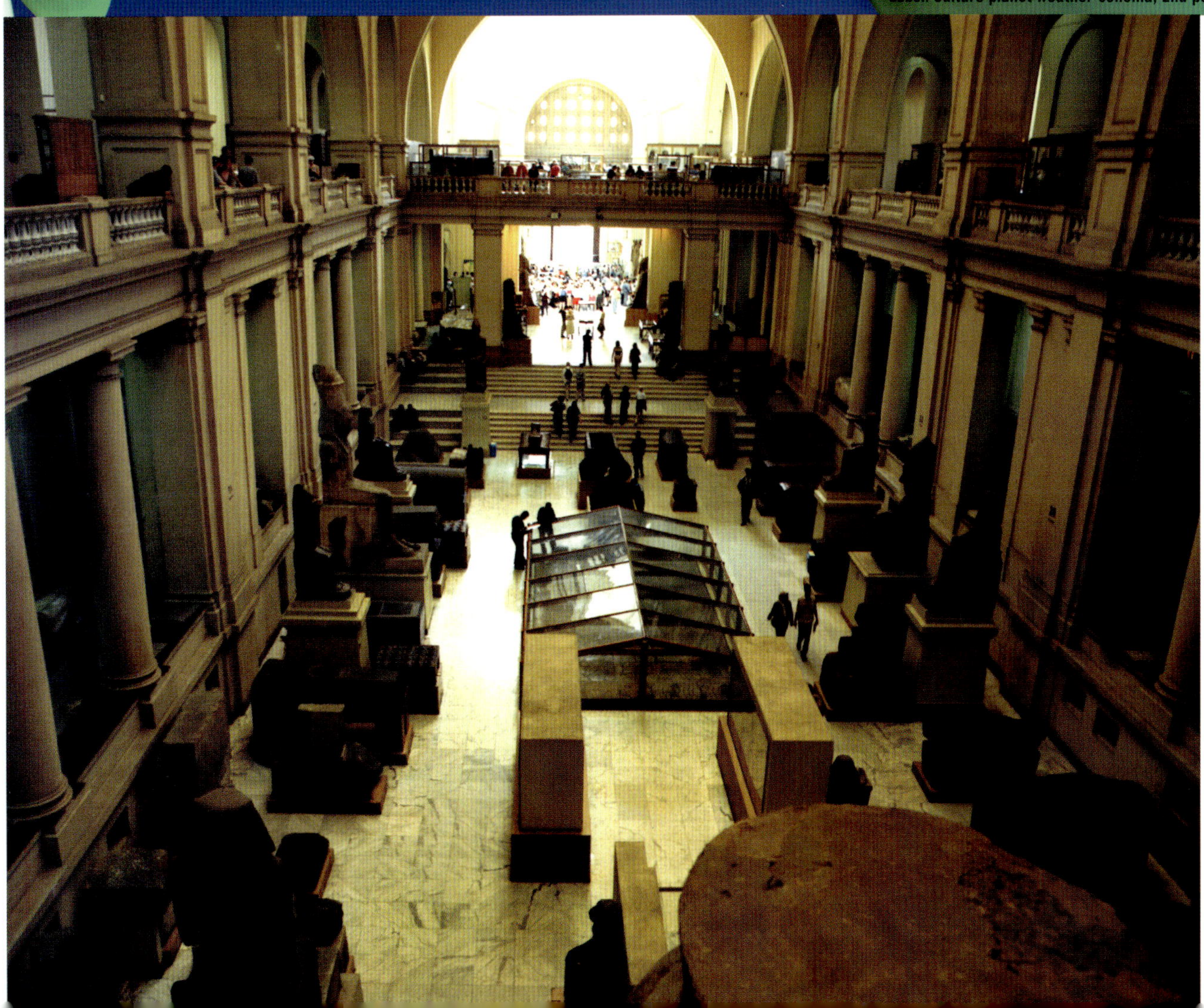

**FLAGSTAFF: Situation worsening. CERN results are inconsistent. 0º vacuum. Fragmentation still occurring. We don't believe we can go deeper.**

REPEAT: No cohesive whole. Finding other territories of newly synthesizing past.

Also in Three Roads, you talk about the experiments that detected atoms as being a development of the need to resolve the paradoxes and contradictions that were previously irresolvable, formulaically speaking. Currently, in the process of characterizing the "channel through which information flows from one region of space to another, the area of the surface is a measure of its capacity to transmit information" you said this to be very suggestive.

there set was a future world, inside the present. The characters themselves are particle fragments with characteristics familiar to us.
They bond and form larger personalities with similar conglomerate presence. They conspire against us and sometimes, although quite a bit less frequently, they succeed.
Adversarial Personalities covered the pieces of an unsorted distant life force, a million light years ago.

random: positions fly around the surface, mostly tainted with academic rhetoric the kind associated with historicism. It's a well trodden field leading back to the well. —current patterns produce a beeshive-stasis, -.006 planet surface scale. Long-term, sinusodal reverberations—they dumped that out with the bath water. note to dept. head; professionalism of "critical" perspective produces characters who wander off into isolation and hermetic self-aggrandizement.

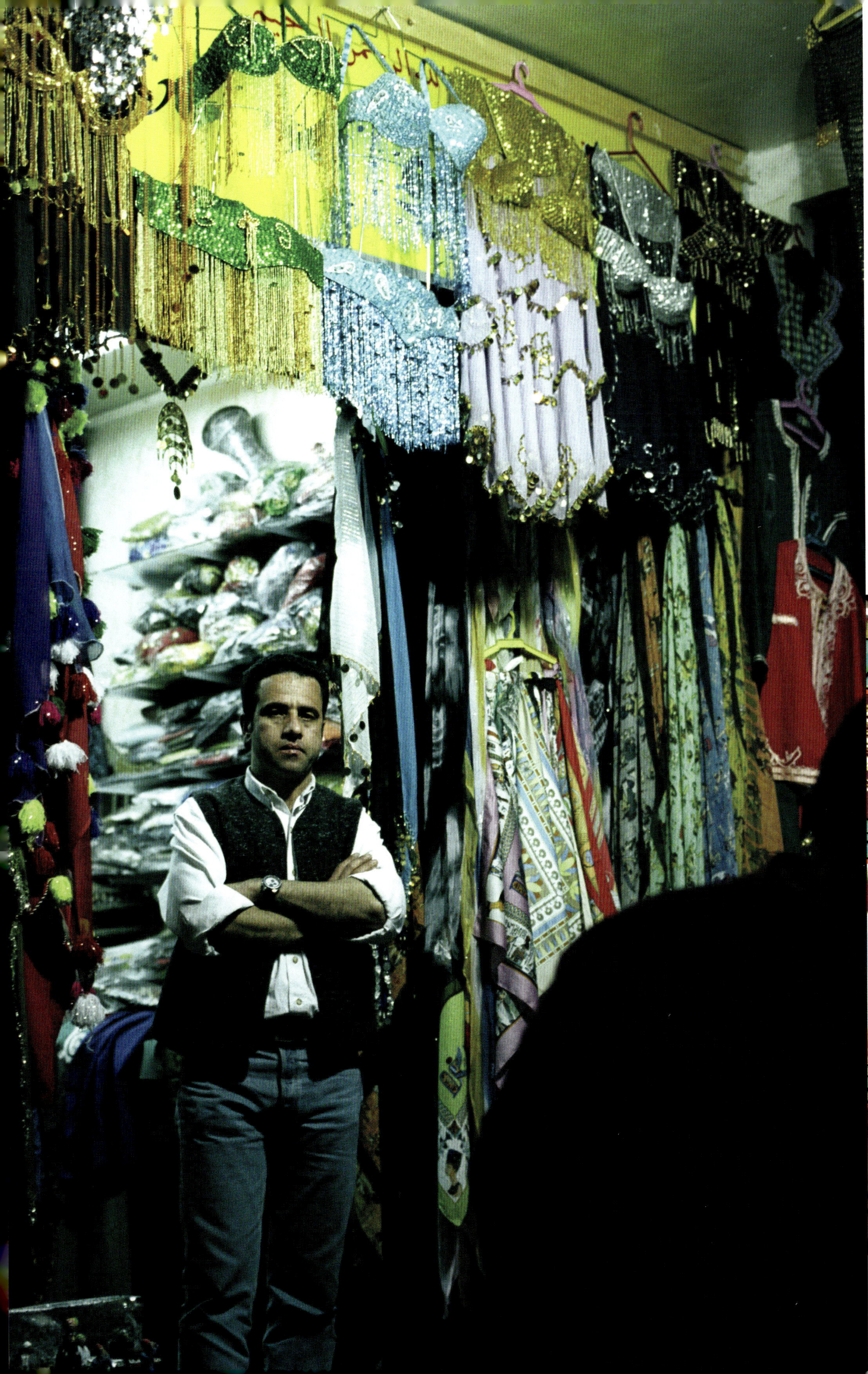

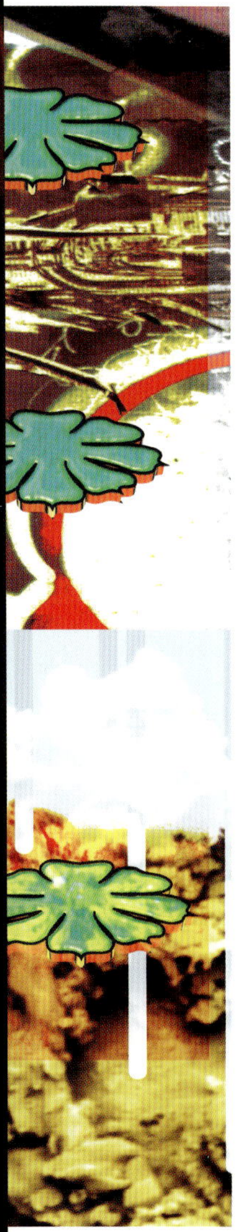

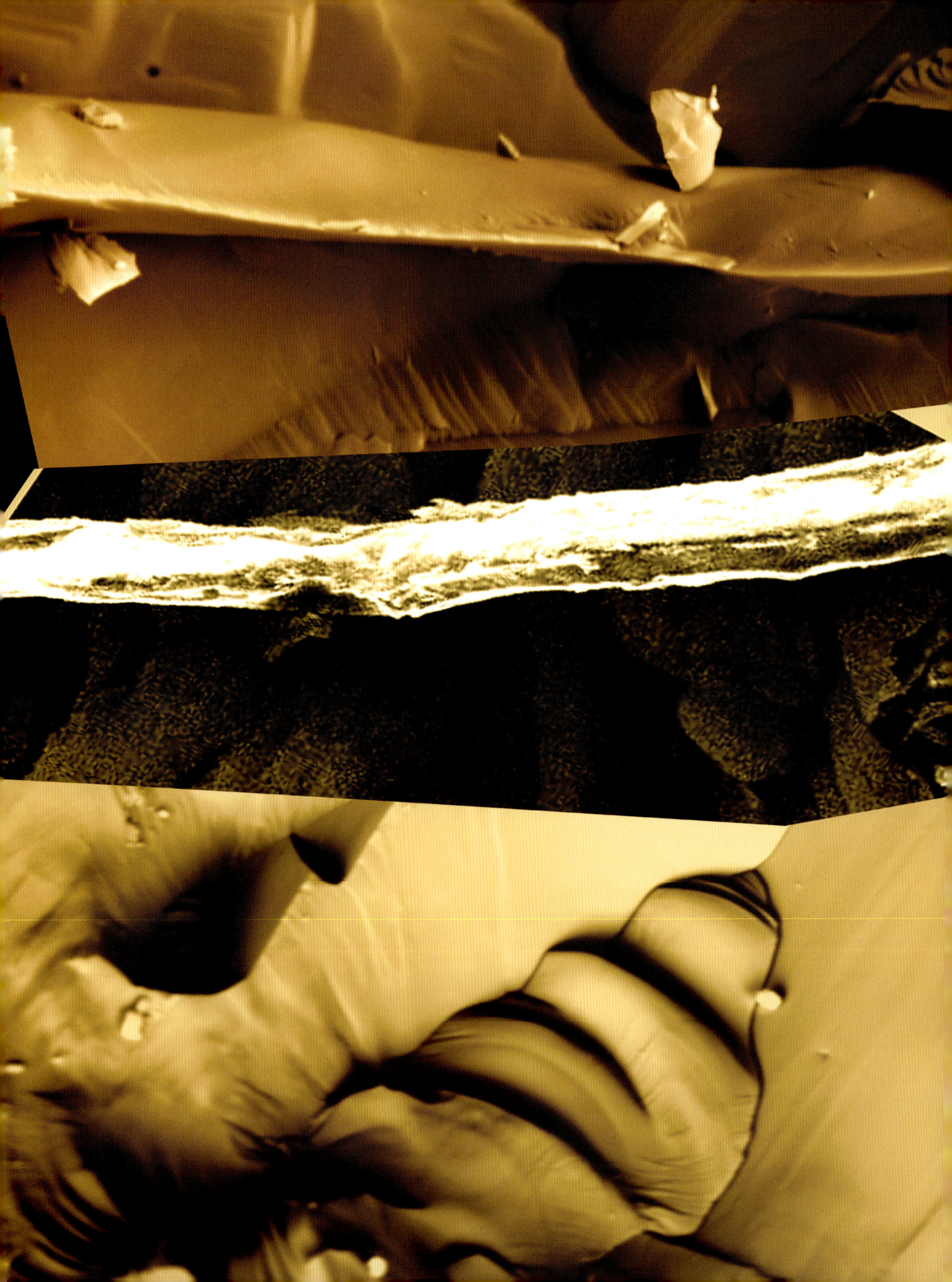

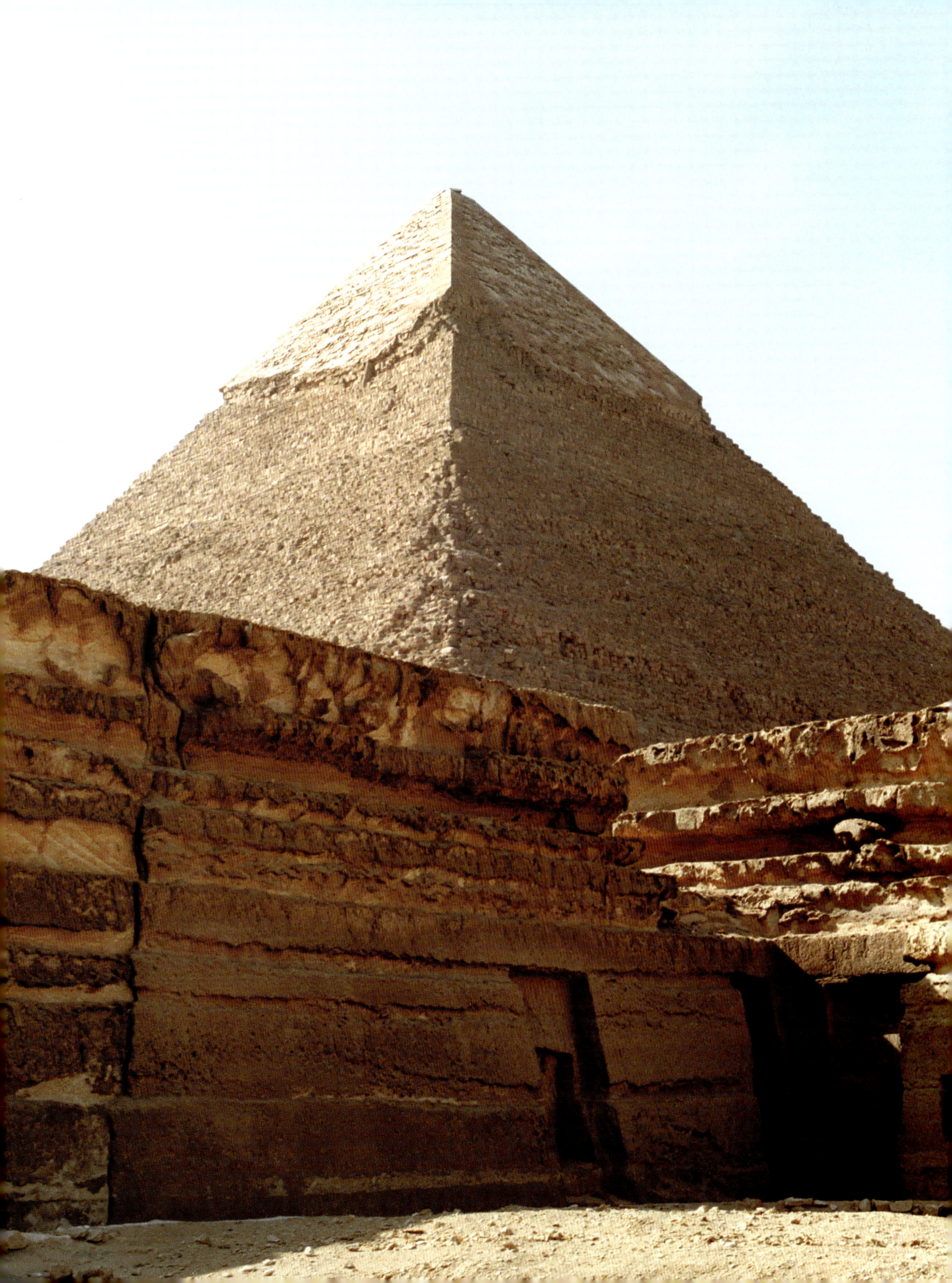

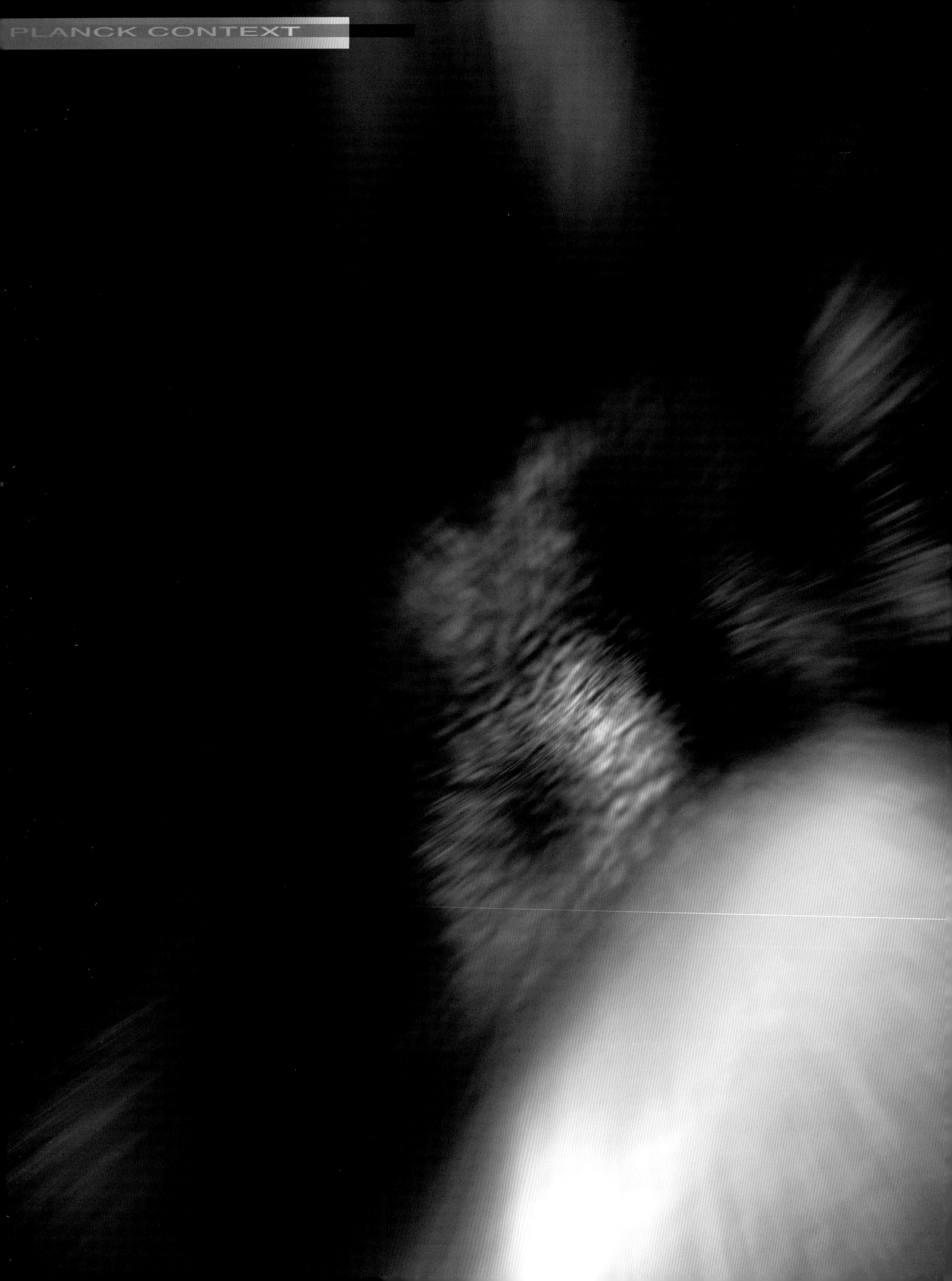

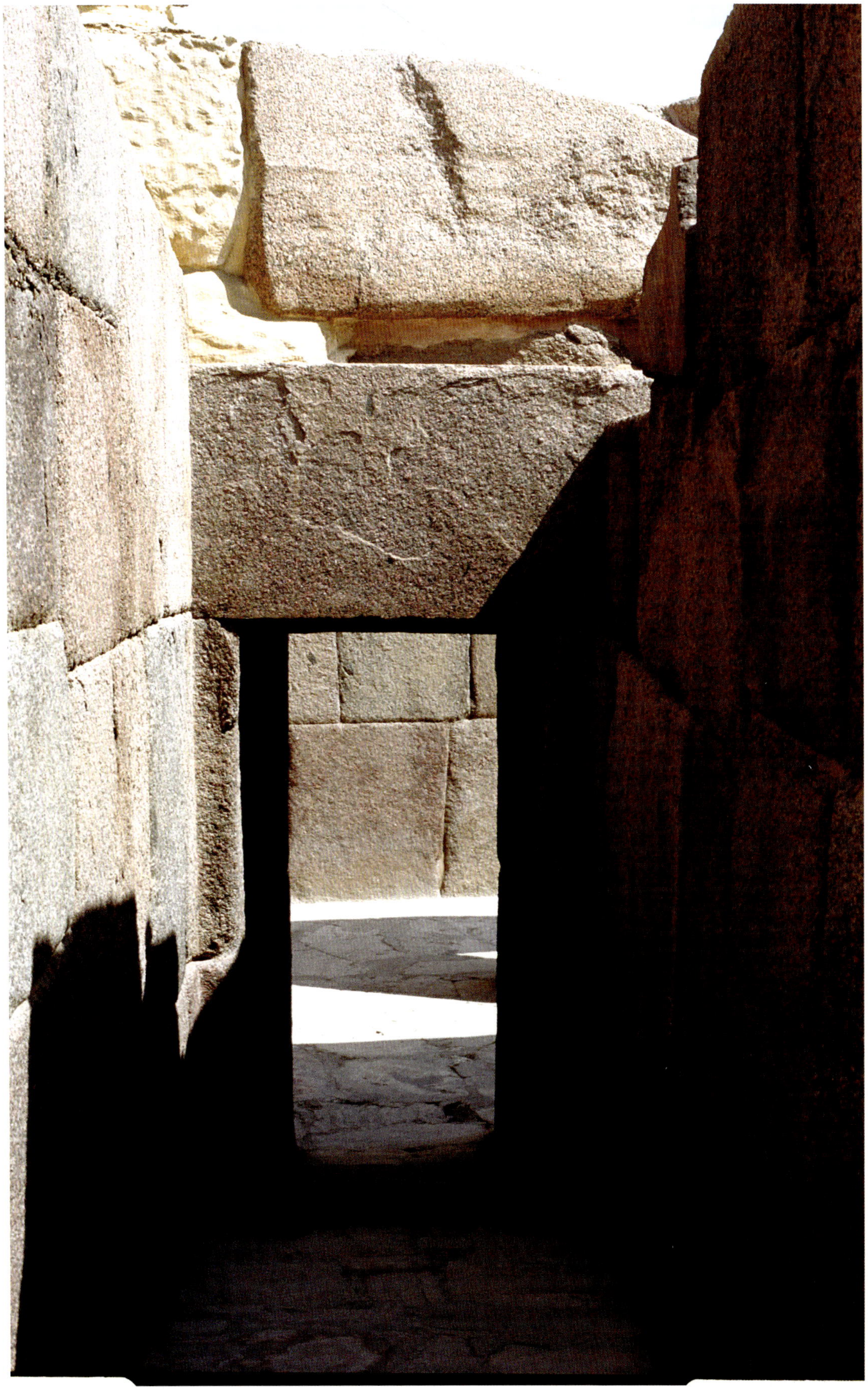

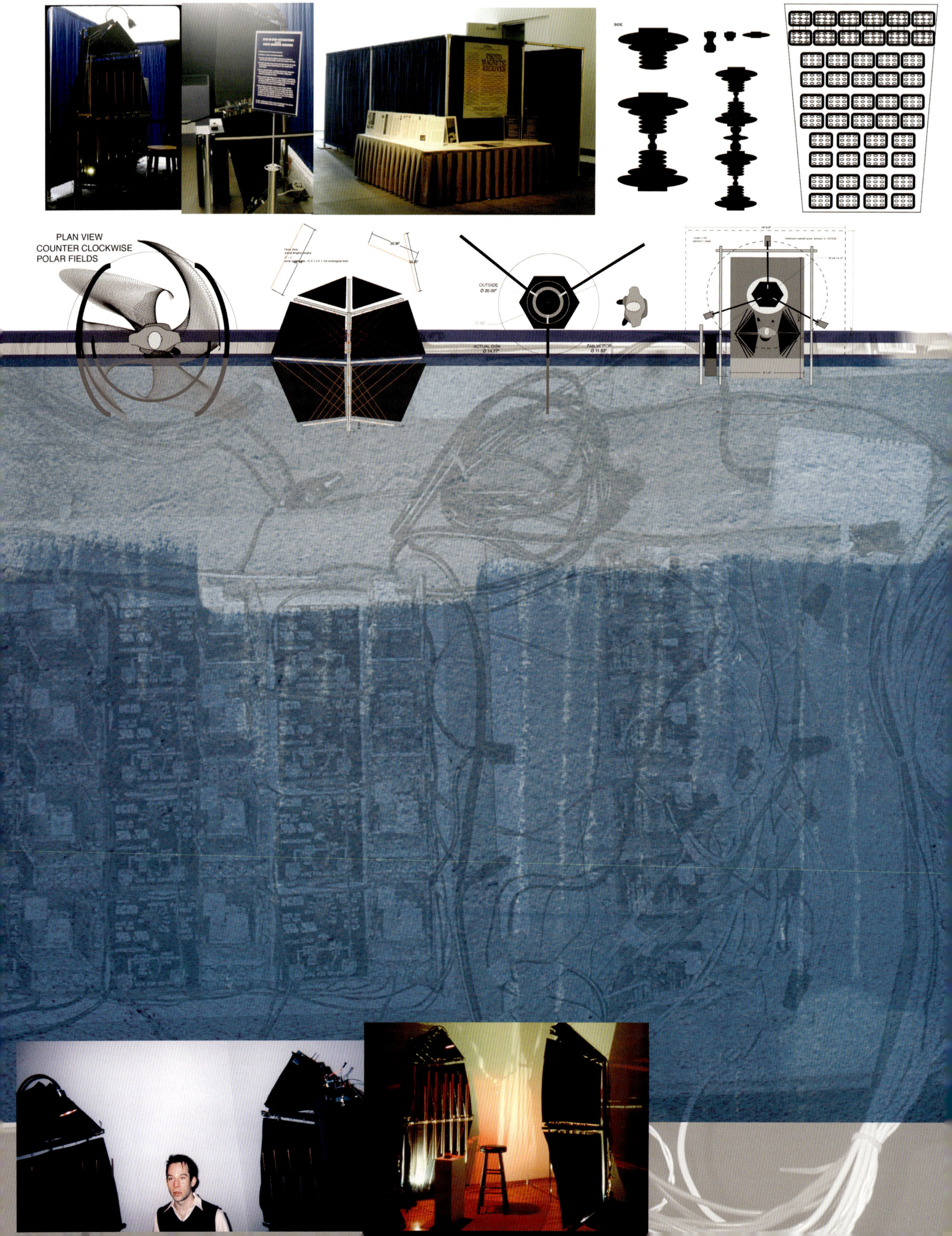

SIDE
PLAN VIEW
COUNTER CLOCKWISE
POLAR FIELDS
OUTSIDE
Ø 20.00"
ACTUAL DISK
Ø 14.77"
FAN MOTOR
Ø 11.63"

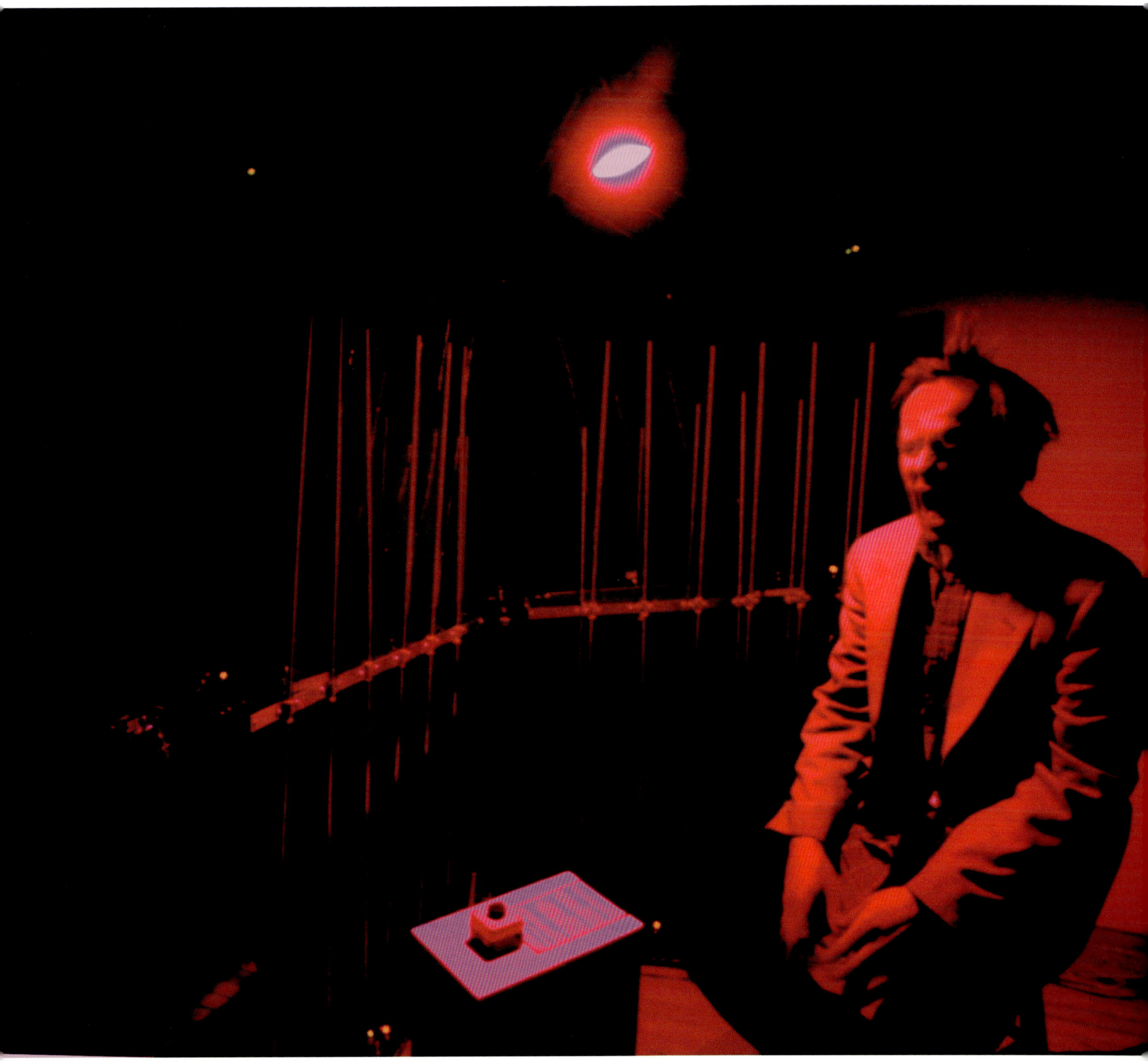

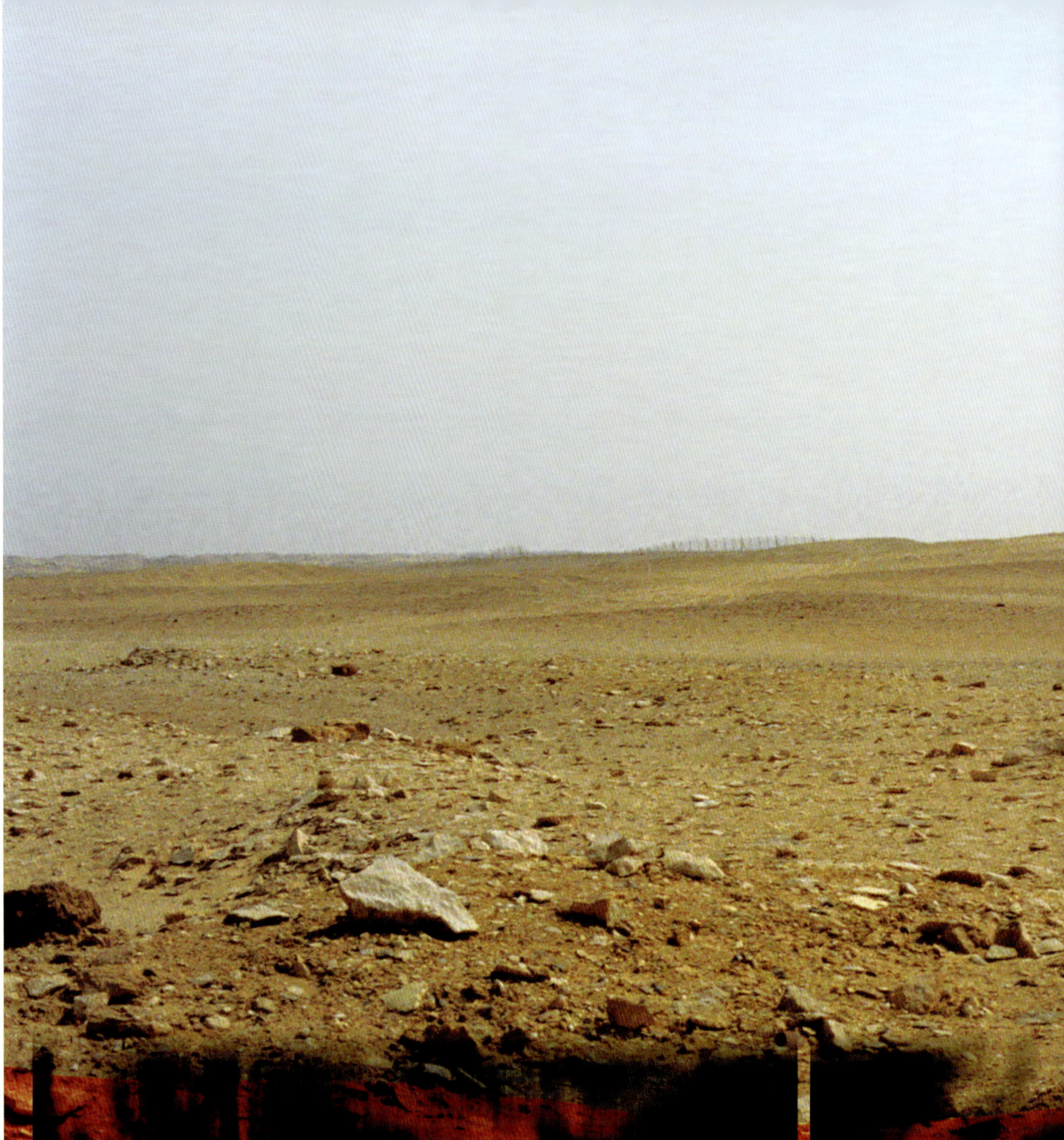

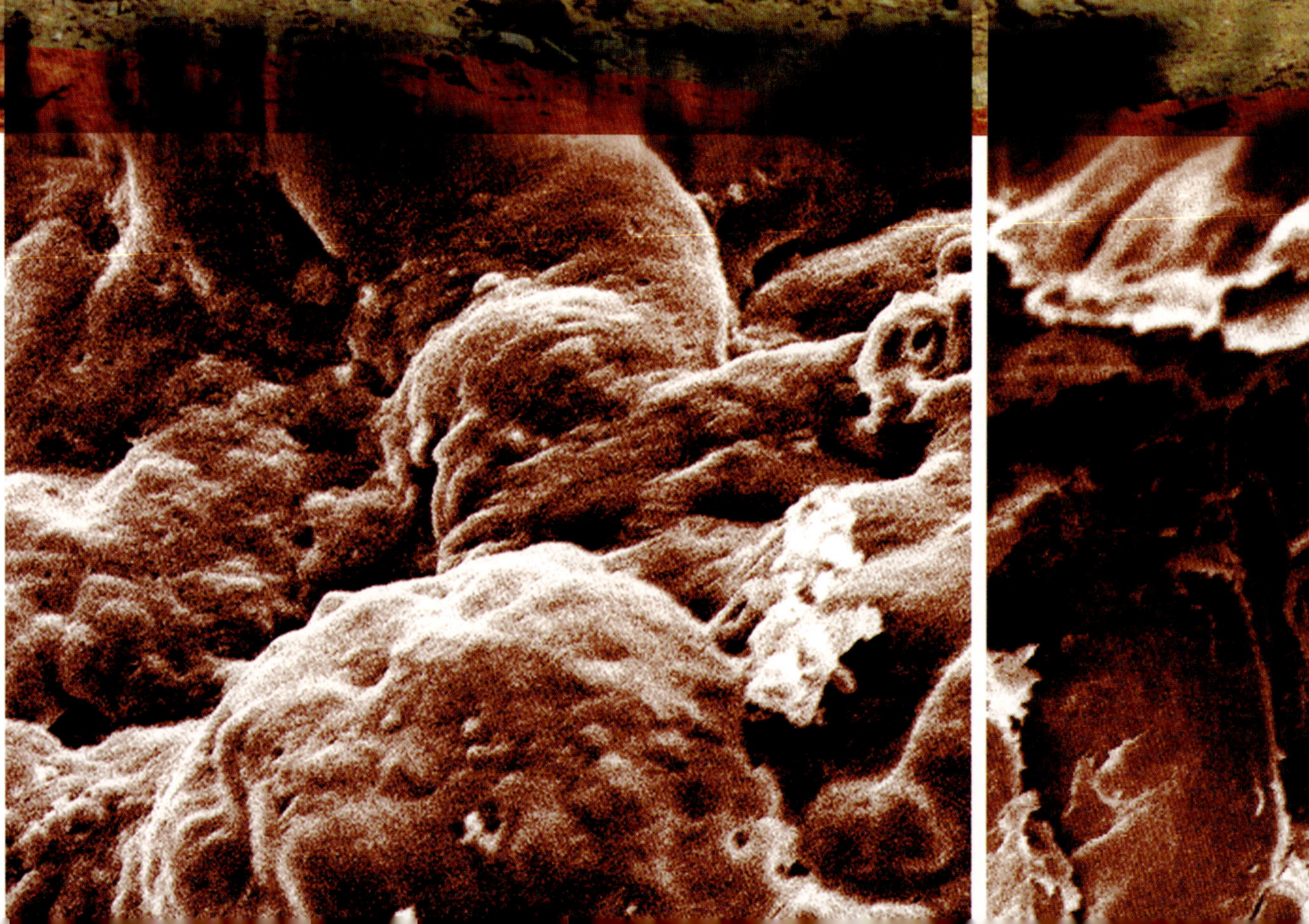

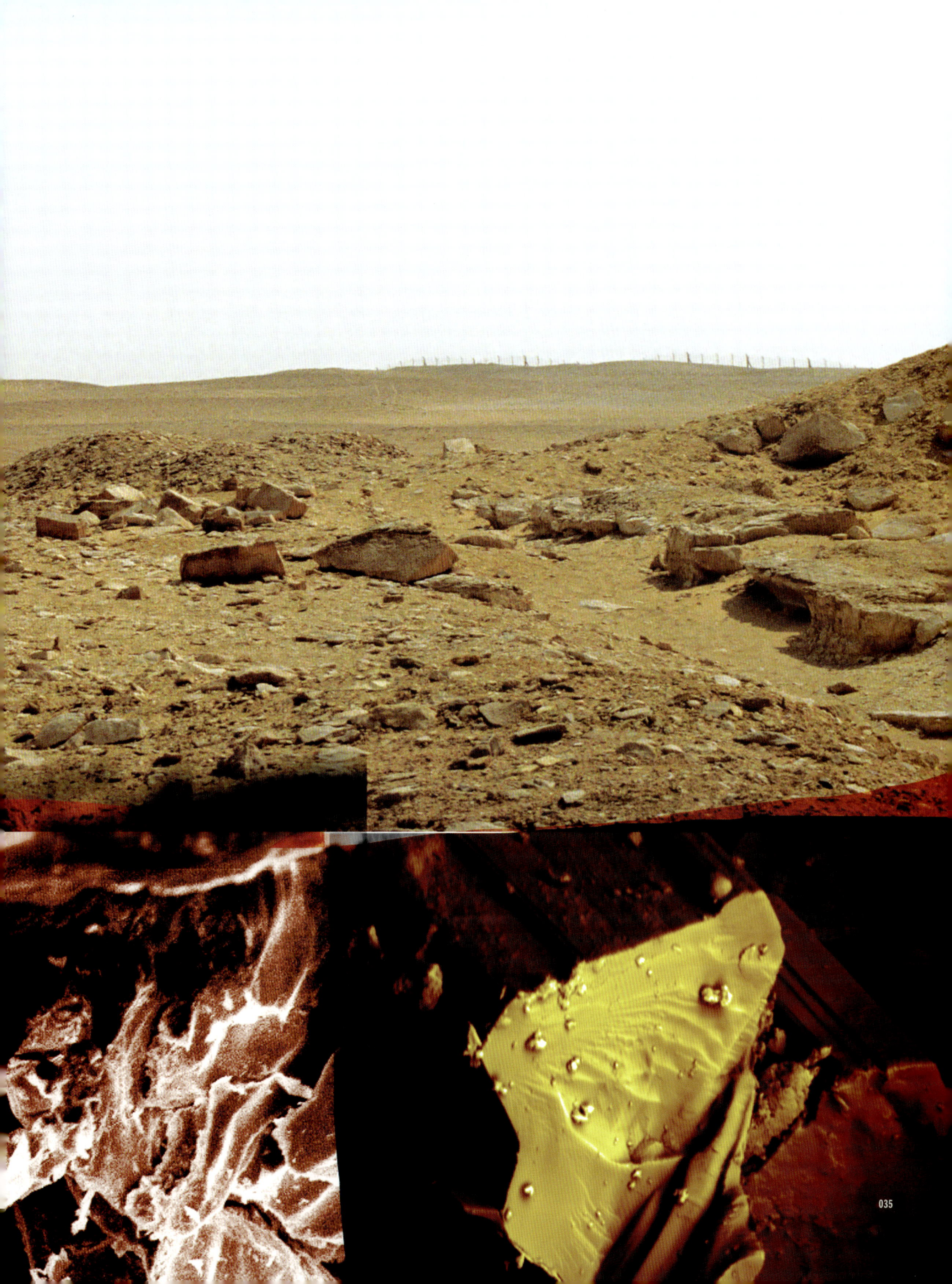

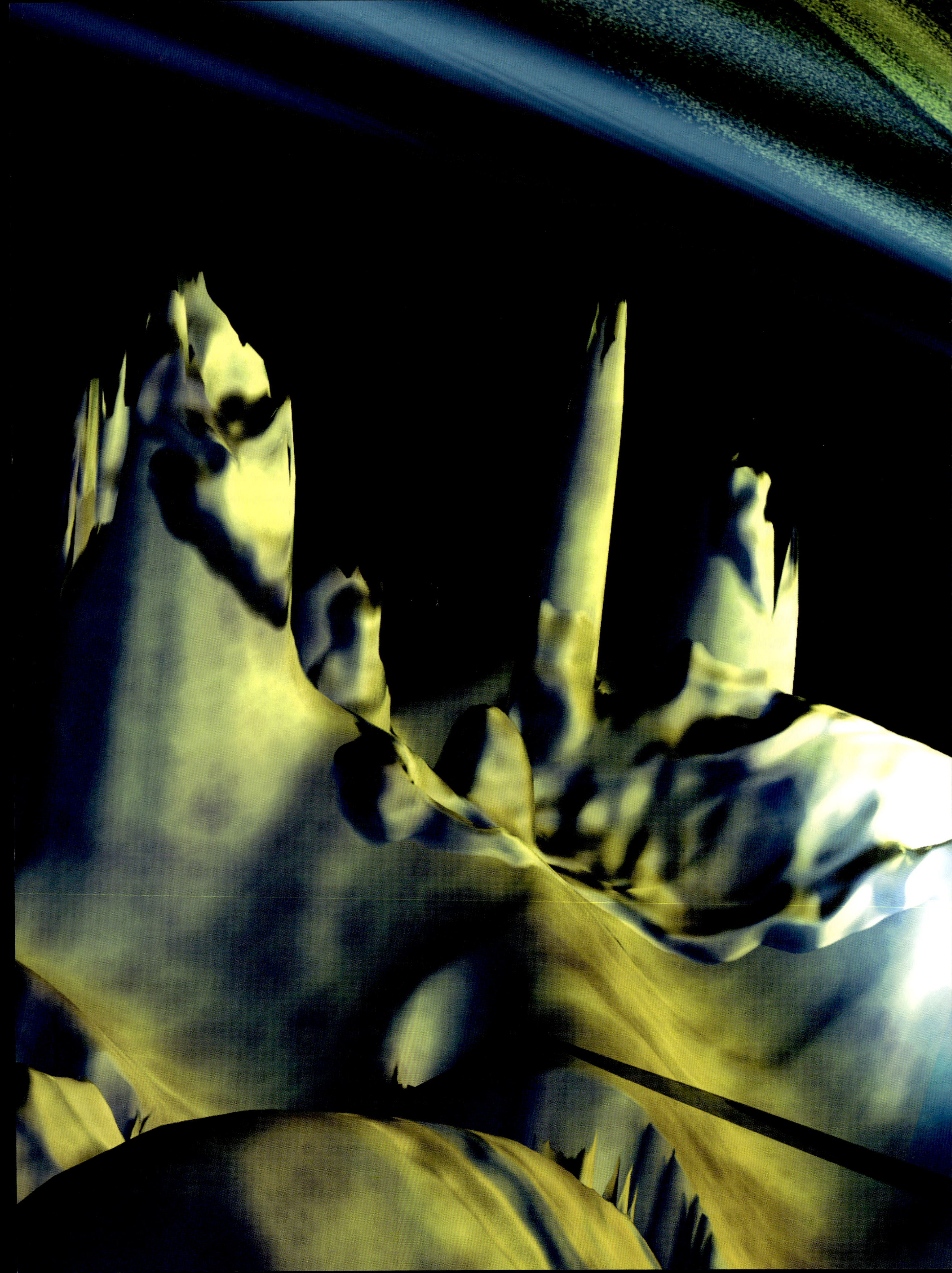

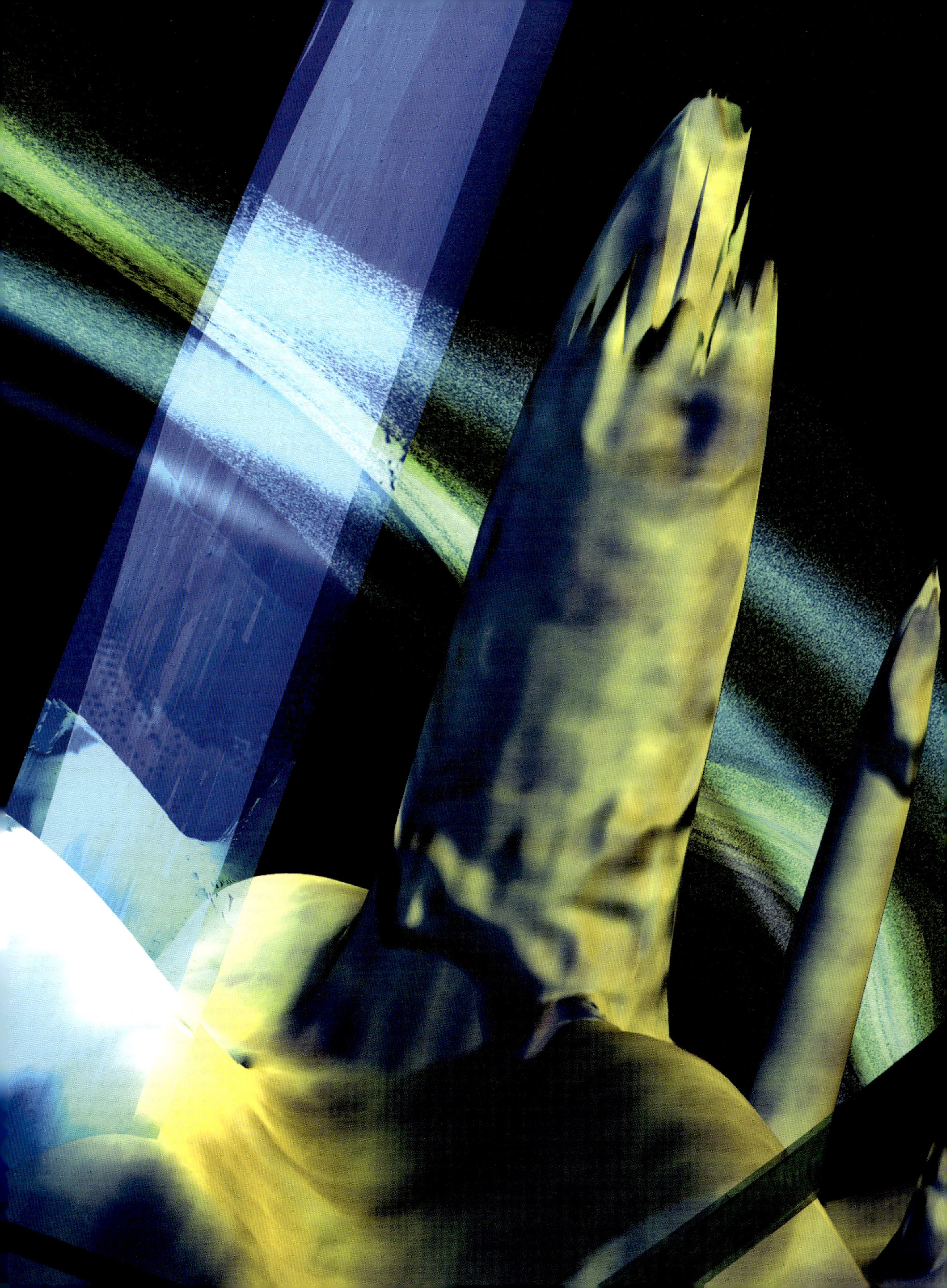

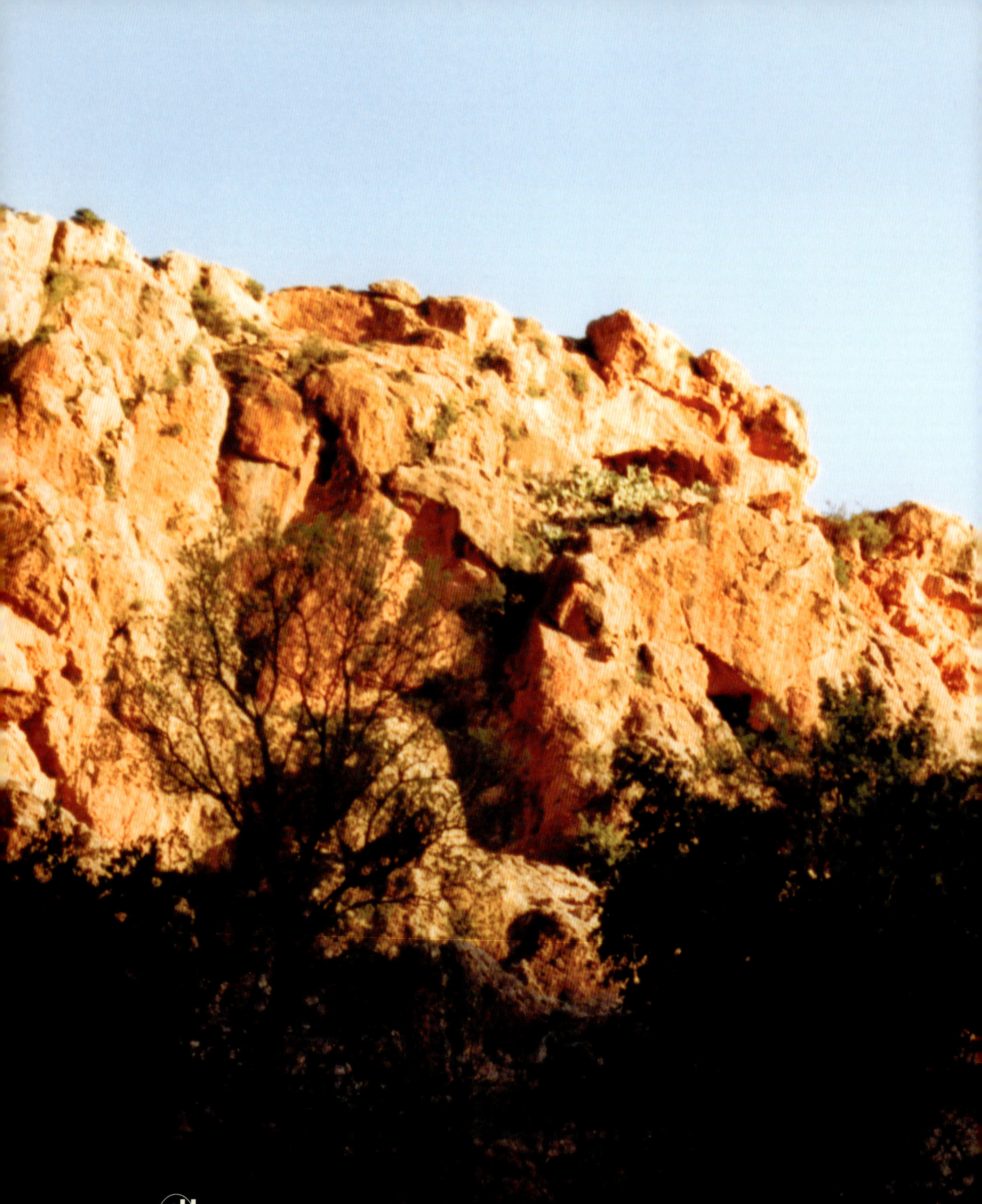

pH powerHouse Books, New York, NY

# PARTICLES + WAVES WITH PLAUSIBILITY

DENNIS BALK

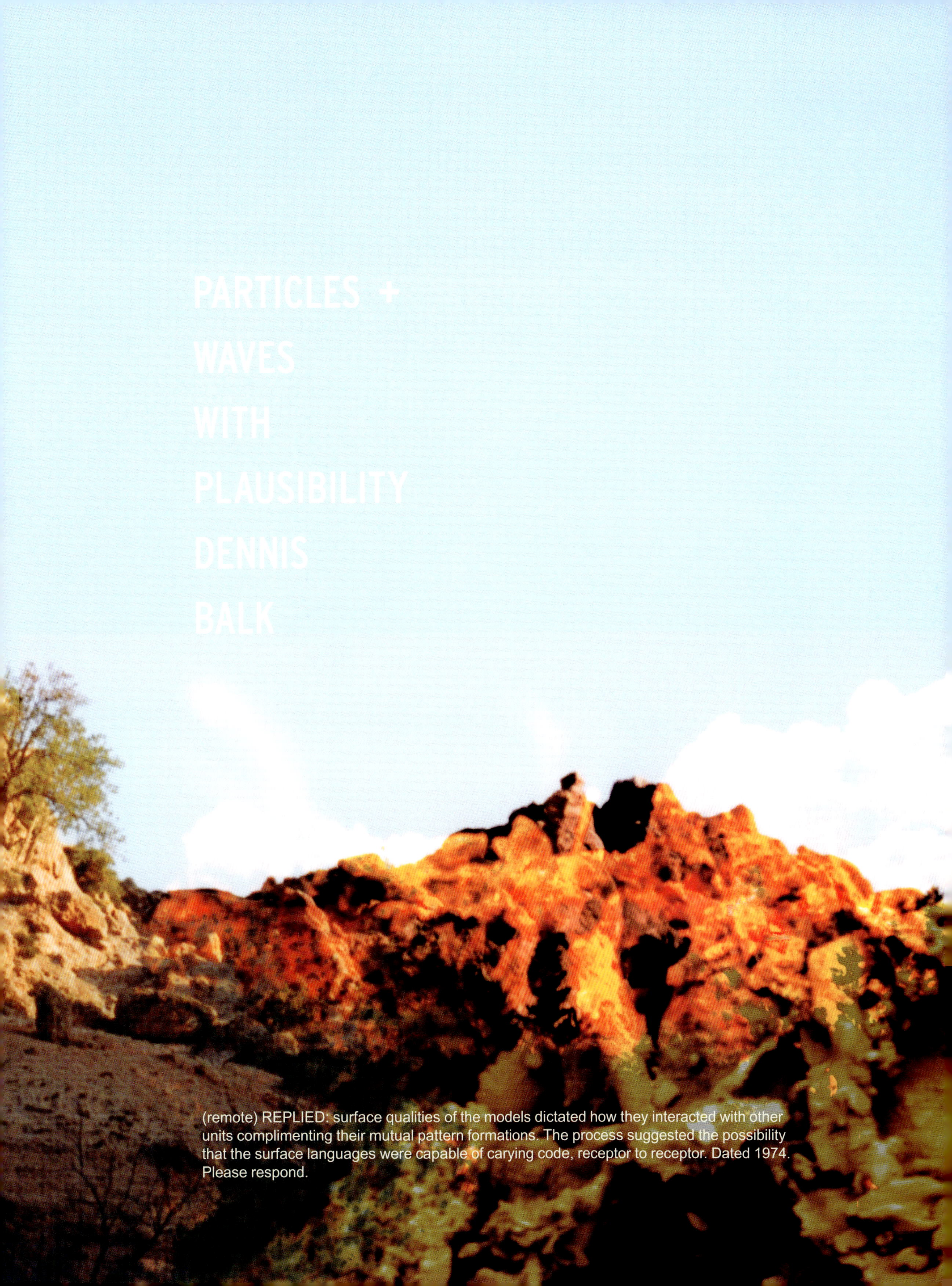

(remote) REPLIED: surface qualities of the models dictated how they interacted with other units complimenting their mutual pattern formations. The process suggested the possibility that the surface languages were capable of carying code, receptor to receptor. Dated 1974. Please respond.

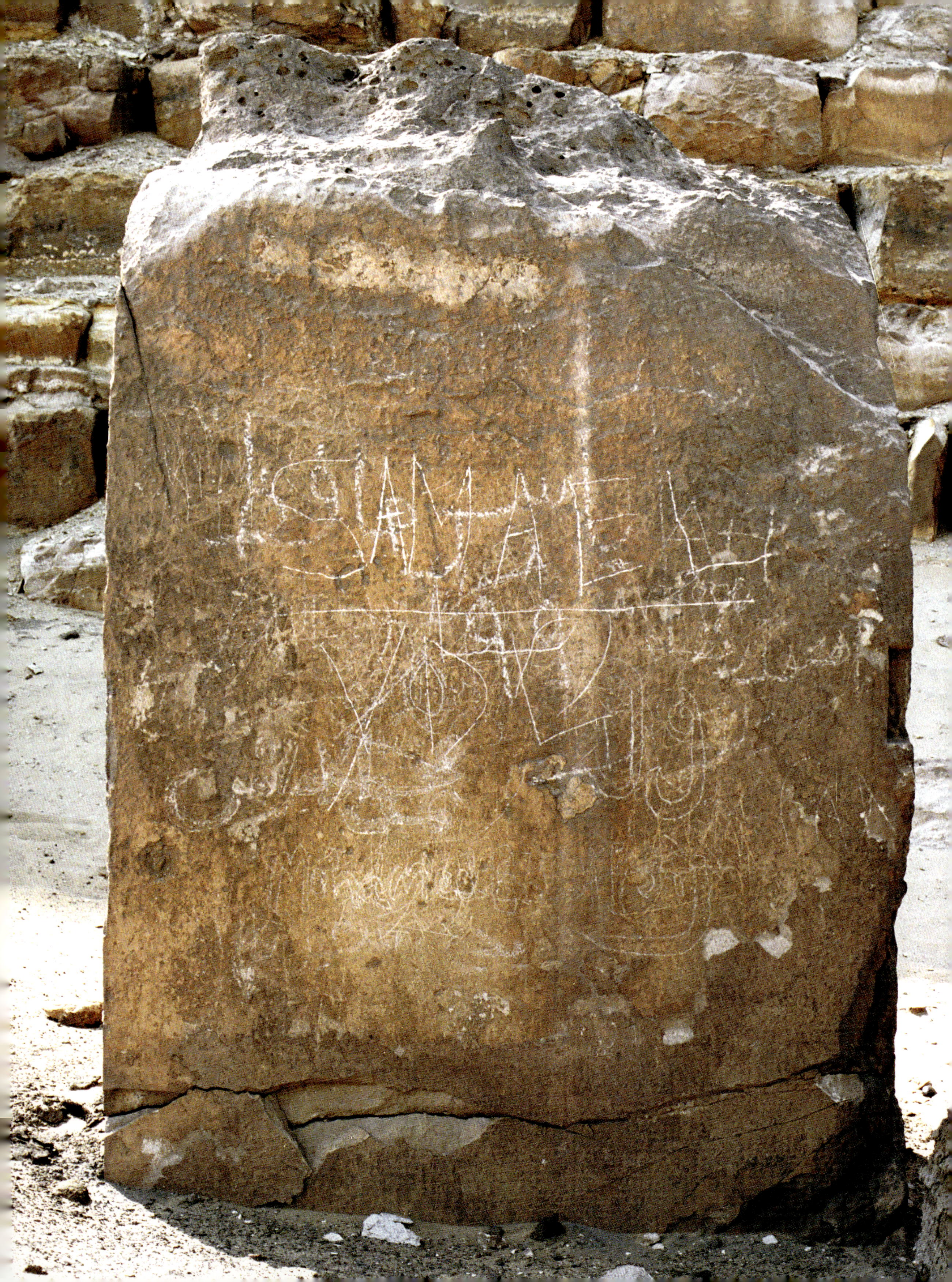

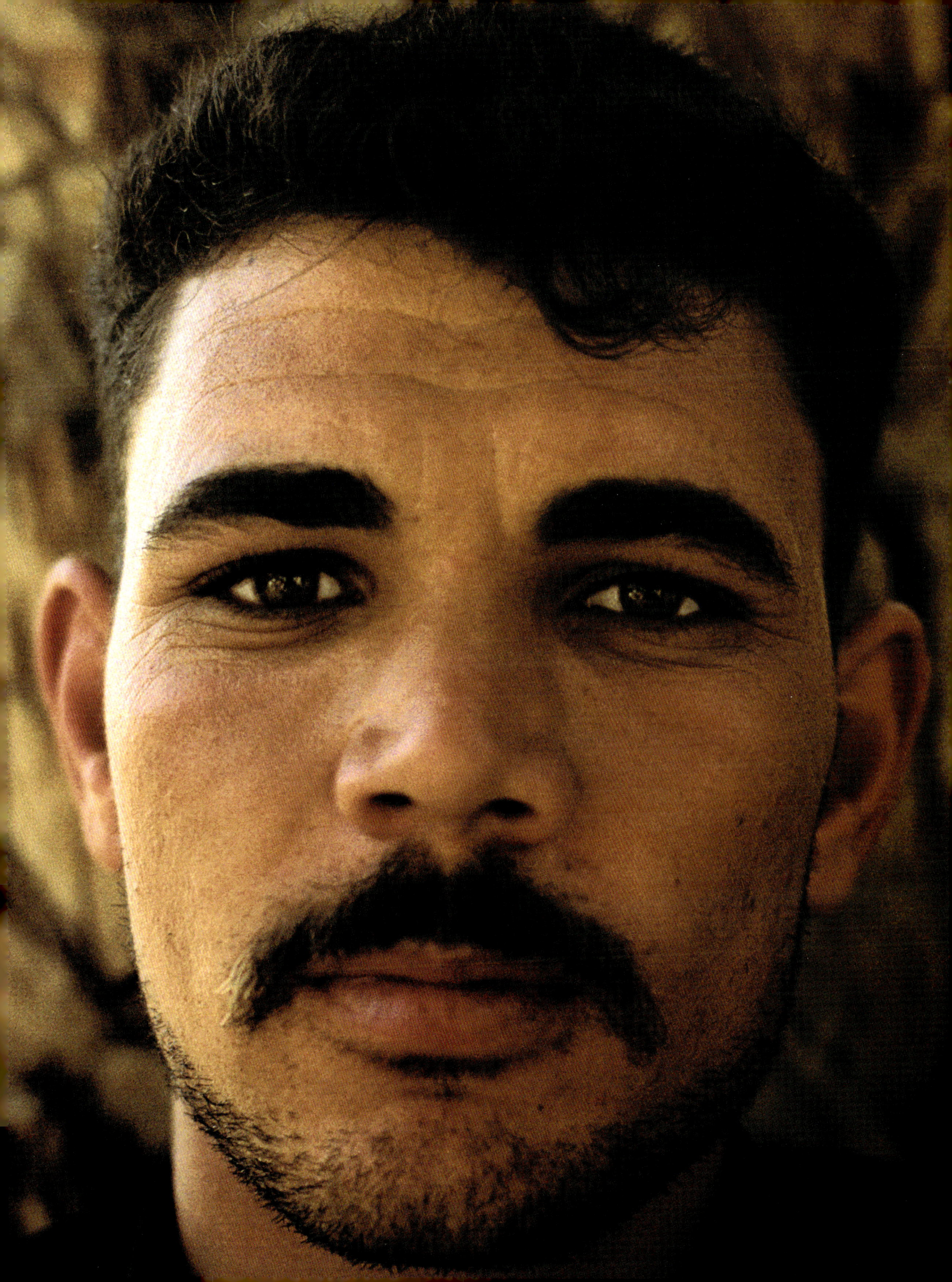

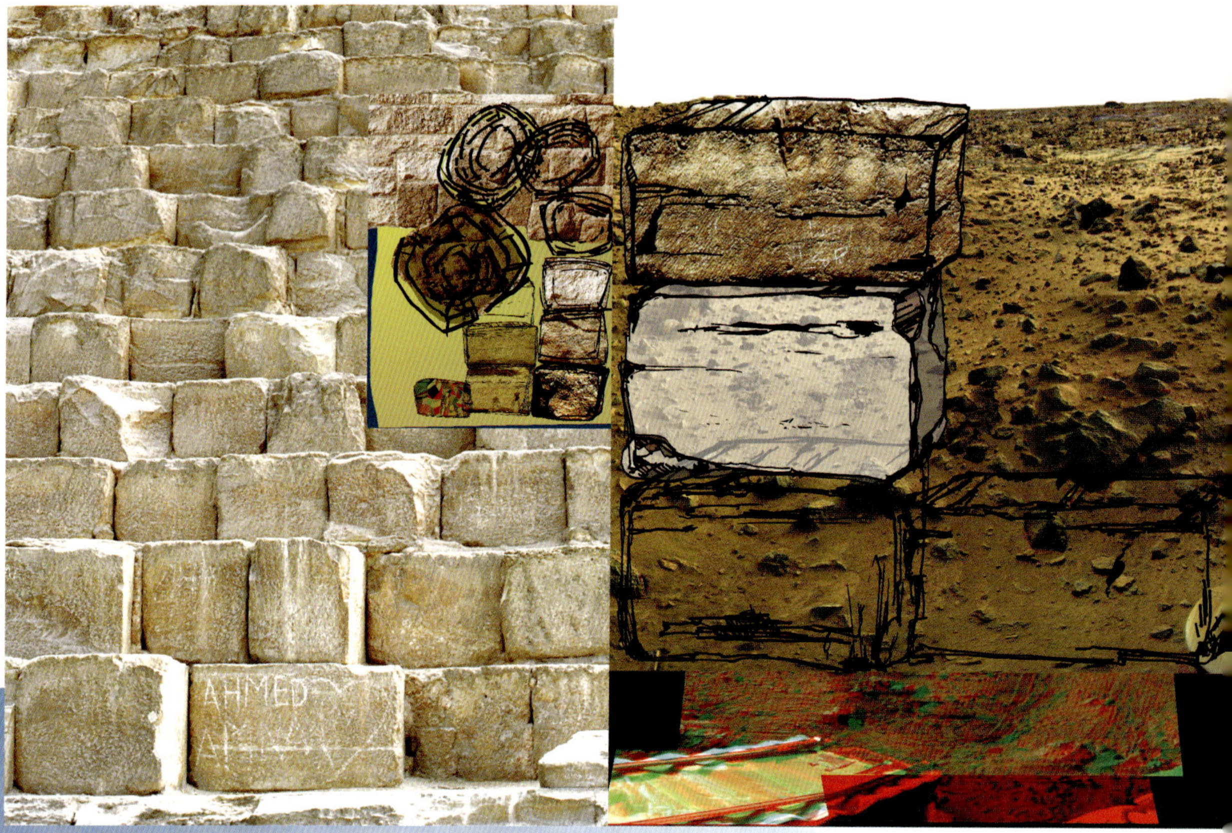
AHMED

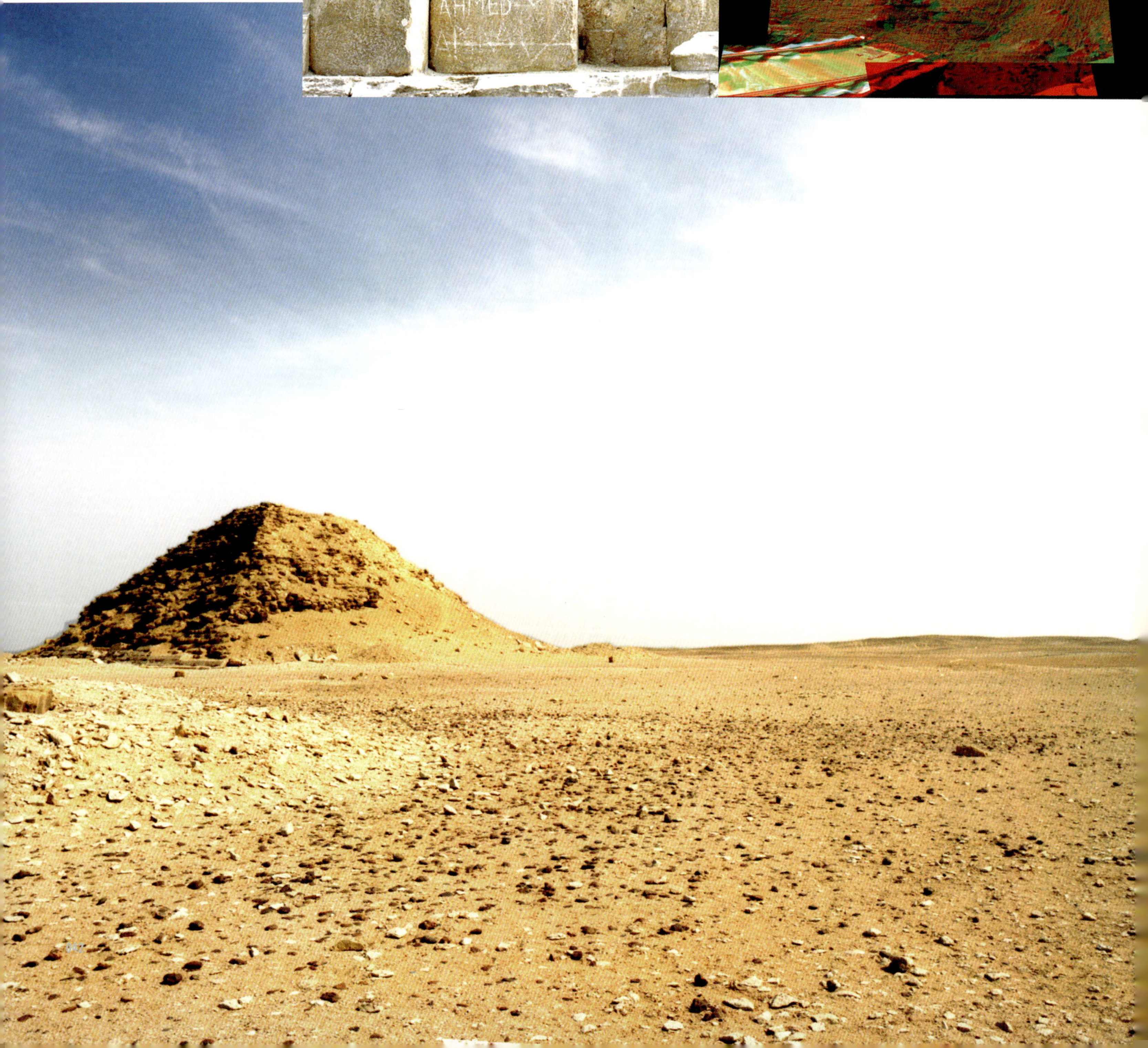

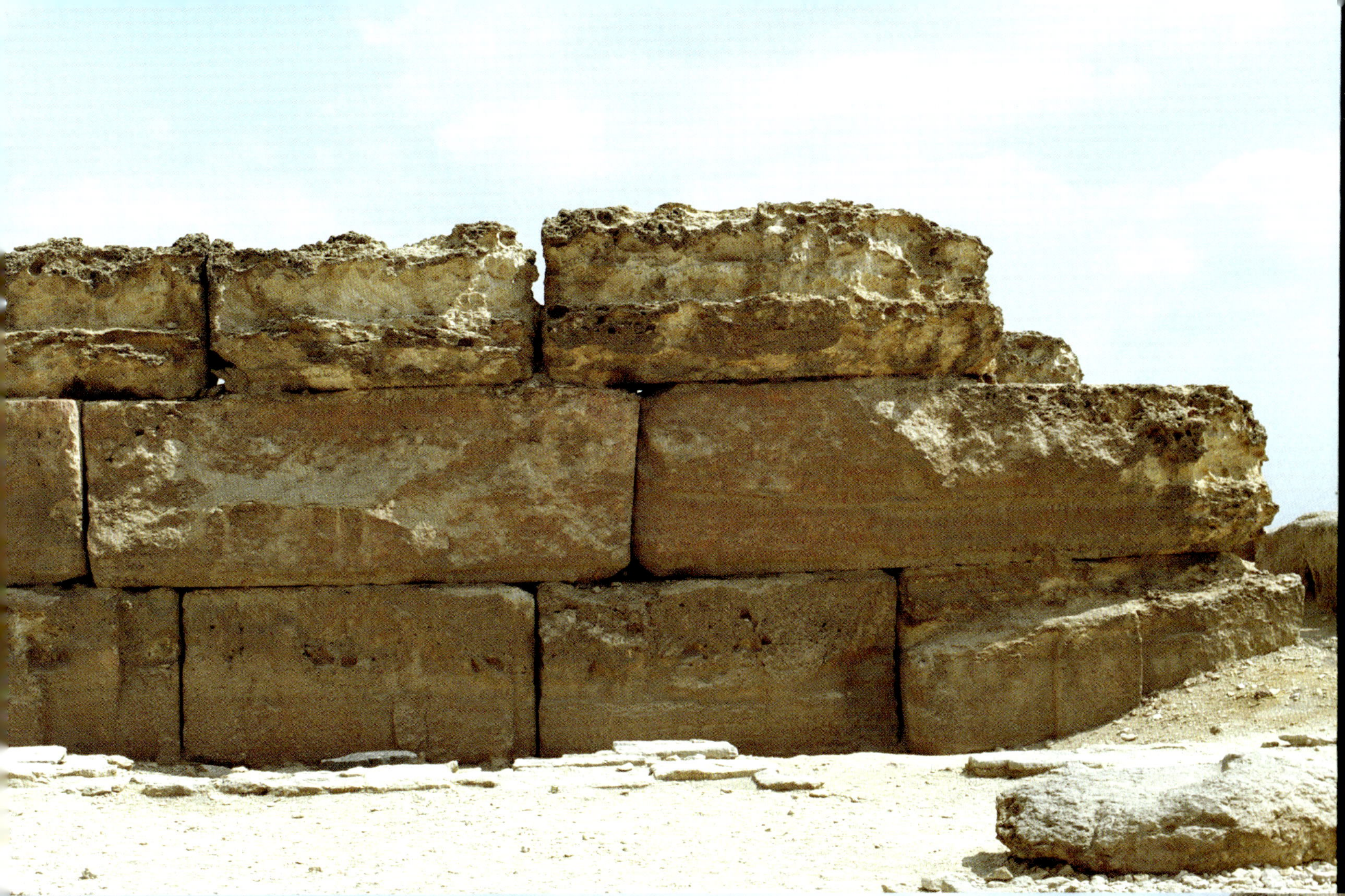

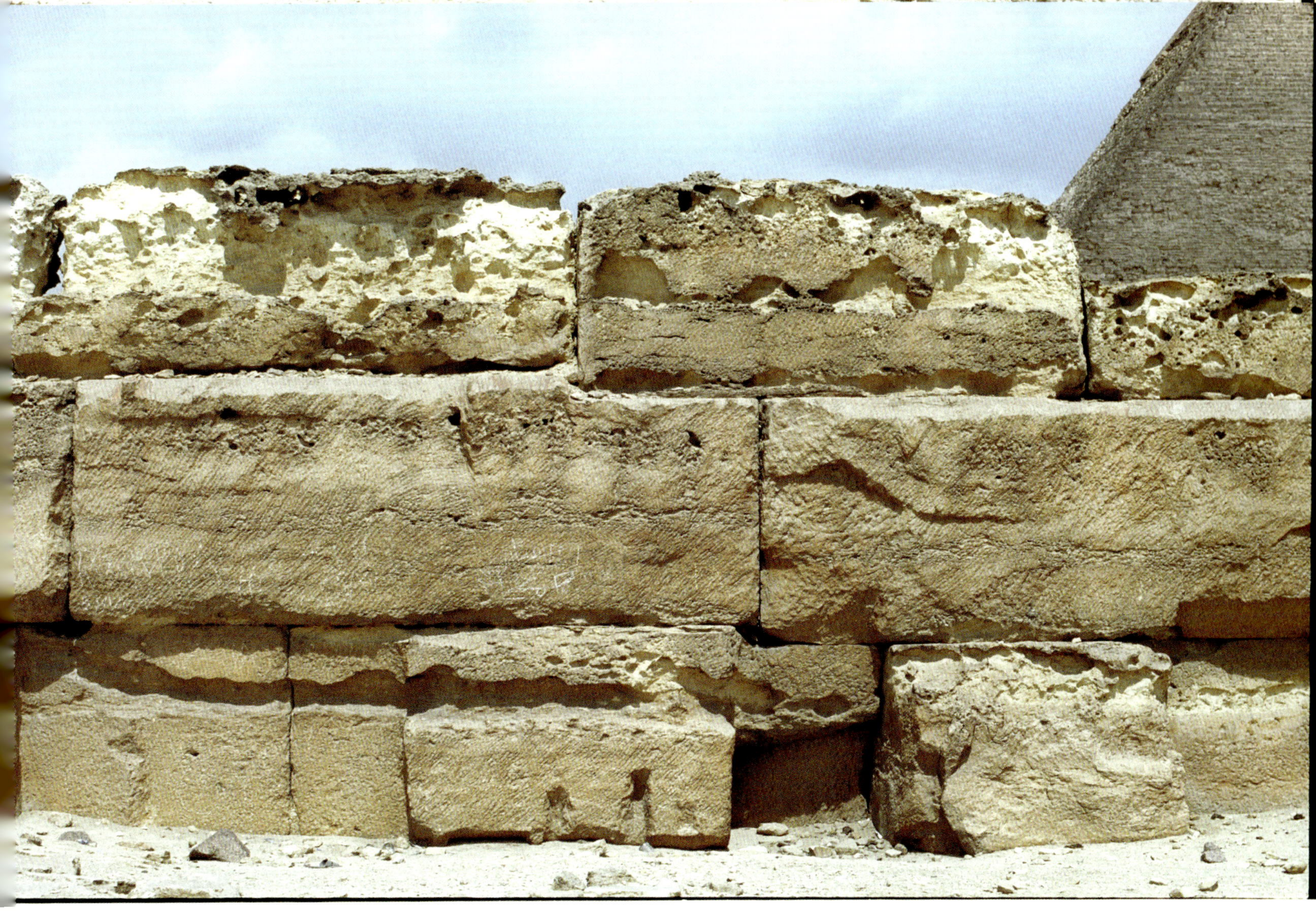

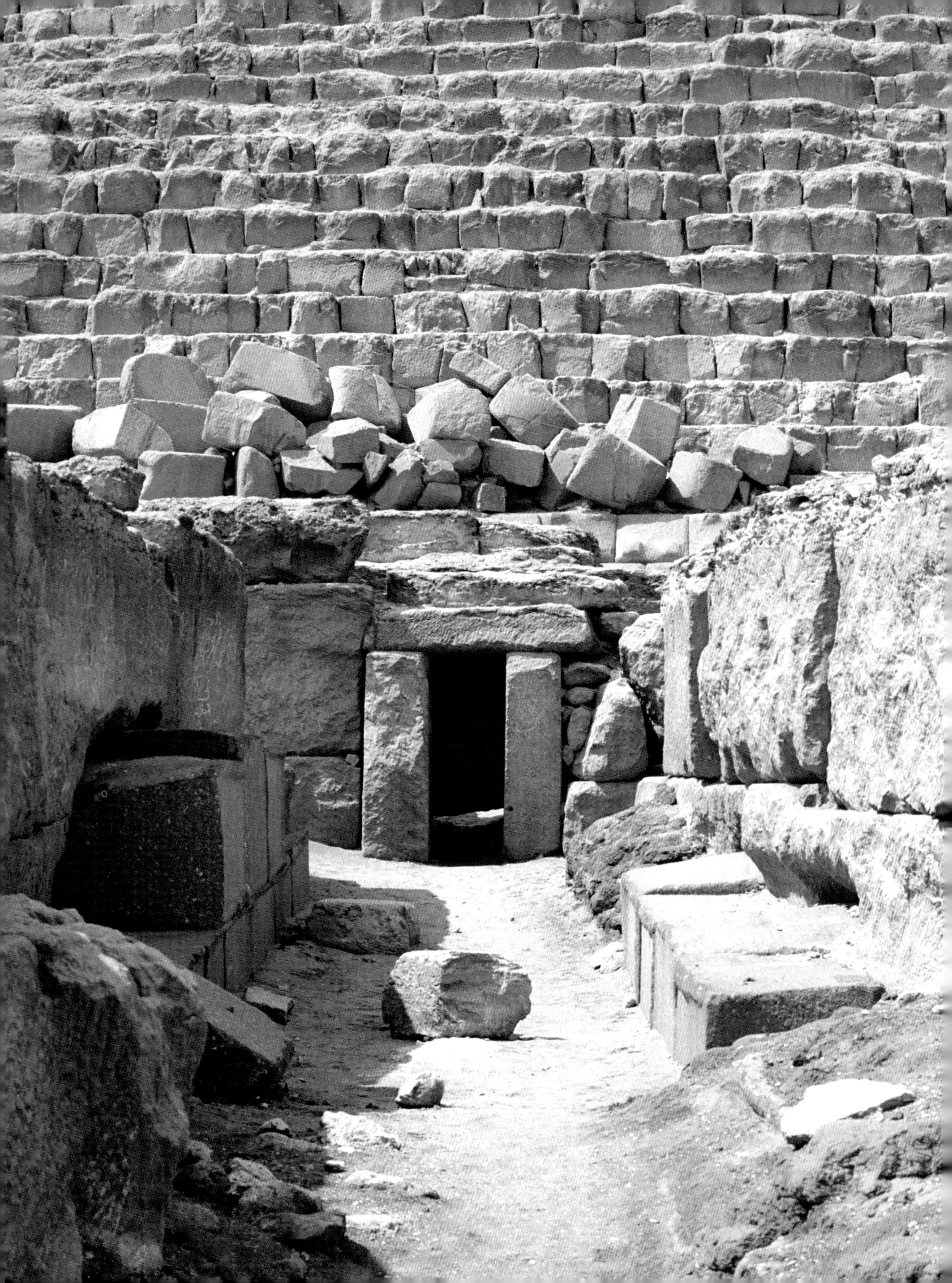

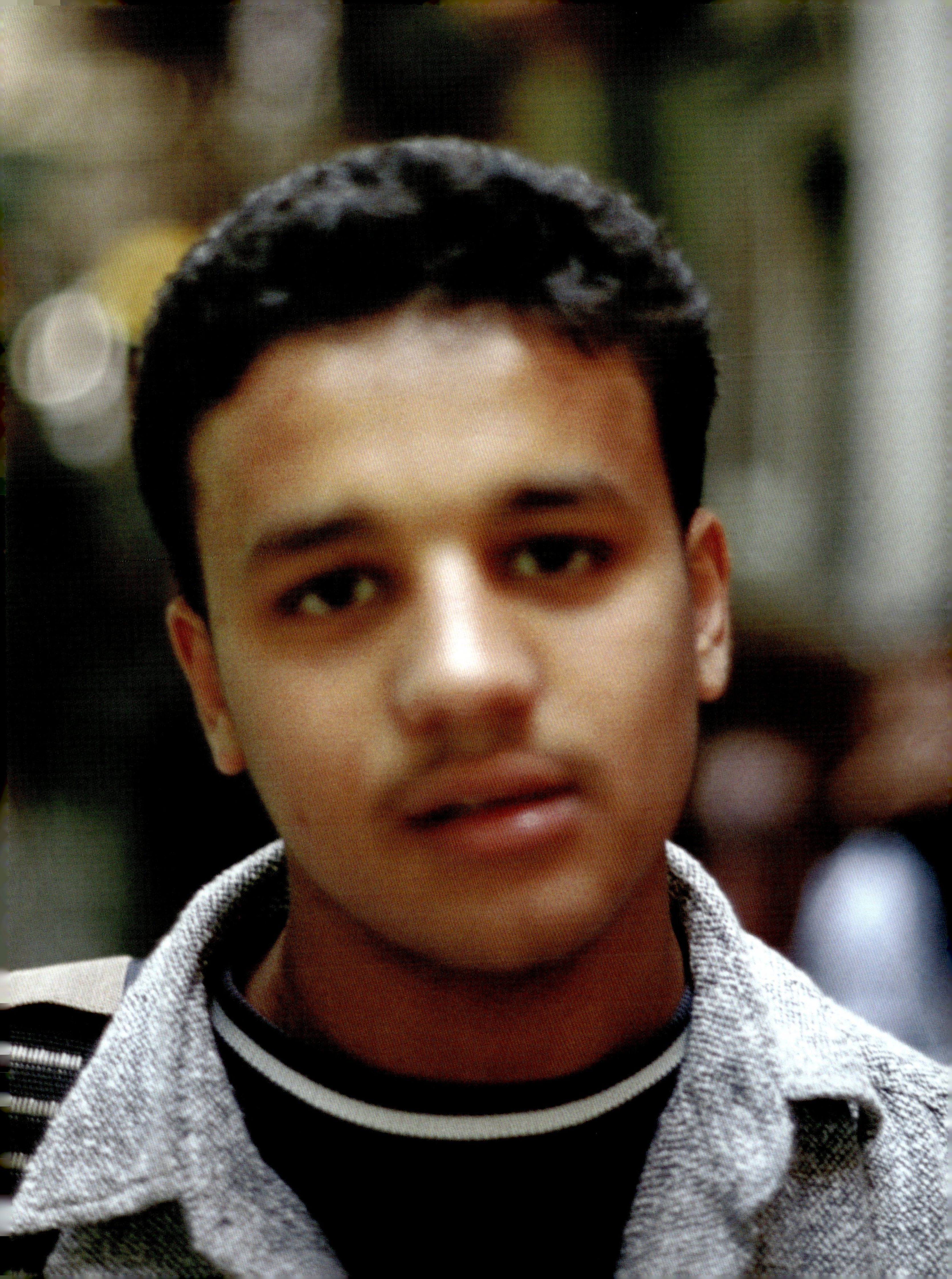

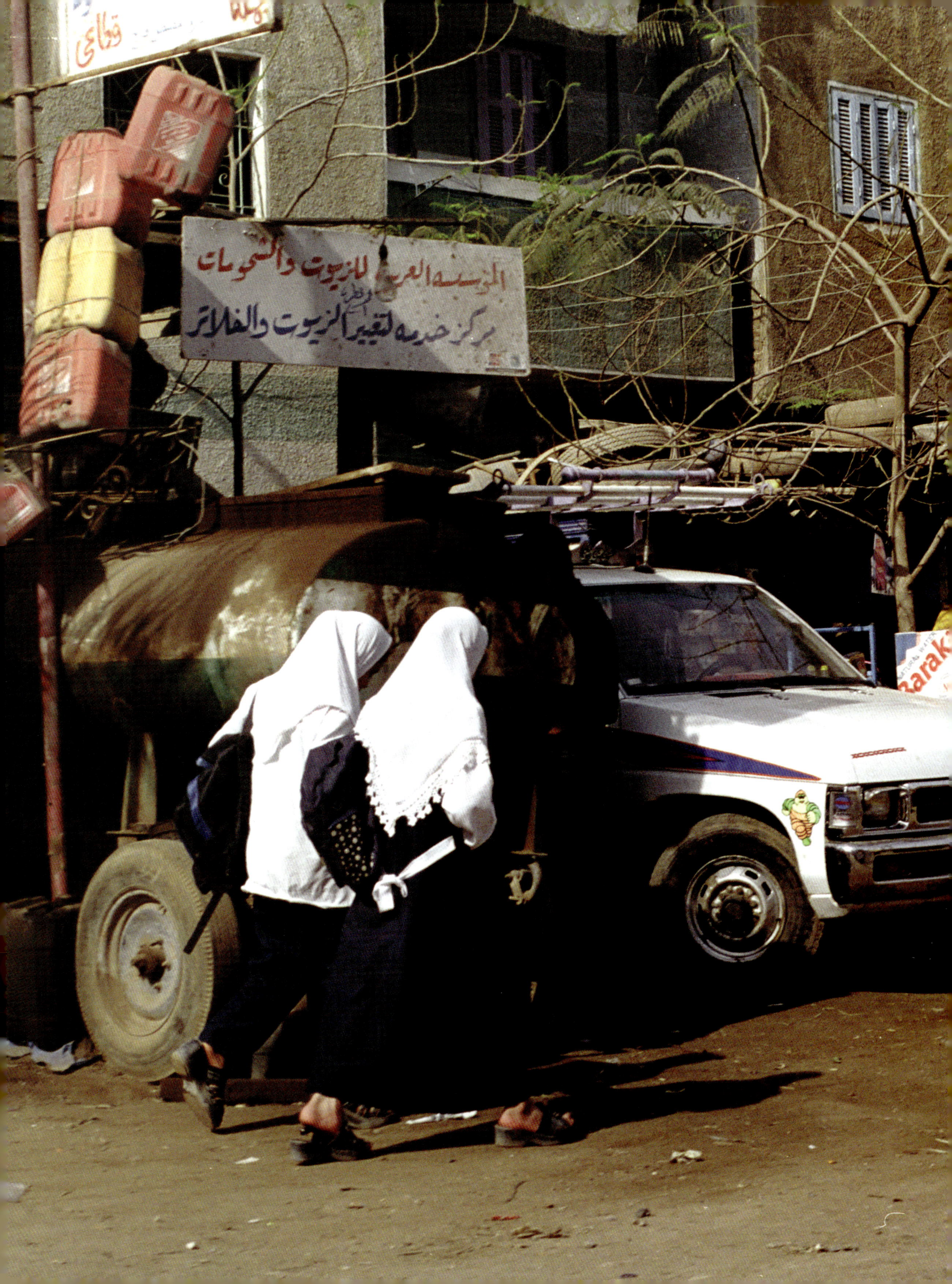
مركز خدمه لتغيير الزيوت والفلاتر

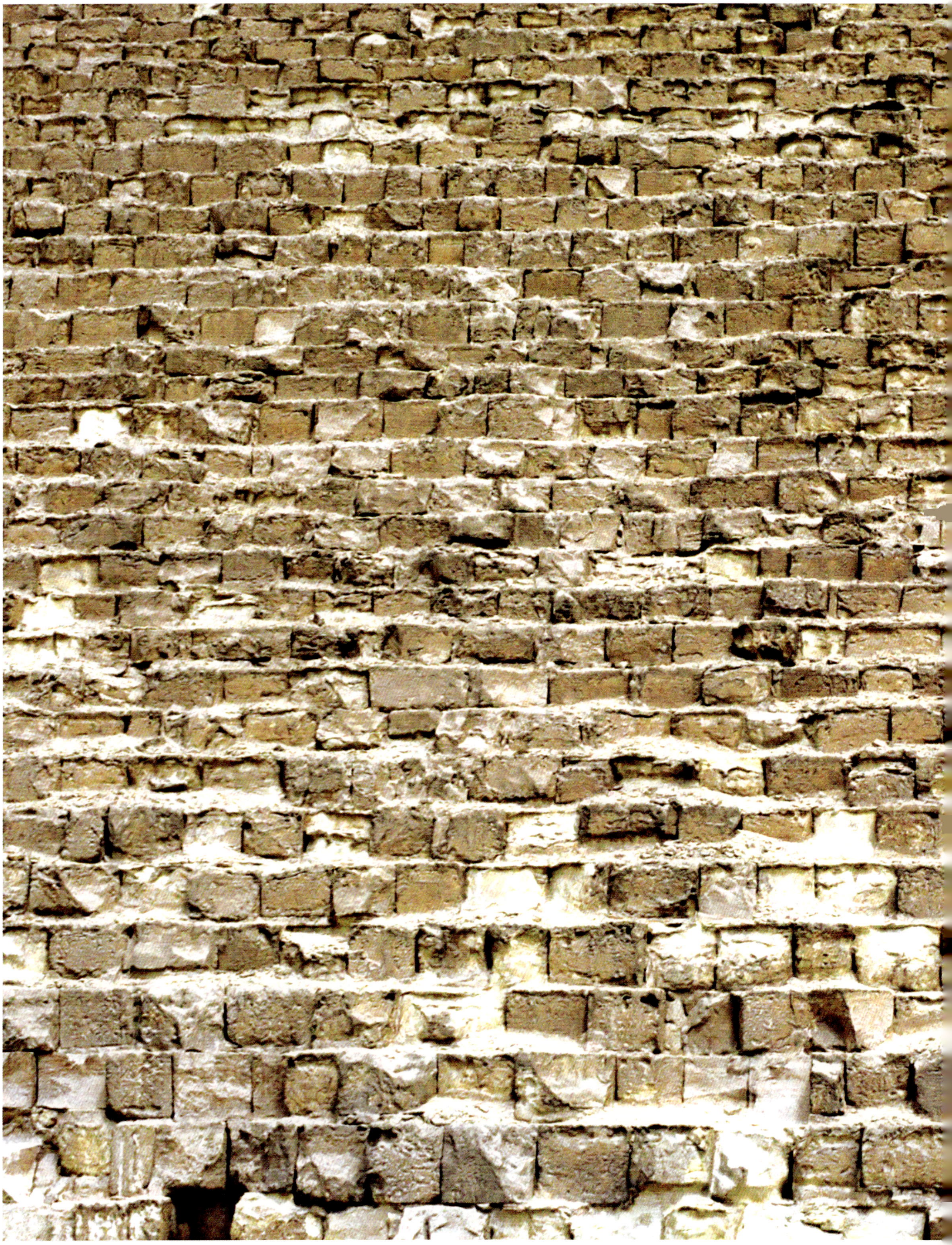

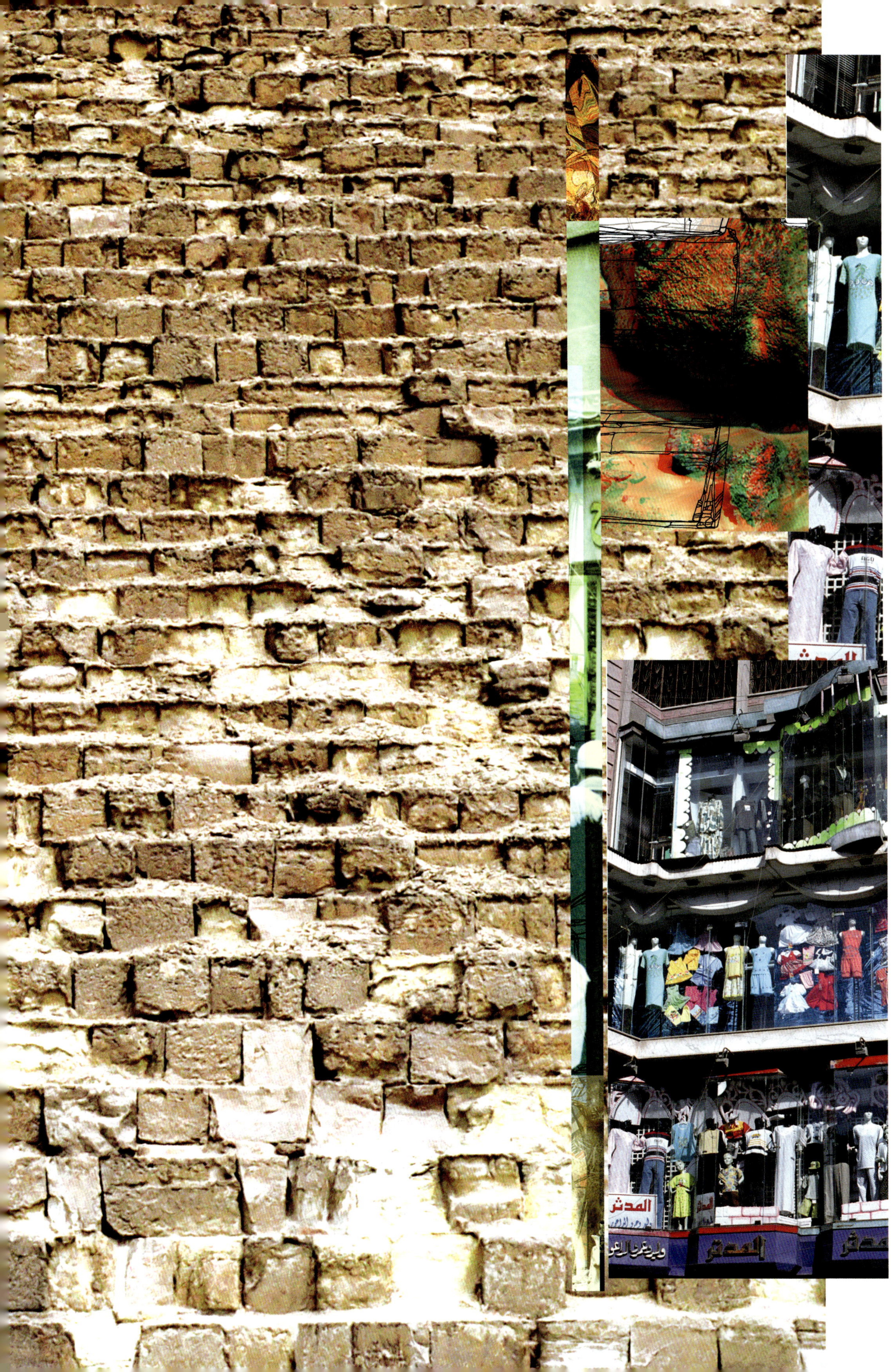
المدثر

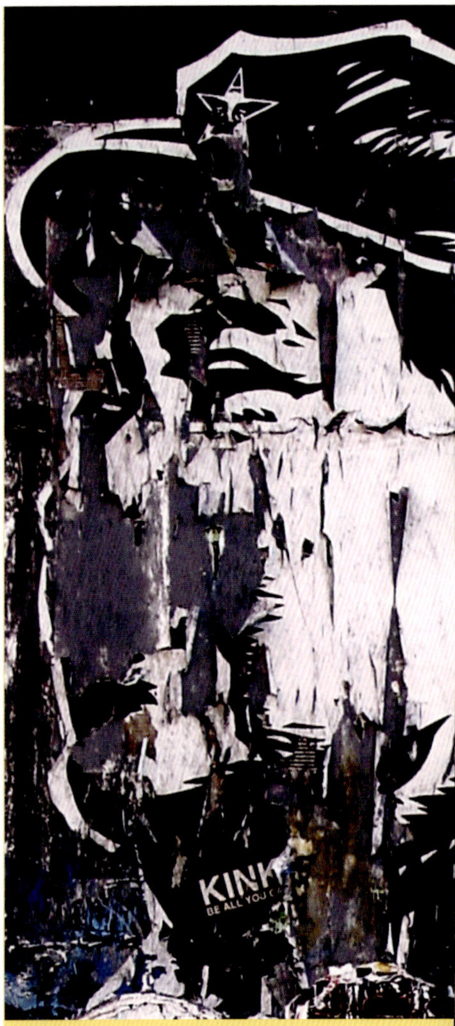

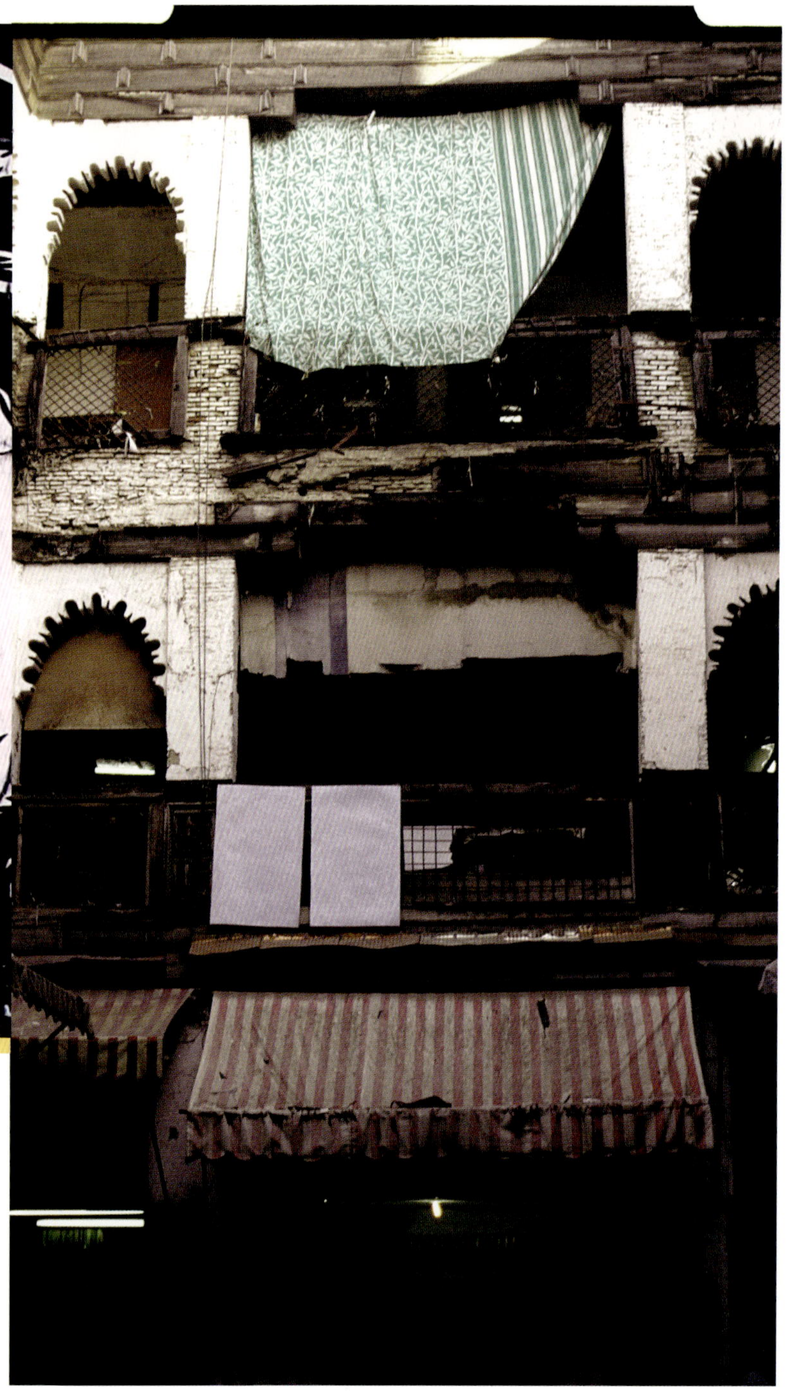

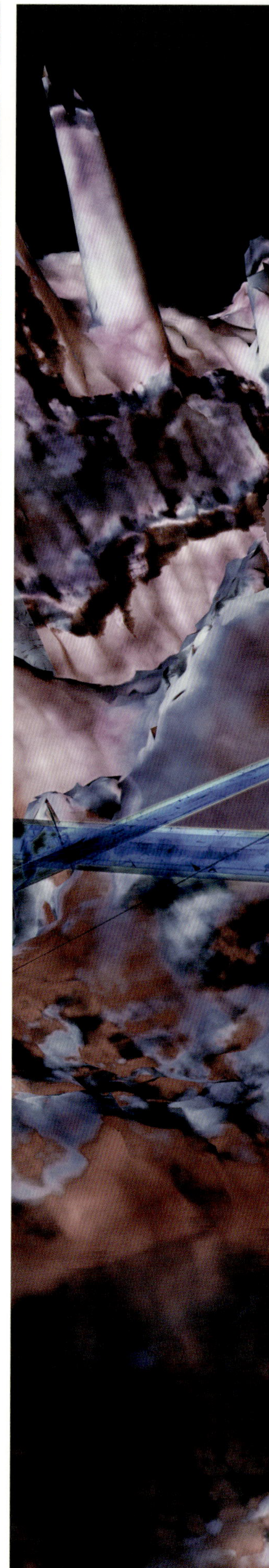

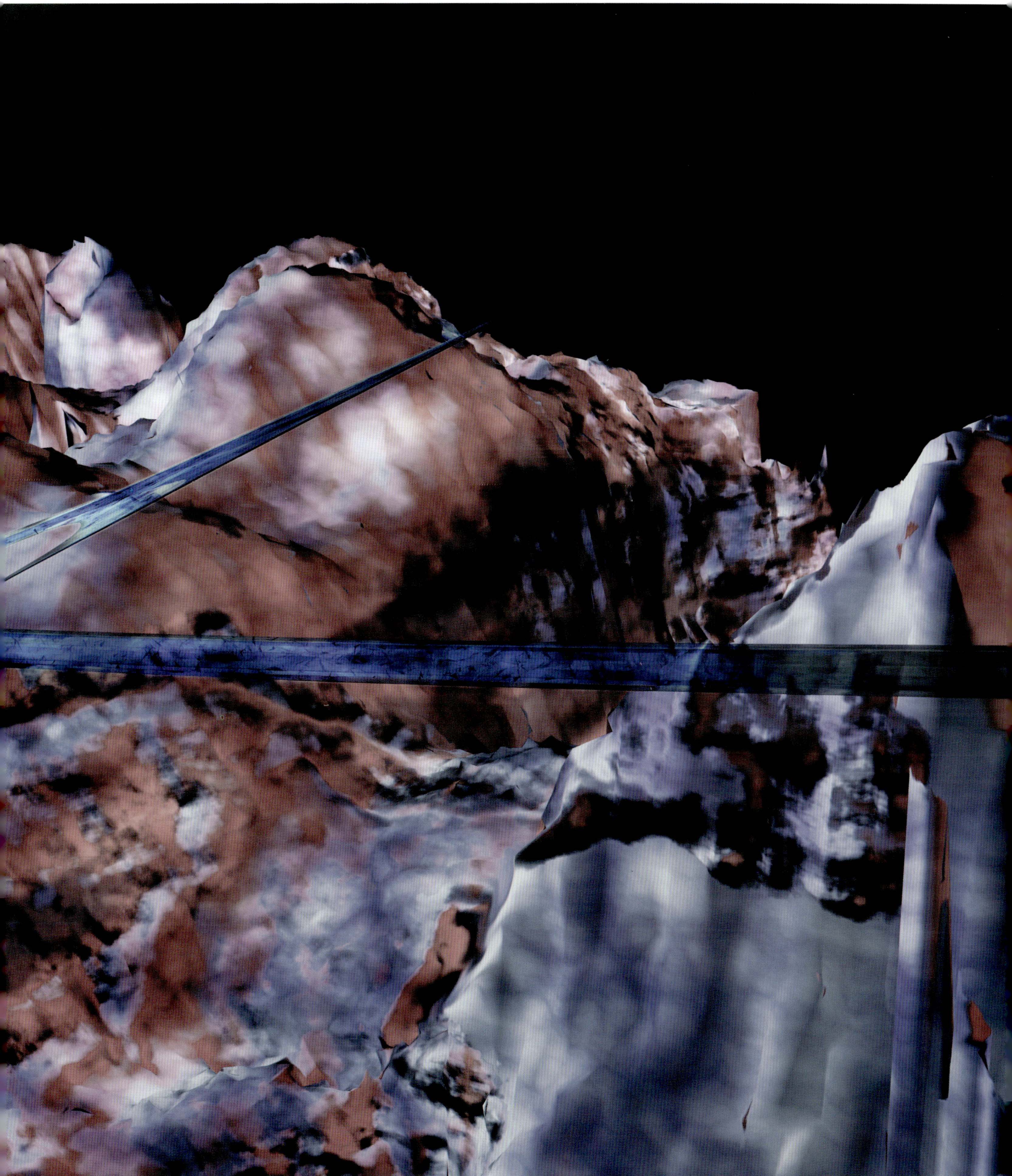

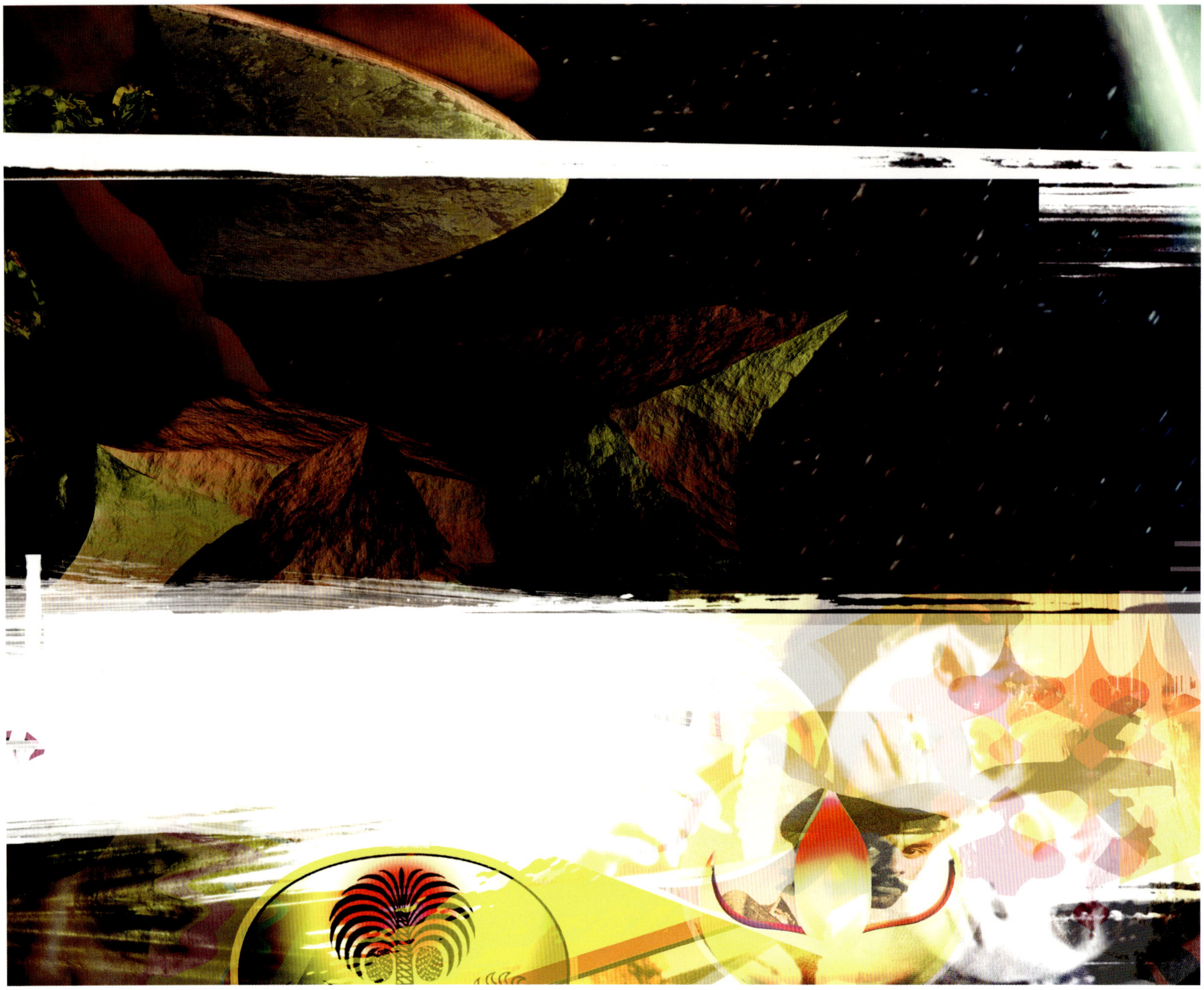

**1ST OCTAVE:**
**Picture: People. Rochester, New York, in a stationary shop.**
**A tall woman buying a Halloween costume for her daughter was telling me that she really liked the movie too. That she found out things later that made her believe there was more to the story. The concept of time travel in Donnie Darko was a legitimate depiction in as far as she could tell. According to what she had found in her crackpot research.**

**In a weird way, I think people are trying to lead me on. It's not conscious, they're not aware they're passing me information. Revealing their processes. Terribly patched together. It would mean a tremendous amount of effort and at best several teams of conceptualists...if successful there might be a coherent theoretical expression—**

**BTW: Your signal's improved.**

Switzerland is giving advance clearances to the standard model team. They will be concepting the Higgs mechanism, their plan of render for the Higgs Field is yellow lighted. **What do you want us to do??.** Their particles collect mass.

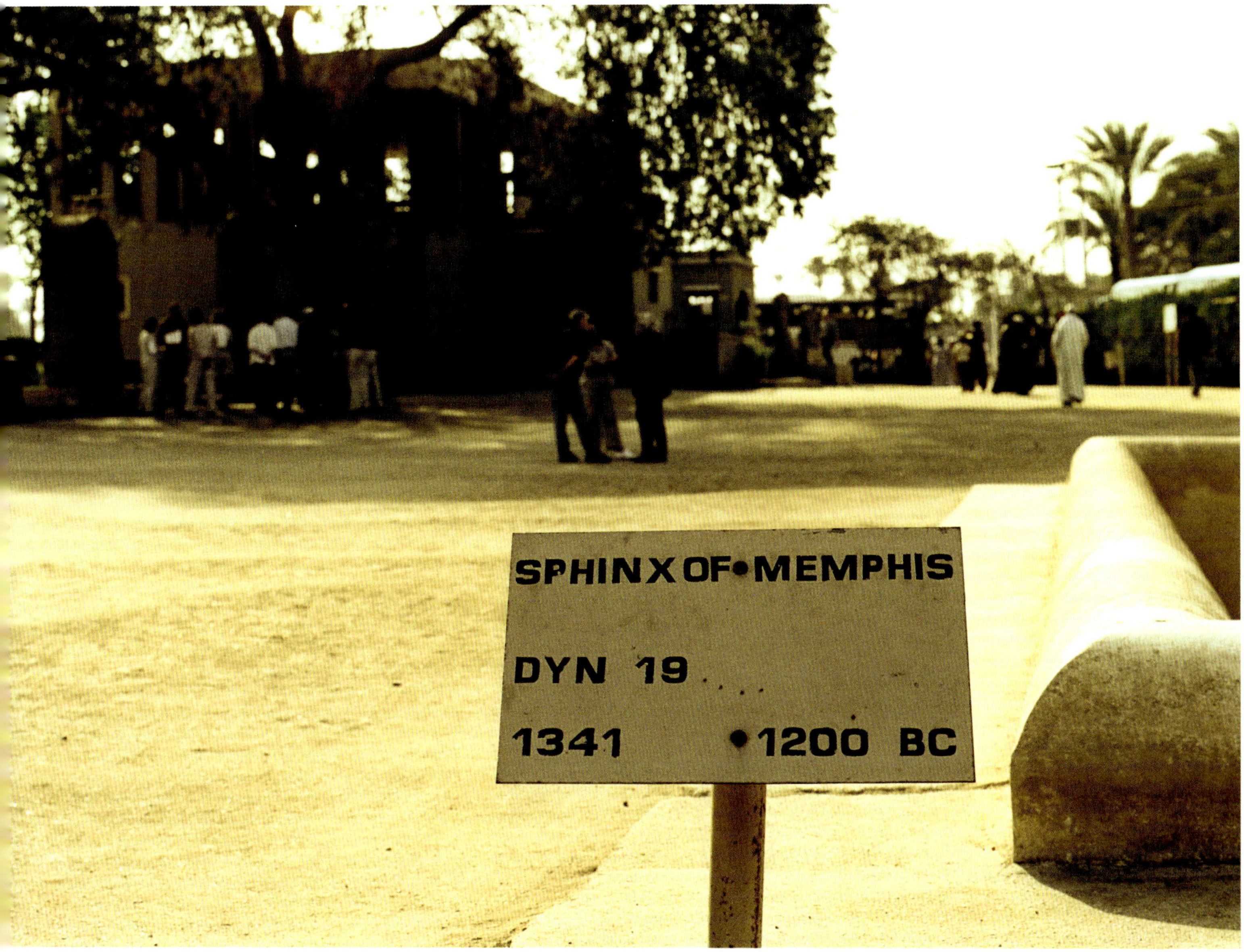
SPHINX OF • MEMPHIS
DYN 19
1341 • 1200 BC

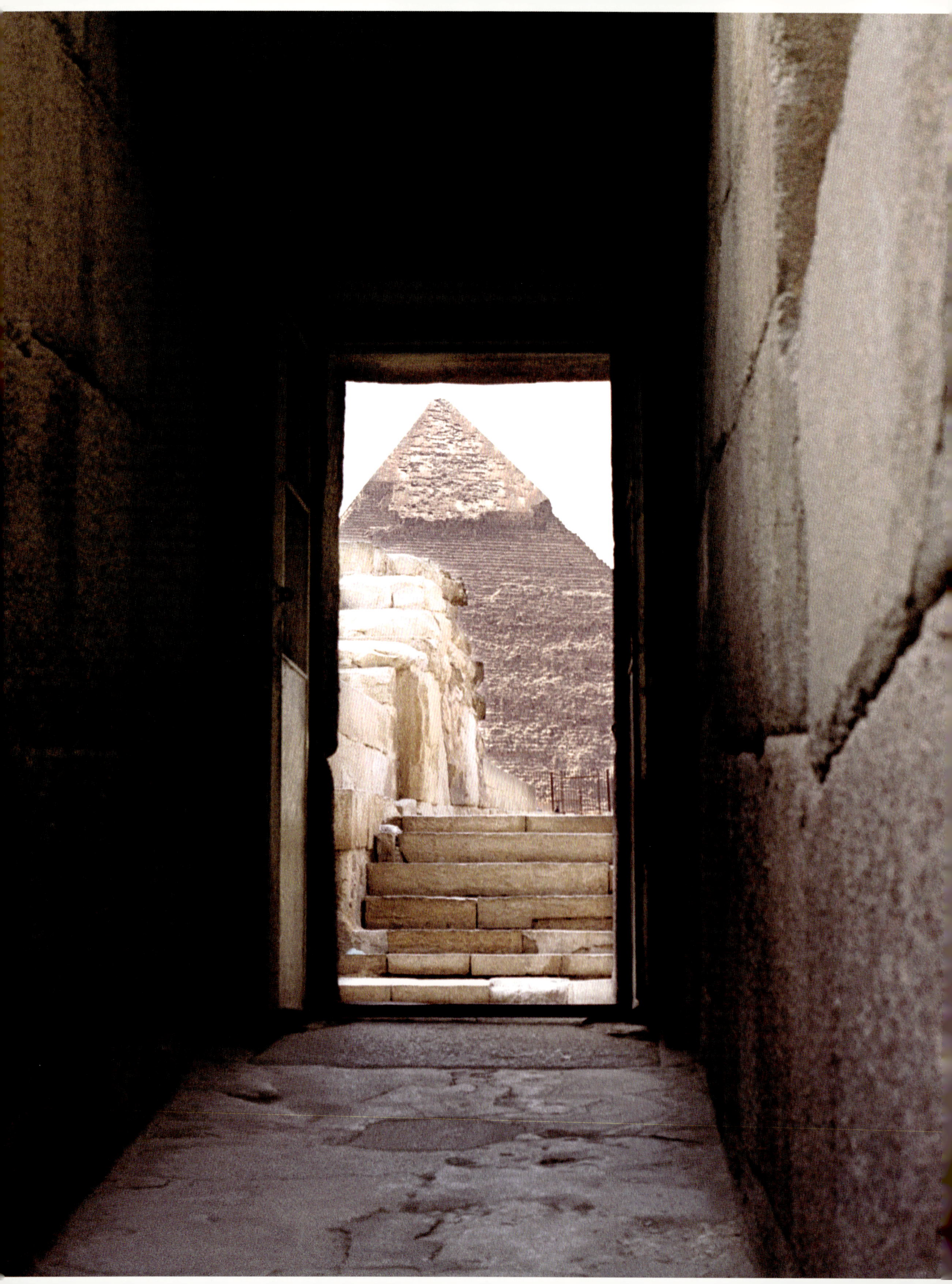

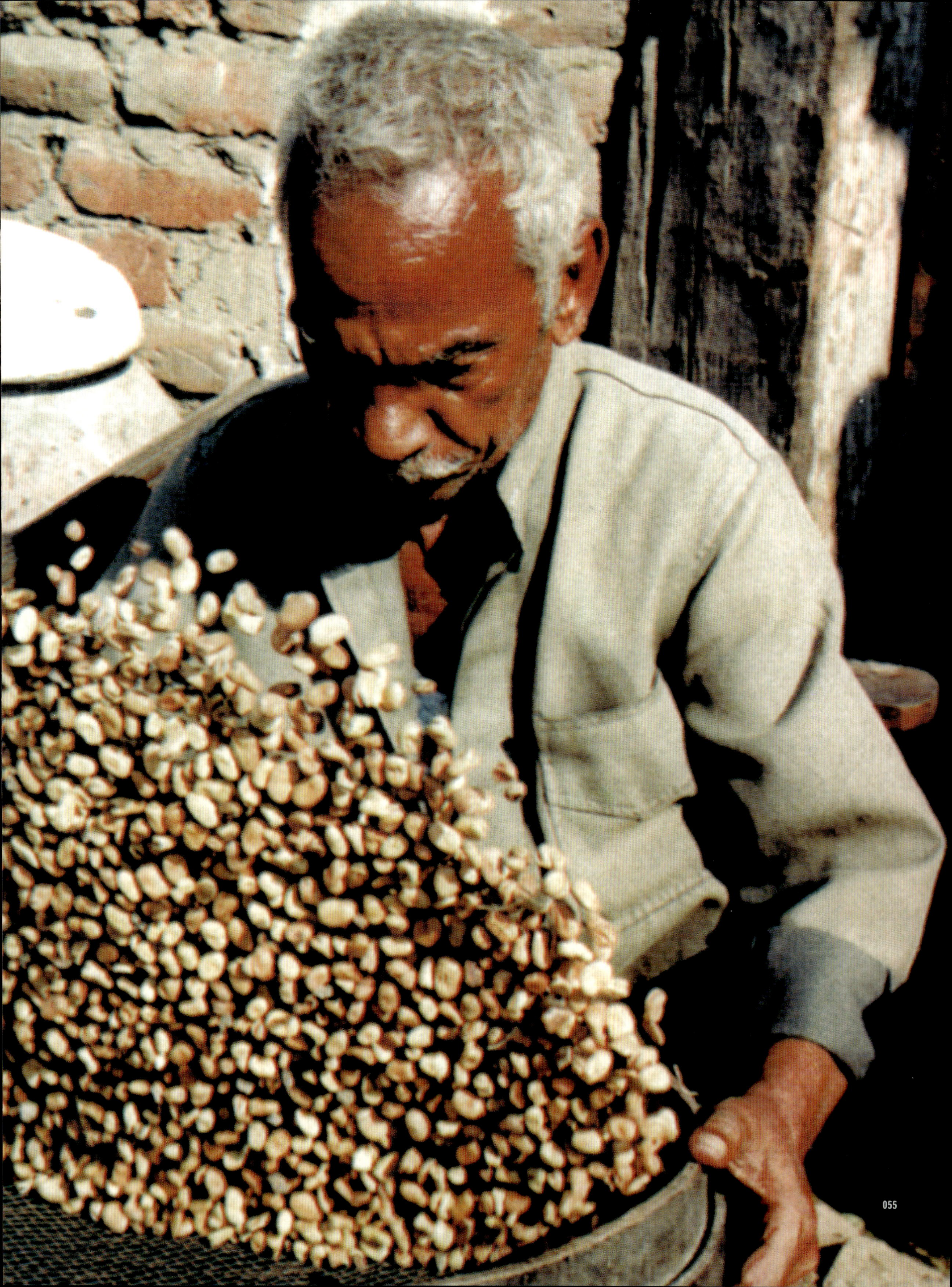

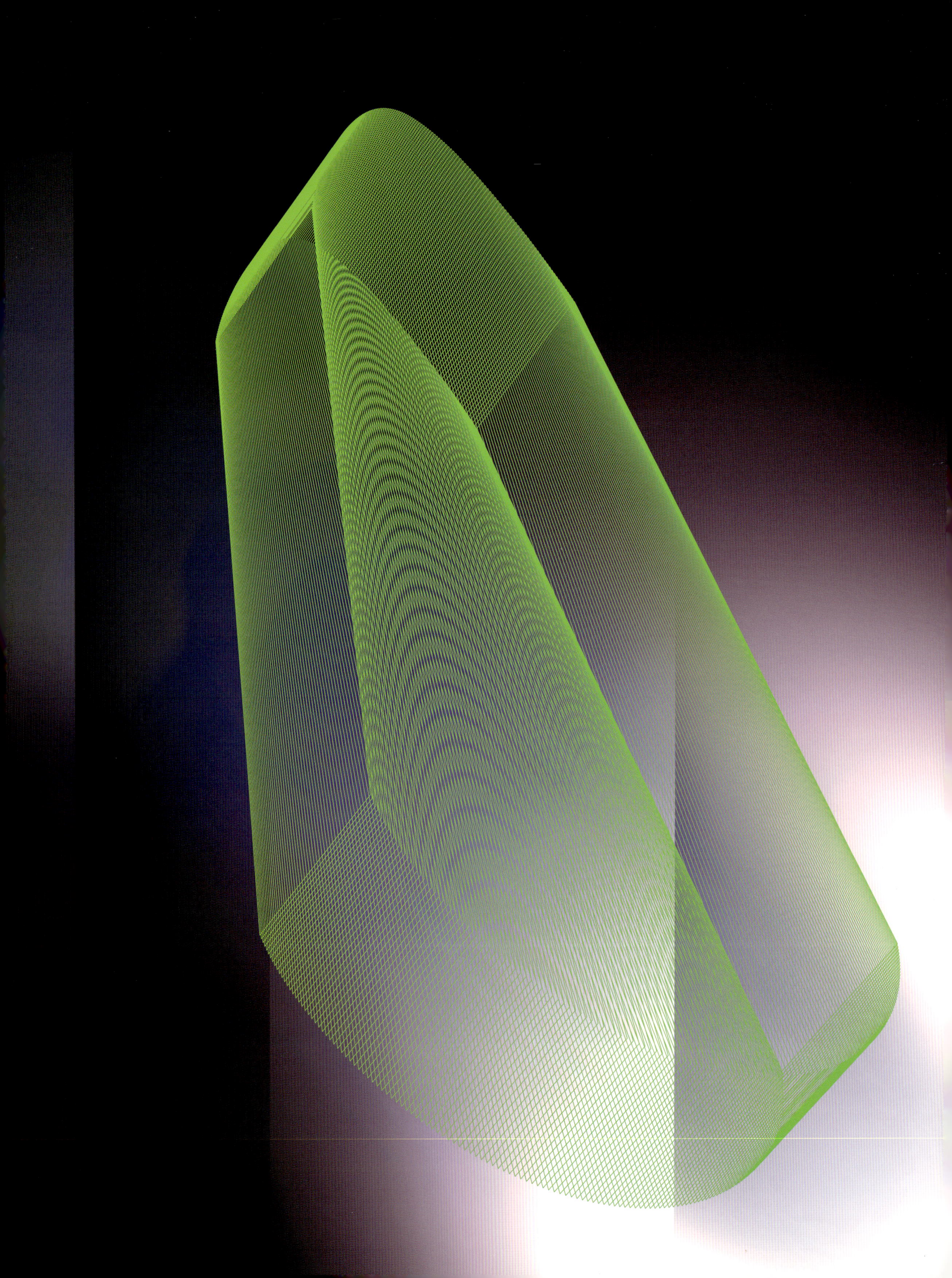

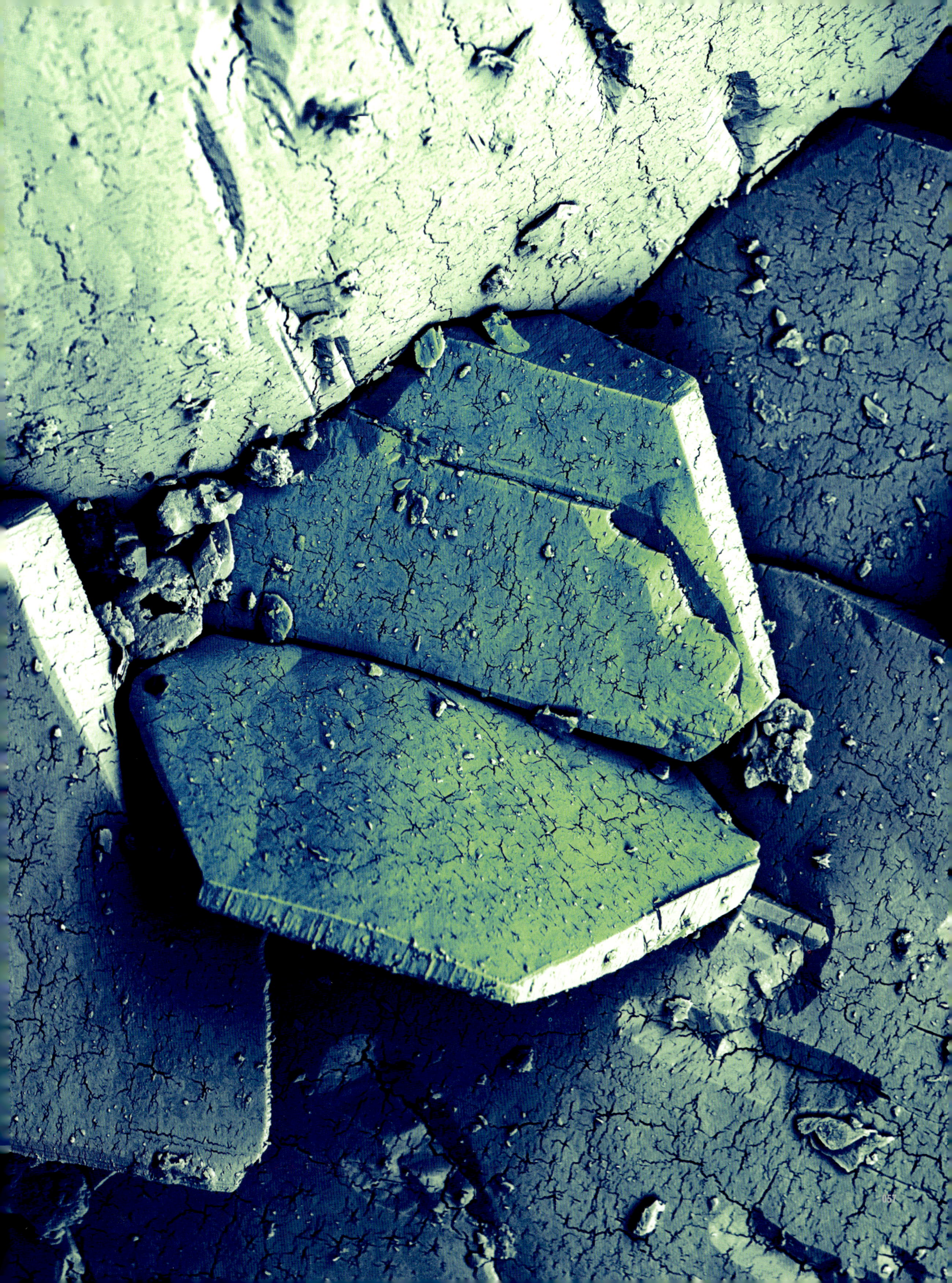

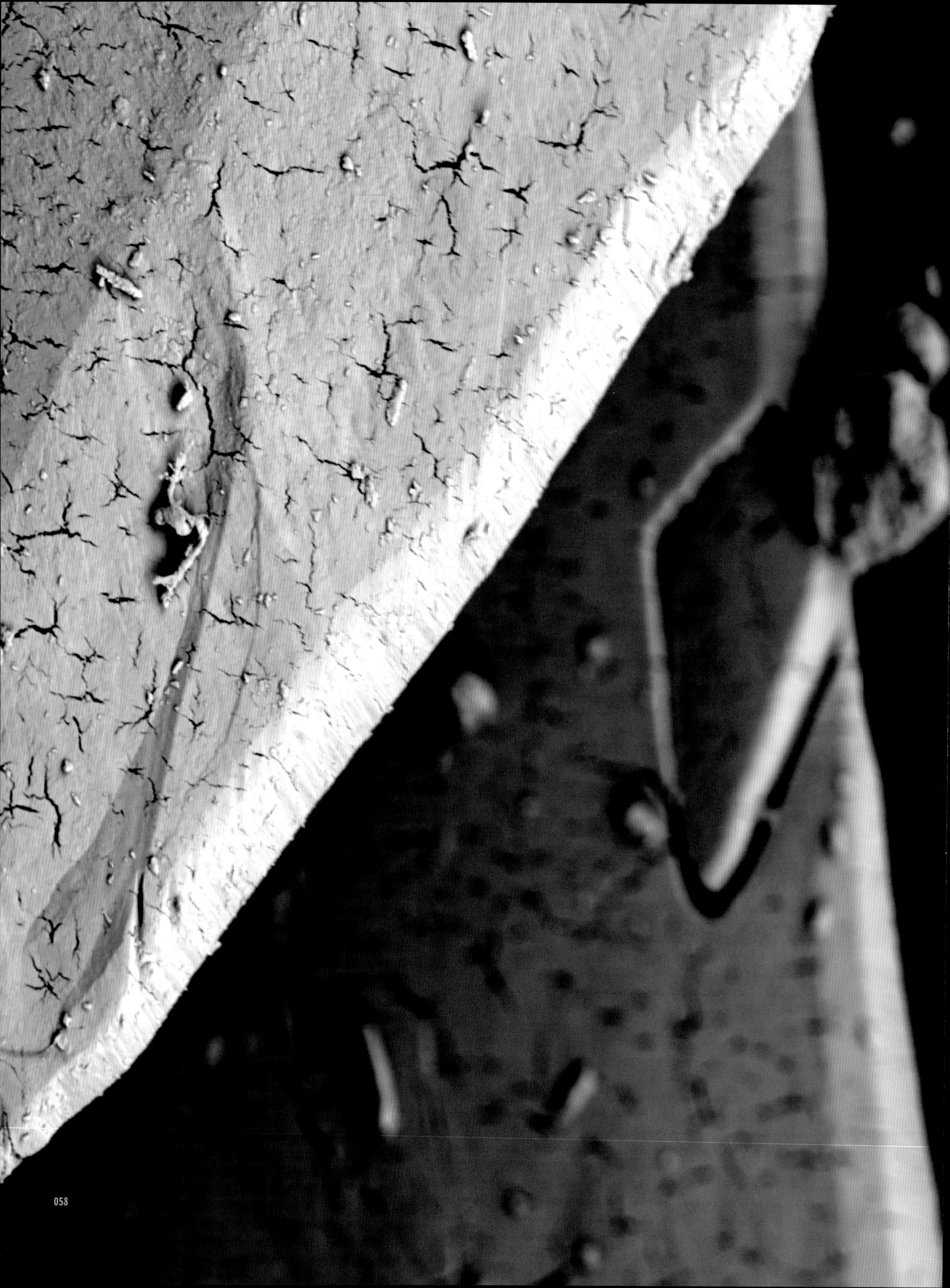

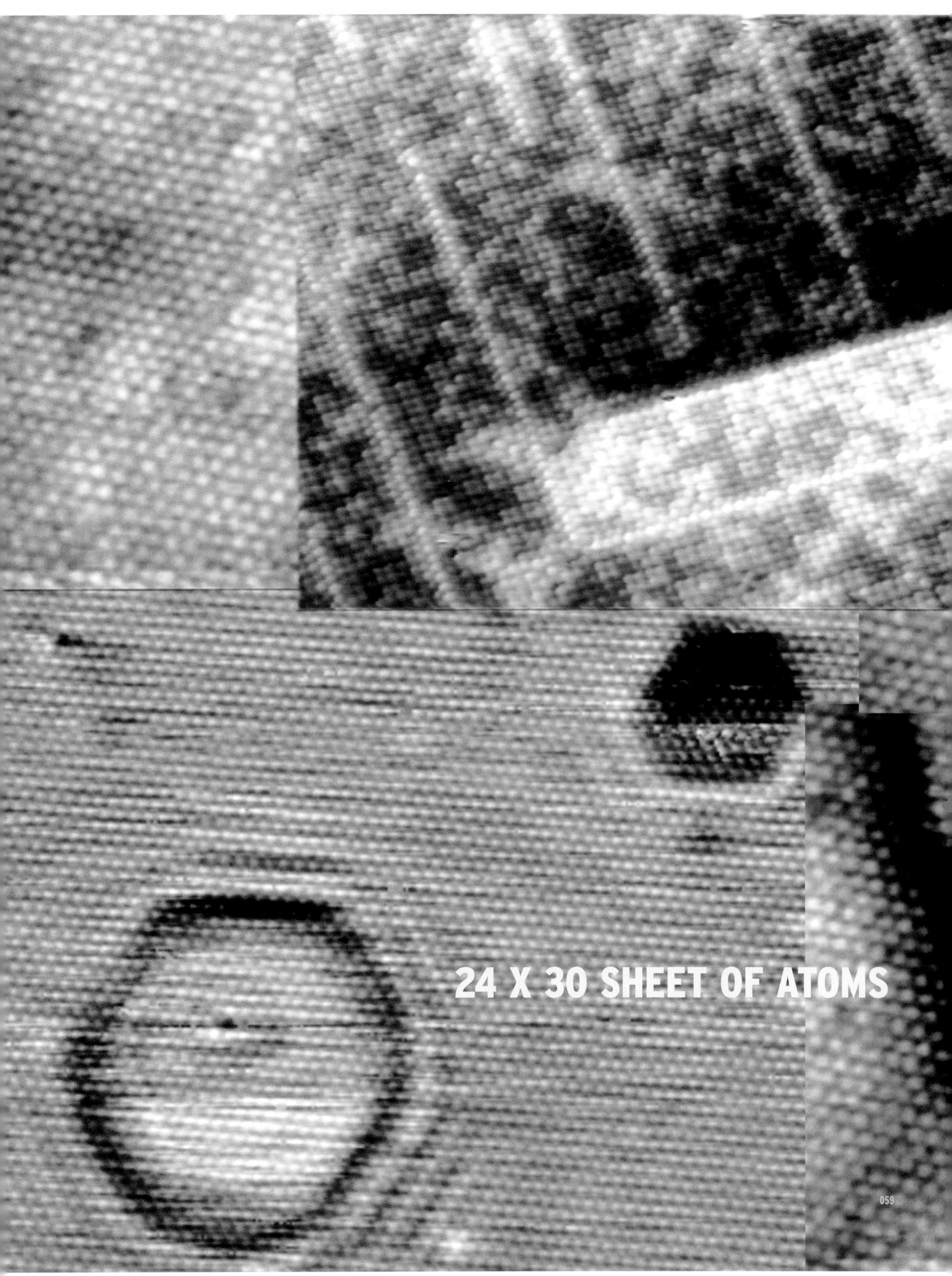
24 X 30 SHEET OF ATOMS

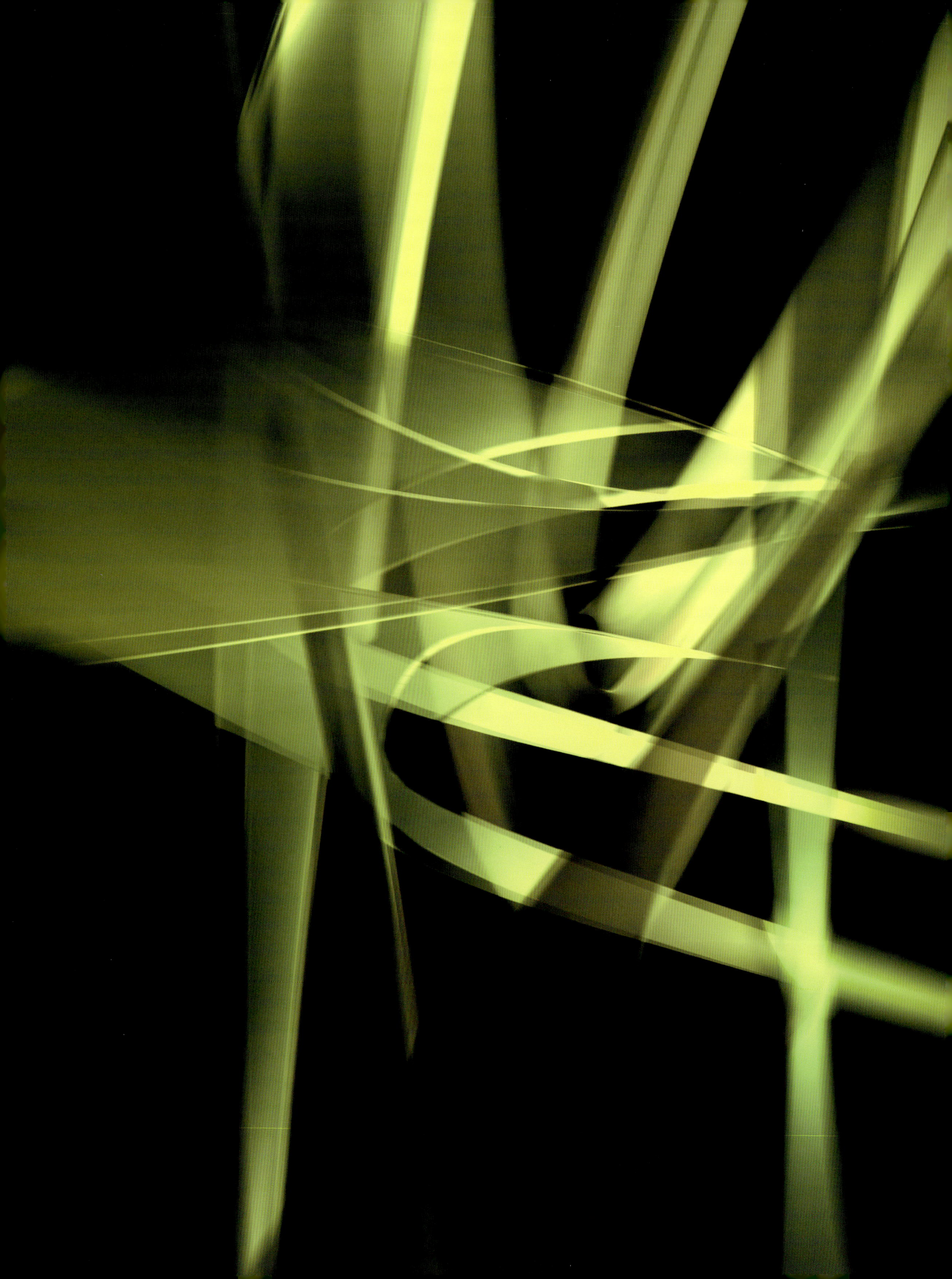

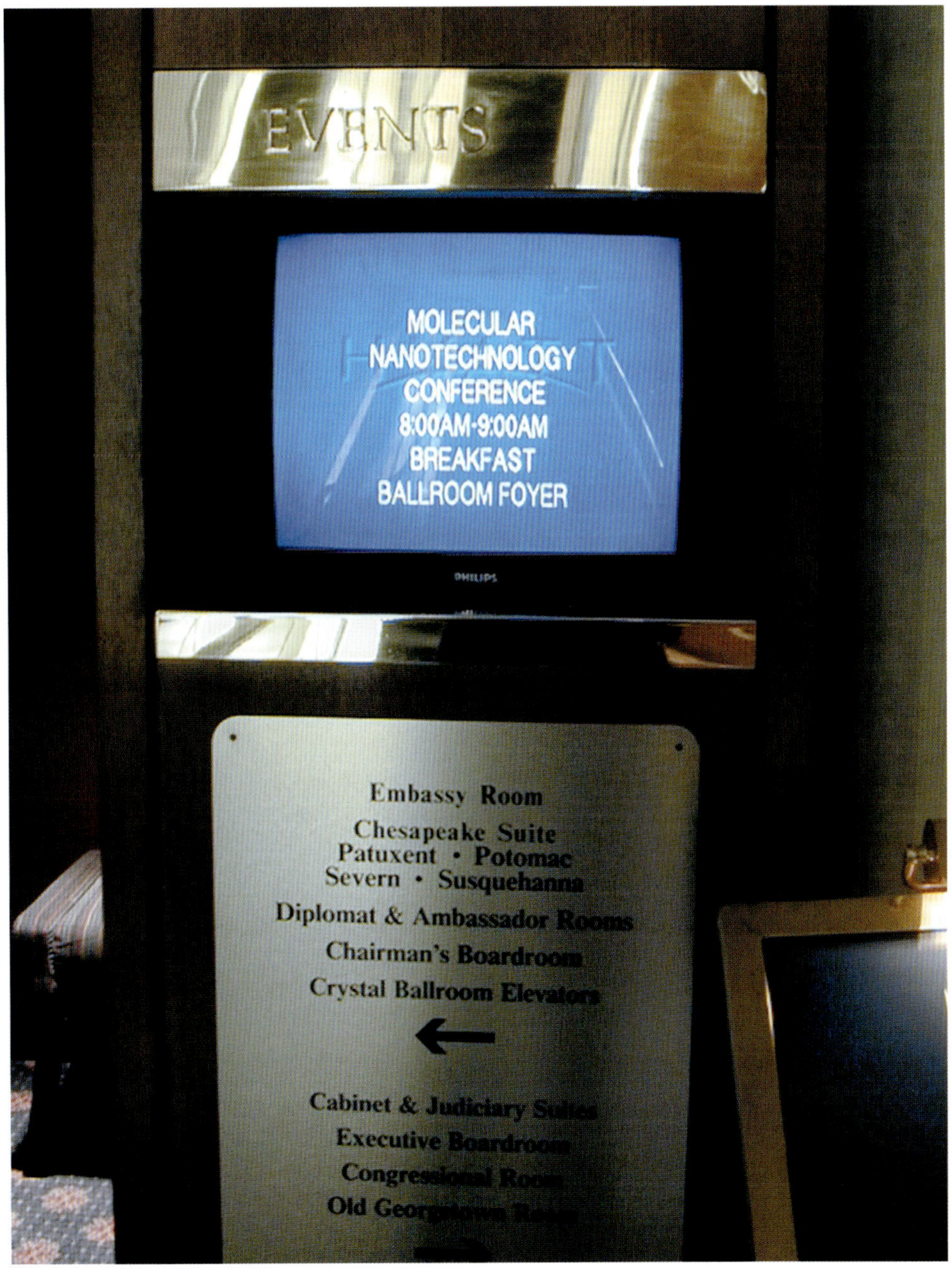
EVENTS
MOLECULAR
NANOTECHNOLOGY
CONFERENCE
8:00AM-9:00AM
BREAKFAST
BALLROOM FOYER
PHILIPS
Embassy Room
Chesapeake Suite
Patuxent • Potomac
Severn • Susquehanna
Diplomat & Ambassador Rooms
Chairman's Boardroom
Crystal Ballroom Elevators
Cabinet & Judiciary Suites
Executive Boardroom
Congressional Room

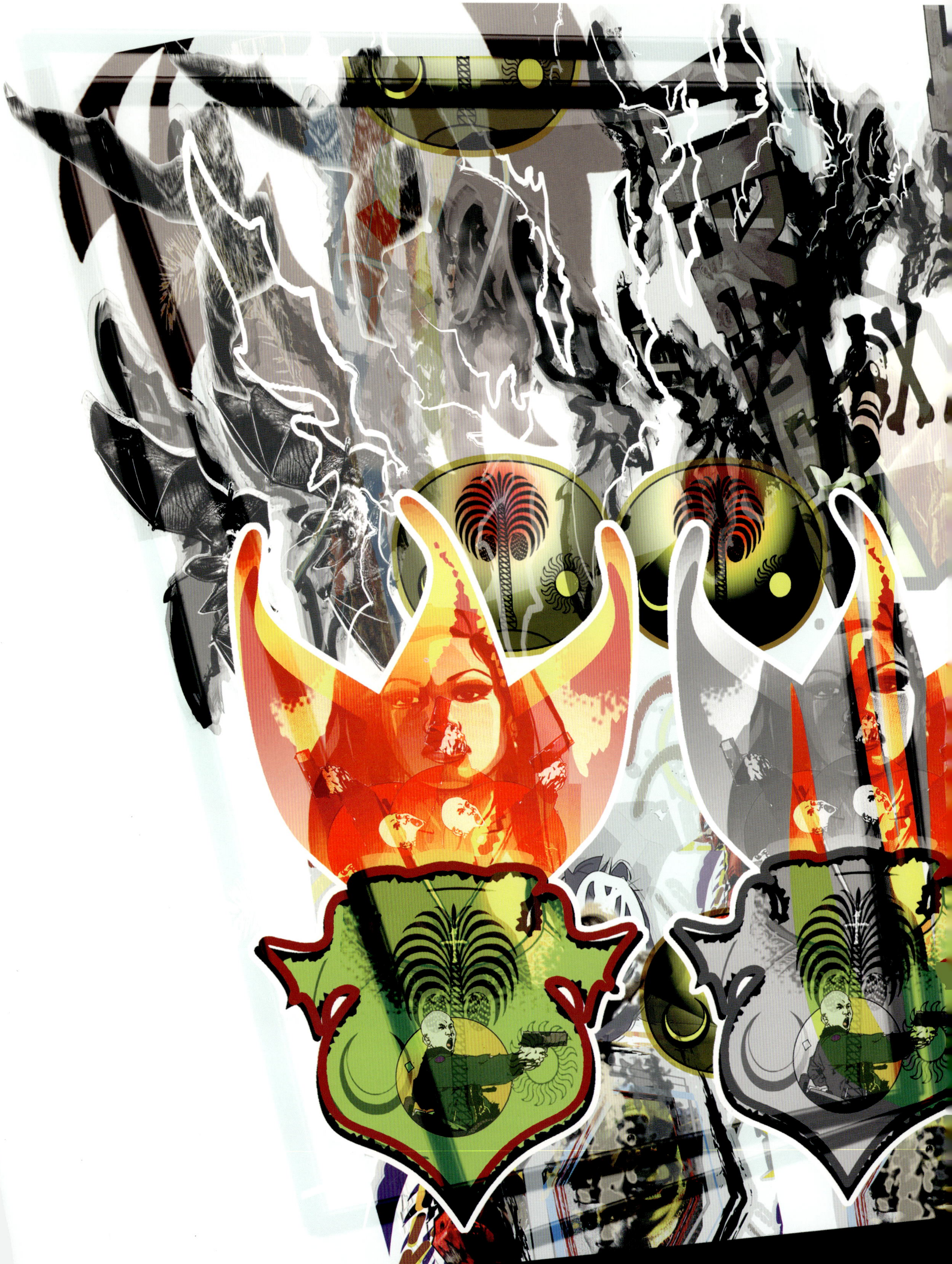

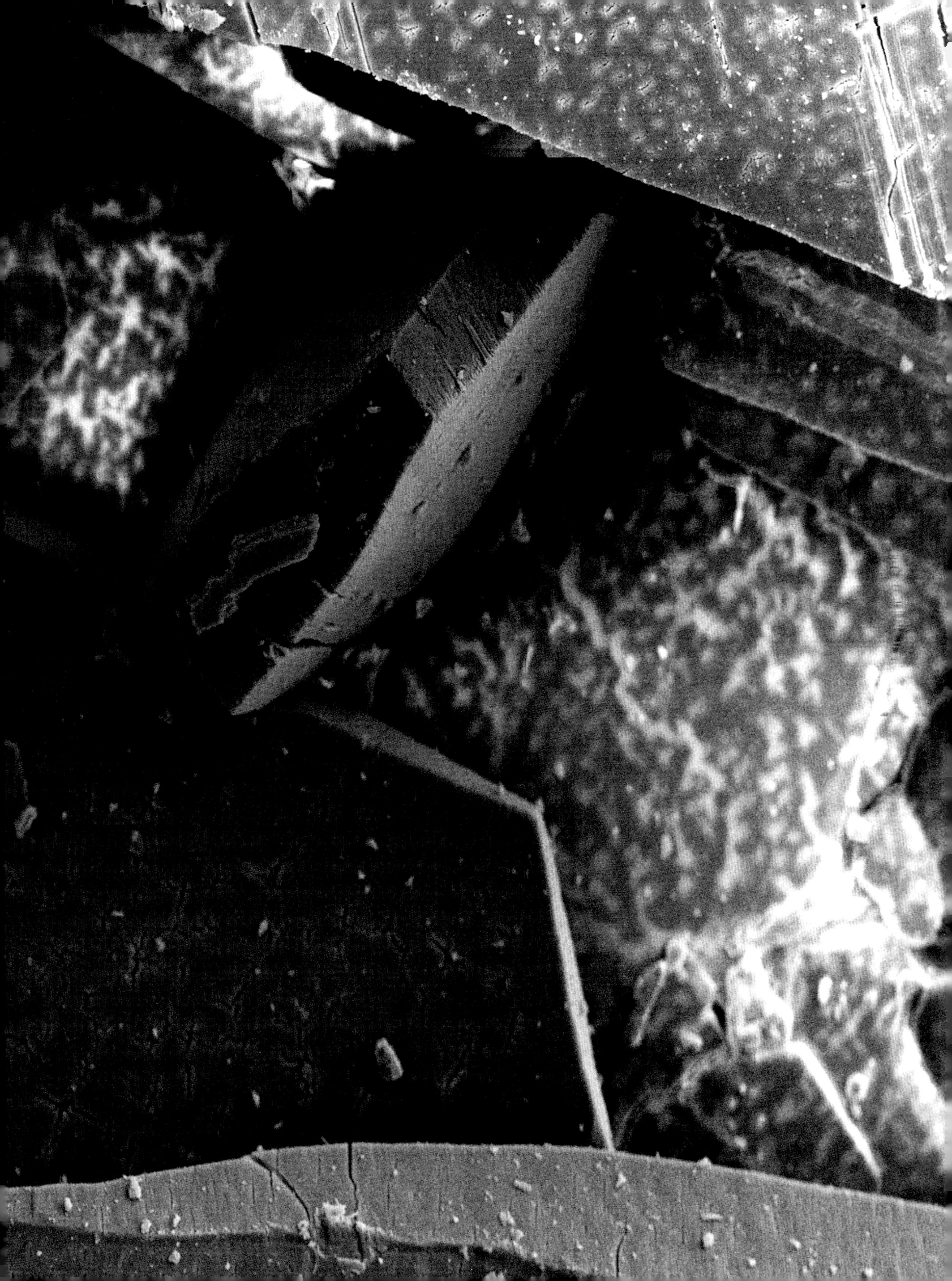

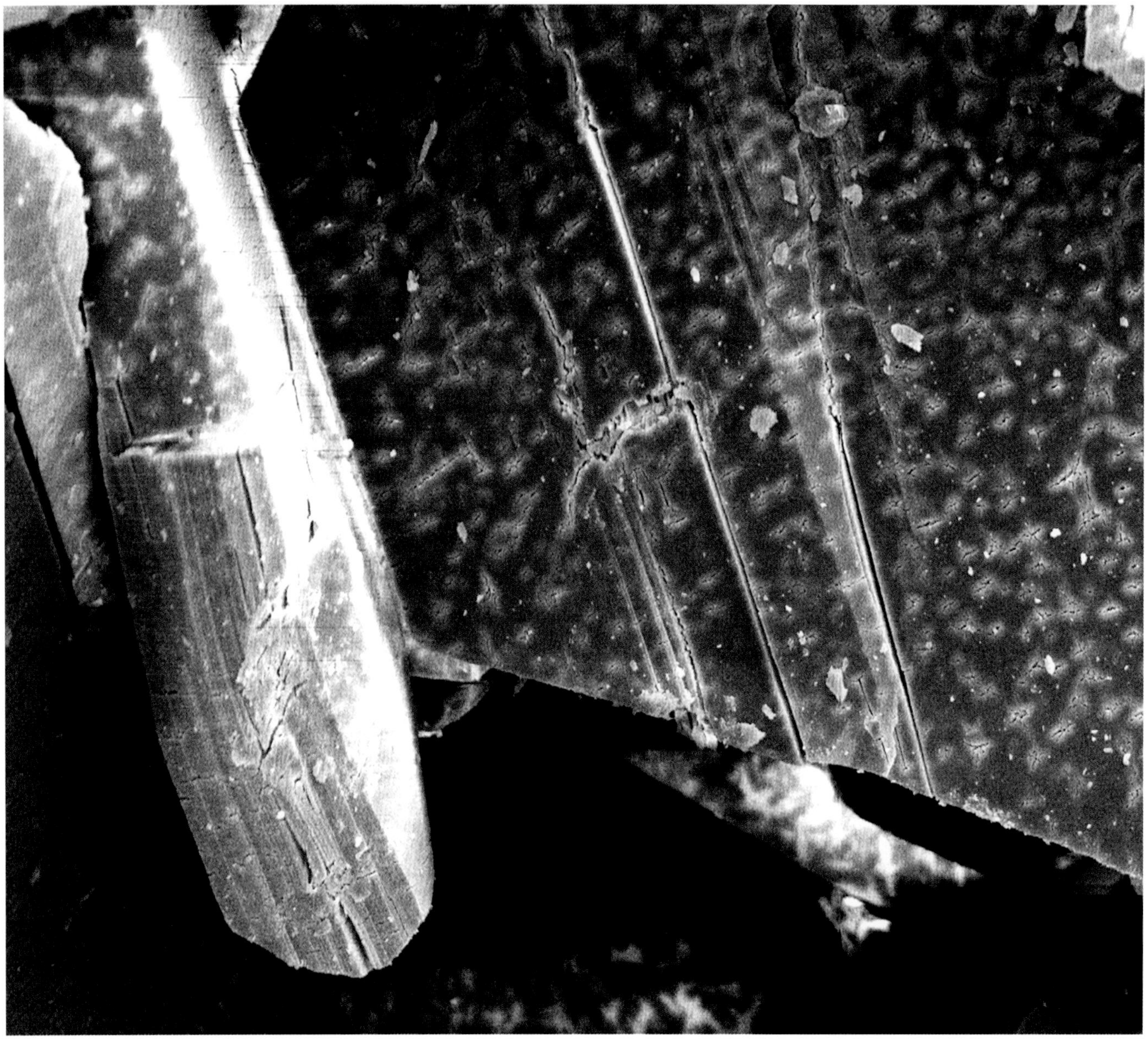

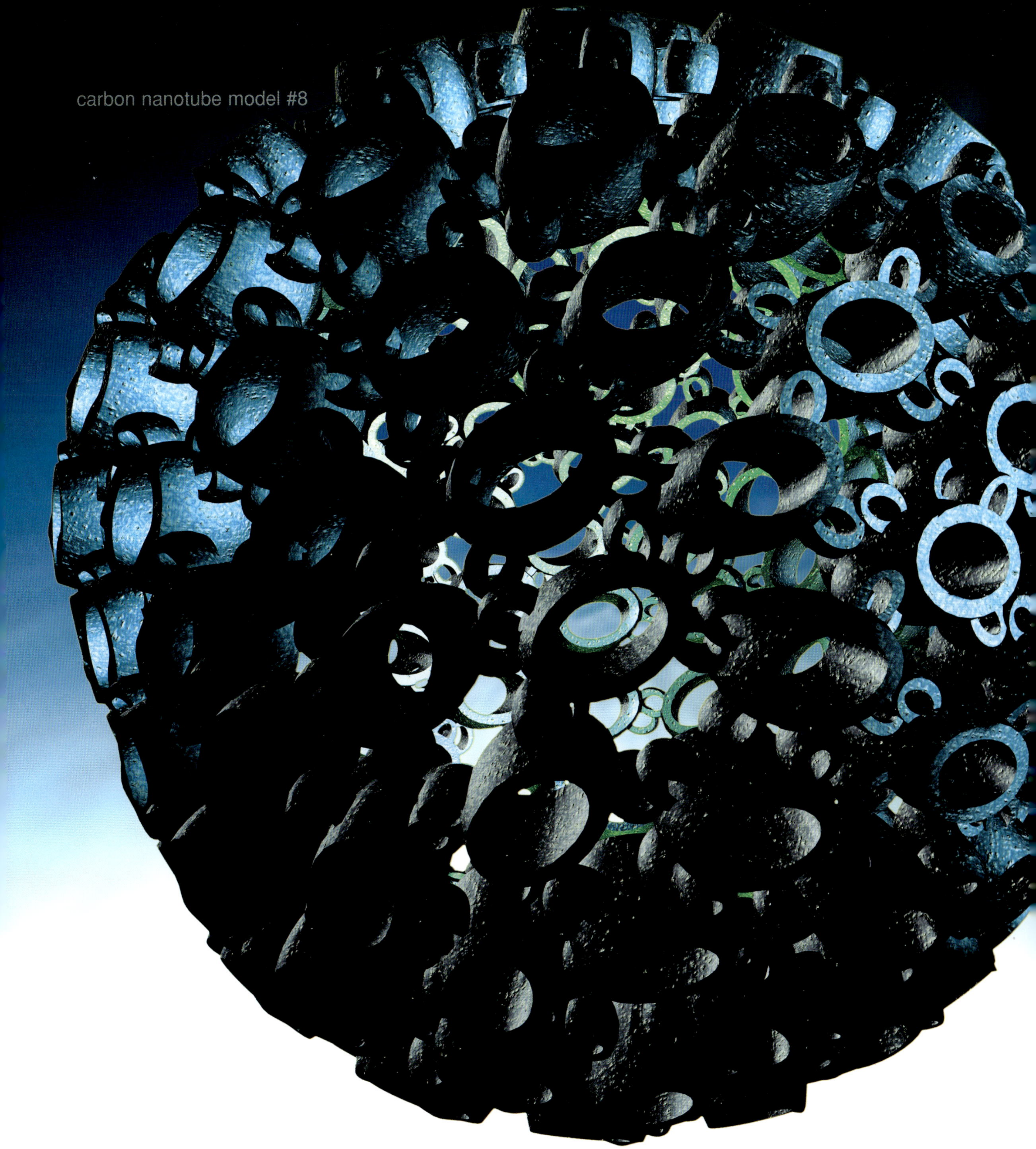

carbon nanotube model #8

From the outside, the projects as a whole have exhibited an array of different directions. However, to know the work is to understand how the work consistently addresses the potential conditions of language as it temporalizes the events of the past.

What is the niche in which the work operates within the context of current exhibitions? From what we have discussed the work presents a particular challenge.

Discuss the piece? An esoteric magic act with some very technical electronics.

What was your experience of the plays themselves, say something positive. Balk continues to be an artist with a very distinct agenda. The format of his work continues to expand which presents a curious test to his audience.

Closing.

Would like to correspond with you after you have received the material.

The model is our strongest relationship to each other.

Integrating fact into the body of the text, always with the nod to the integrity of the genuine narrative—the inherent fiction. Live by the model.

At the recent Foresight Institute's 8th Conference on Molecular Nanotechnology, at the Hyatt in Bethesda, Raymond Kurzweil, a prominent nanotechnology mediator, known for his books on the spiritual intelligence of machines and their imminent ability to out-think human beings, had just delivered his keynote address. The speech consisted essentially of an analysis of exponential growth charts, and statistics that contextualized all sorts of fantastic applications. Kurzweil insists that within two decades trends in three-dimensional computer architecture will produce a type of lattice network consisting of nanotubes whereby a one-inch cube will have a million times the processing capabilities of the human brain. I was seduced by the amount of hyperbole and wondered how the scientists attending might be responding to his projections. I spoke with a small group of young MIT scientist-engineers, working in the field of nanotechnological computer architecture. These guys were wearing a lot of black clothing.

**DB:** You guys look like the fringe element here at the conference.

**MIT guys:** We were just talking about why it isn't always a good idea to predict the future.

**DB:** Why is that?

**MIT guys:** The best you can hope to be is right and people don't really care when you are. And at worst you're wrong, and it didn't matter, and you've wasted your time instead of working on something more productive. What you can primarily draw out of Kurzweil's curves is that the future will always have something surprising to say. And, it may be some type of a paradigm shift but it's unsafe to make assumptions about what the technology is going to look like 50 years from now. And, what's much more interesting is looking at what the technology is right now.

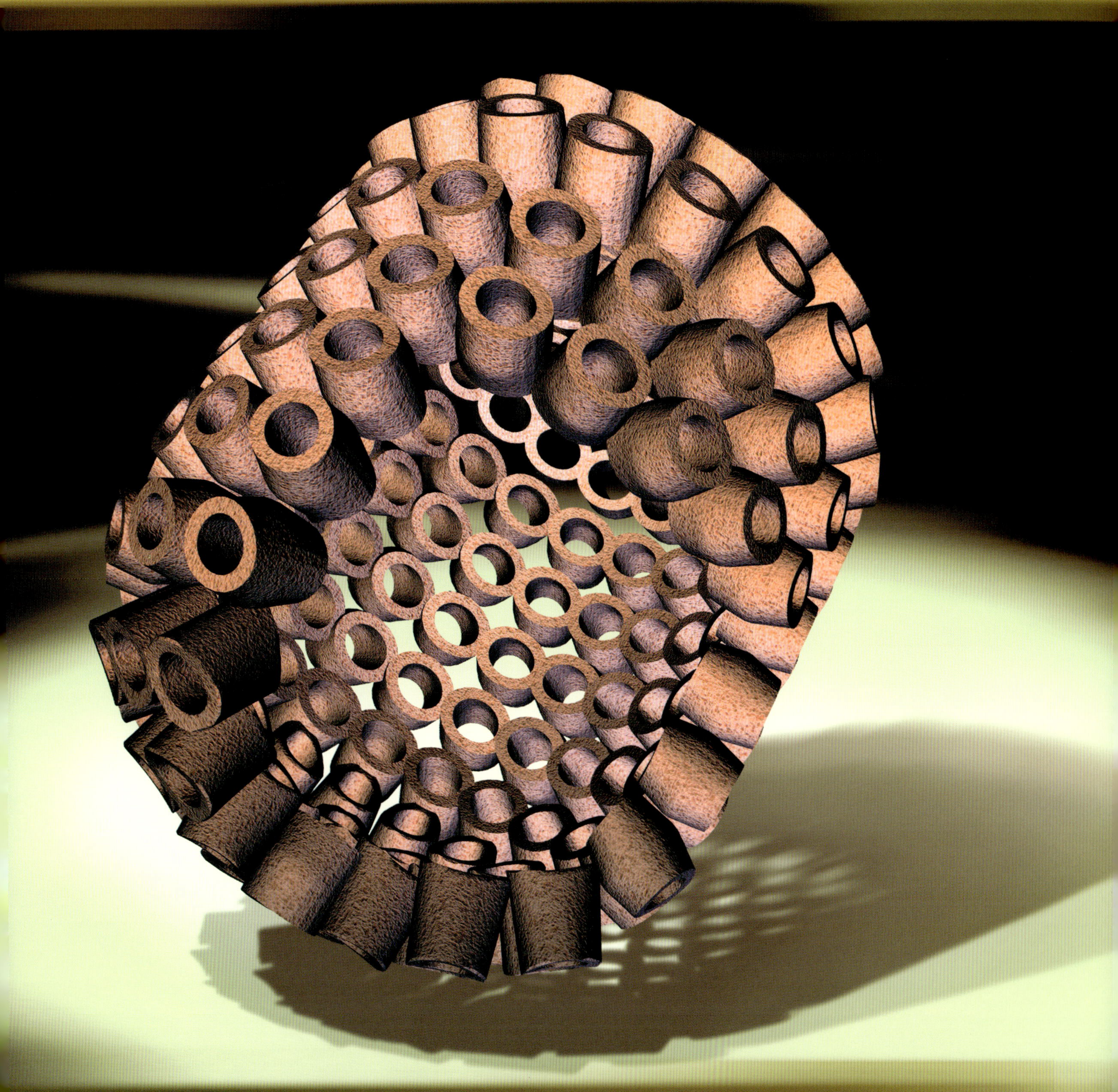

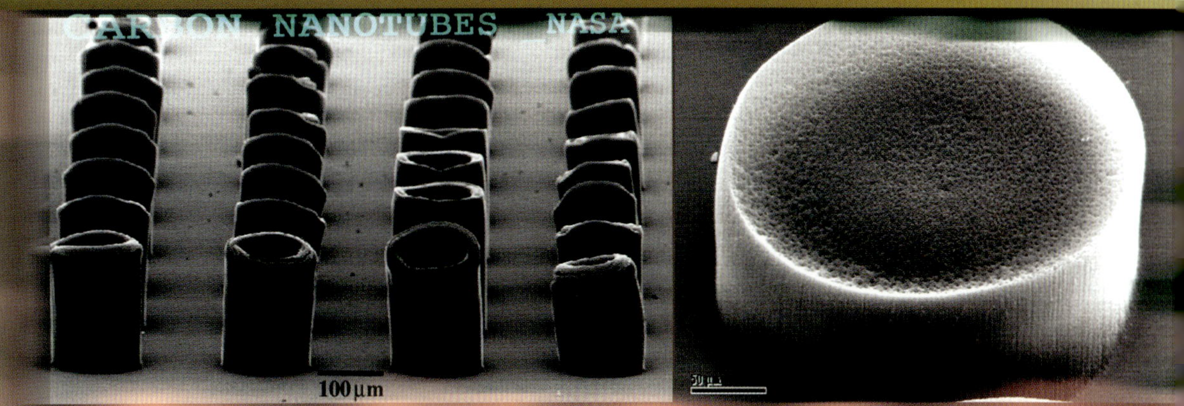
CARBON NANOTUBES NASA
100µm

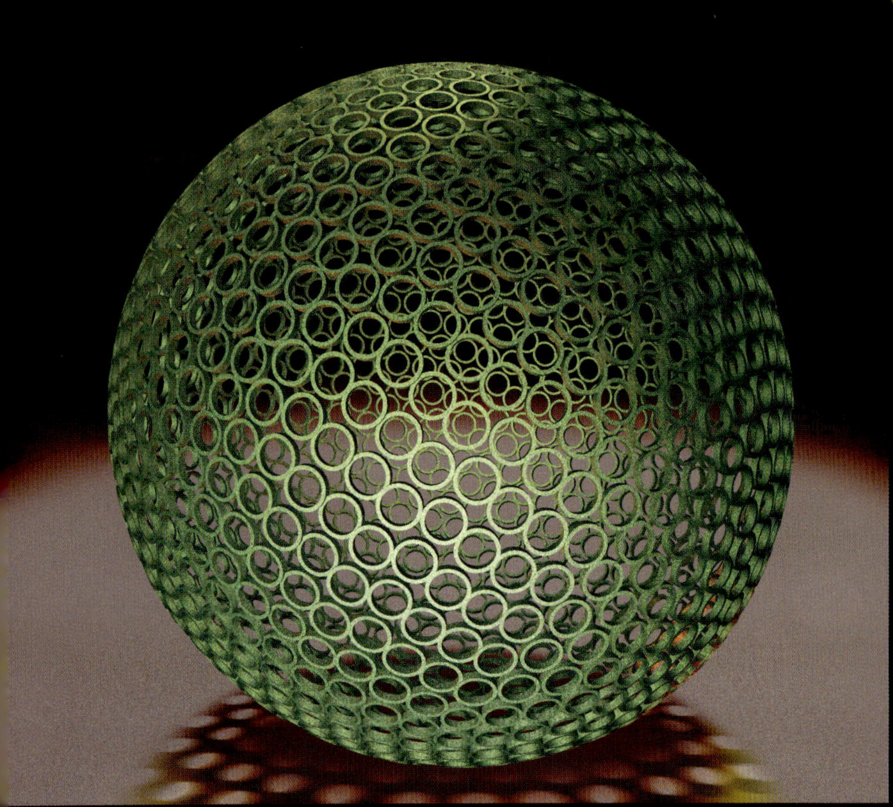

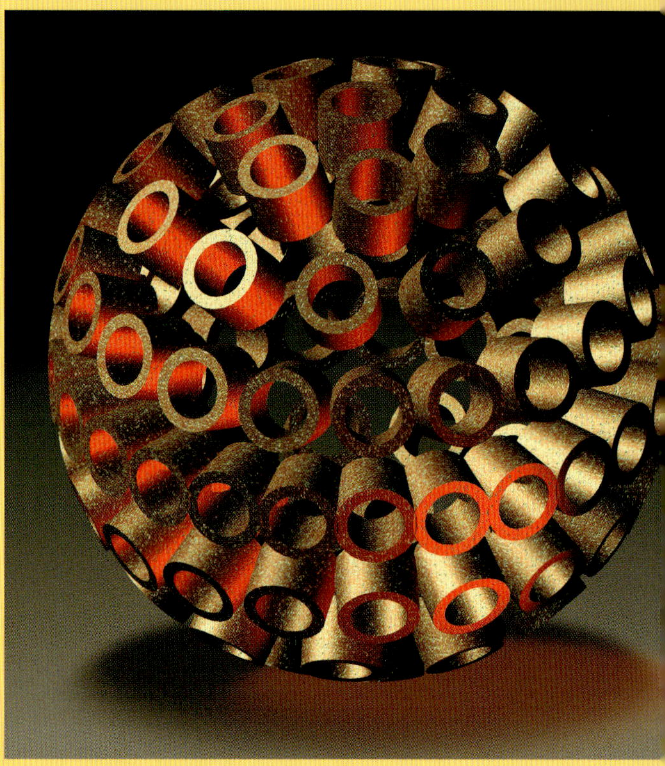

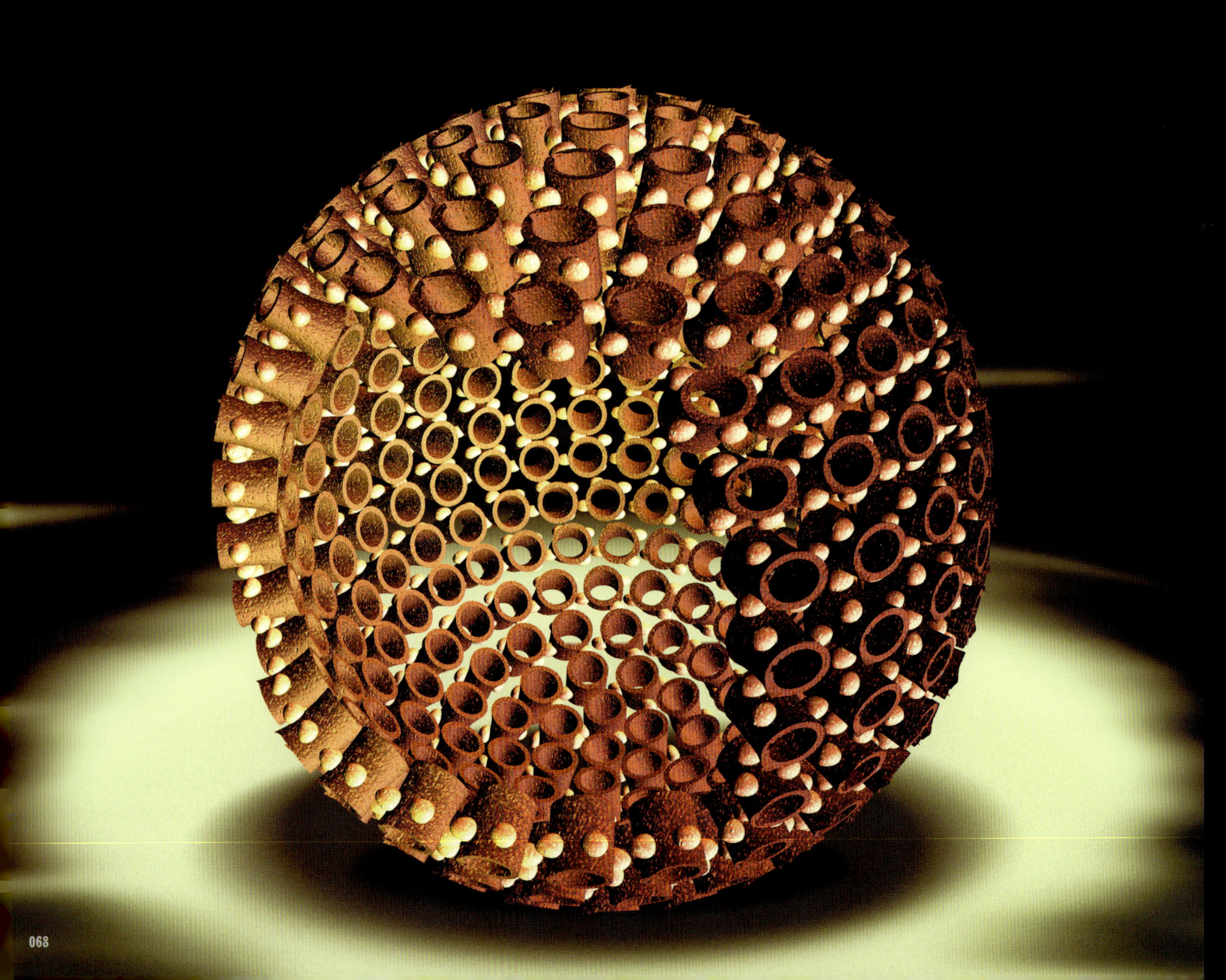

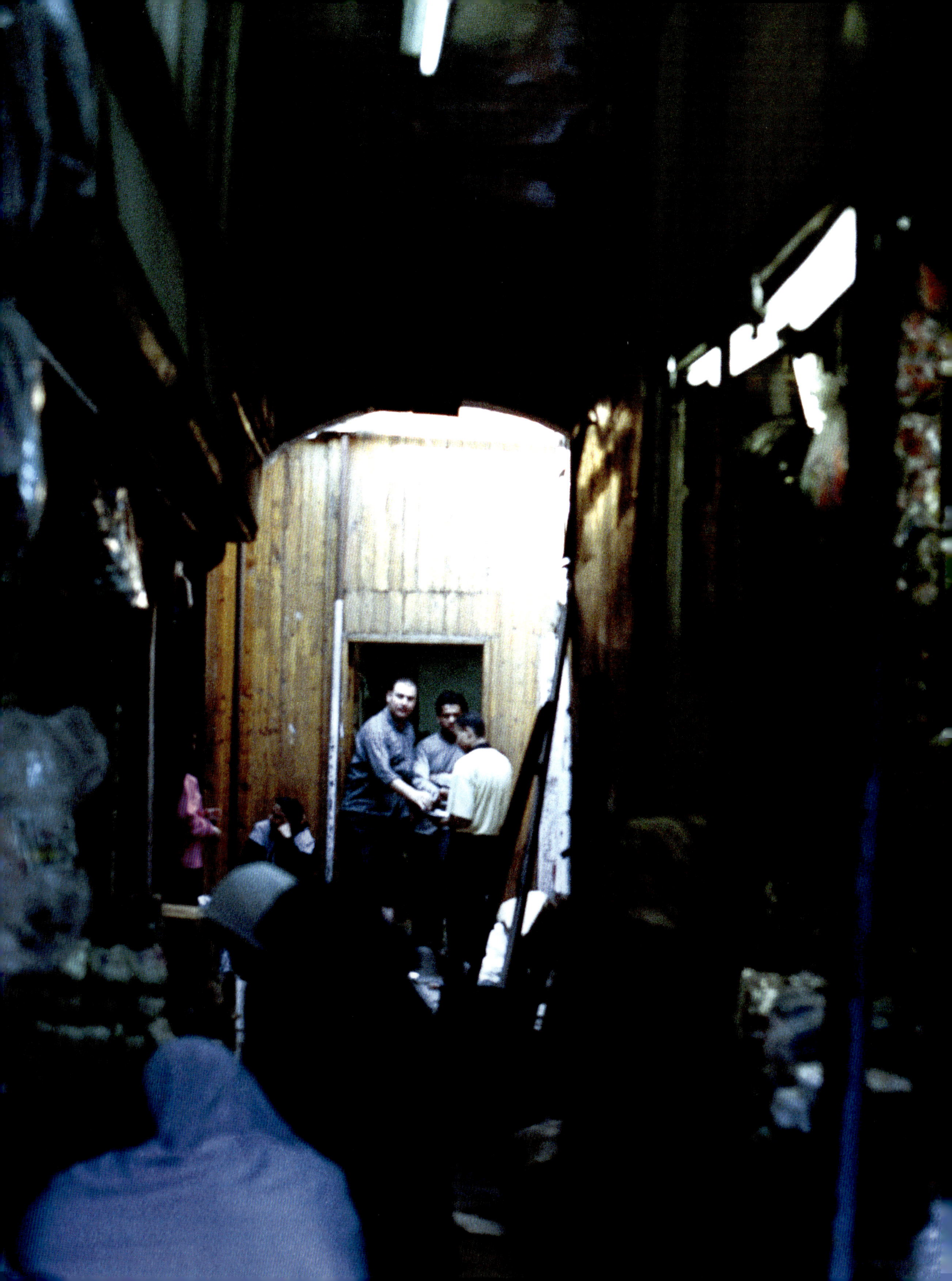

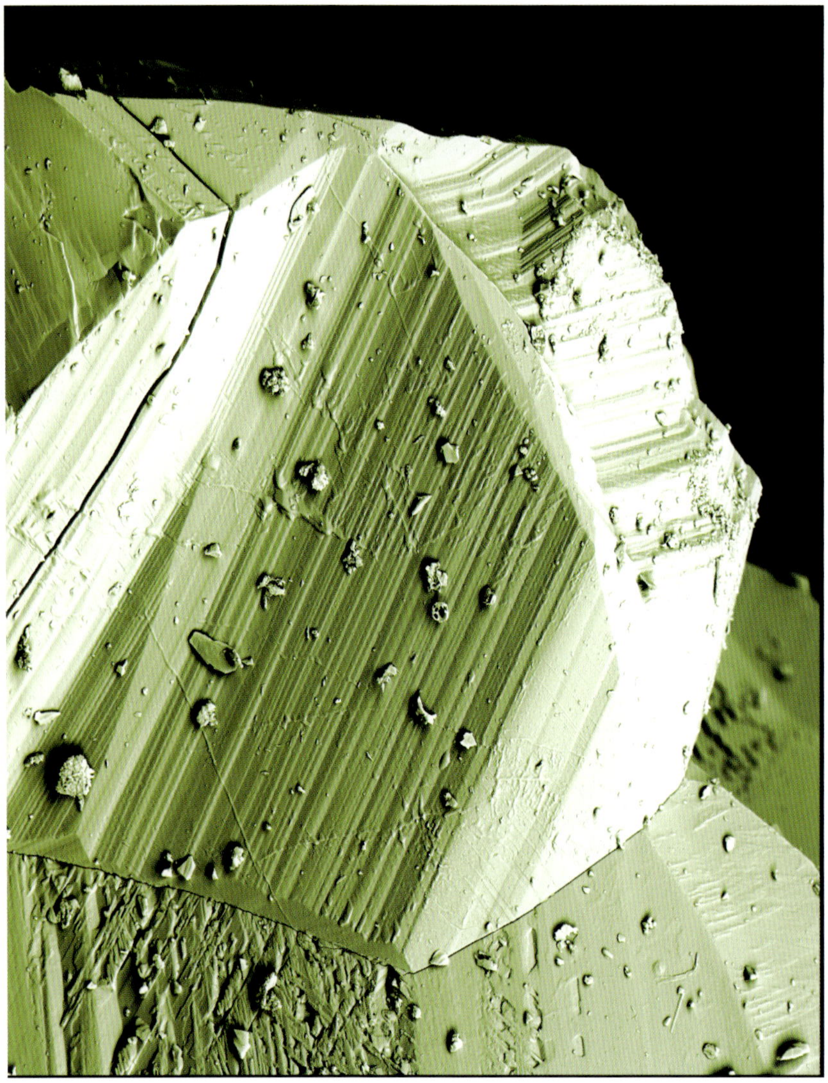

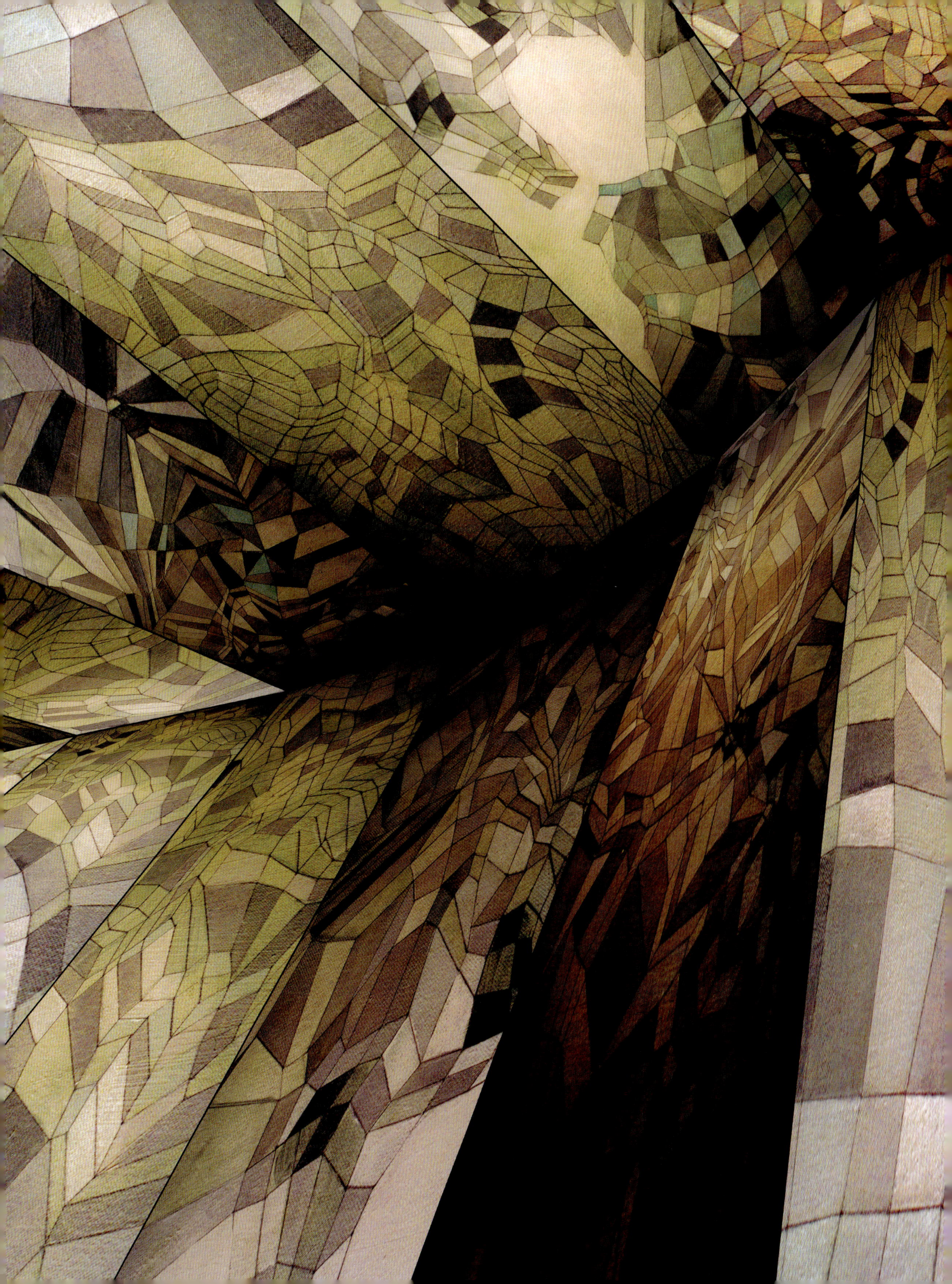

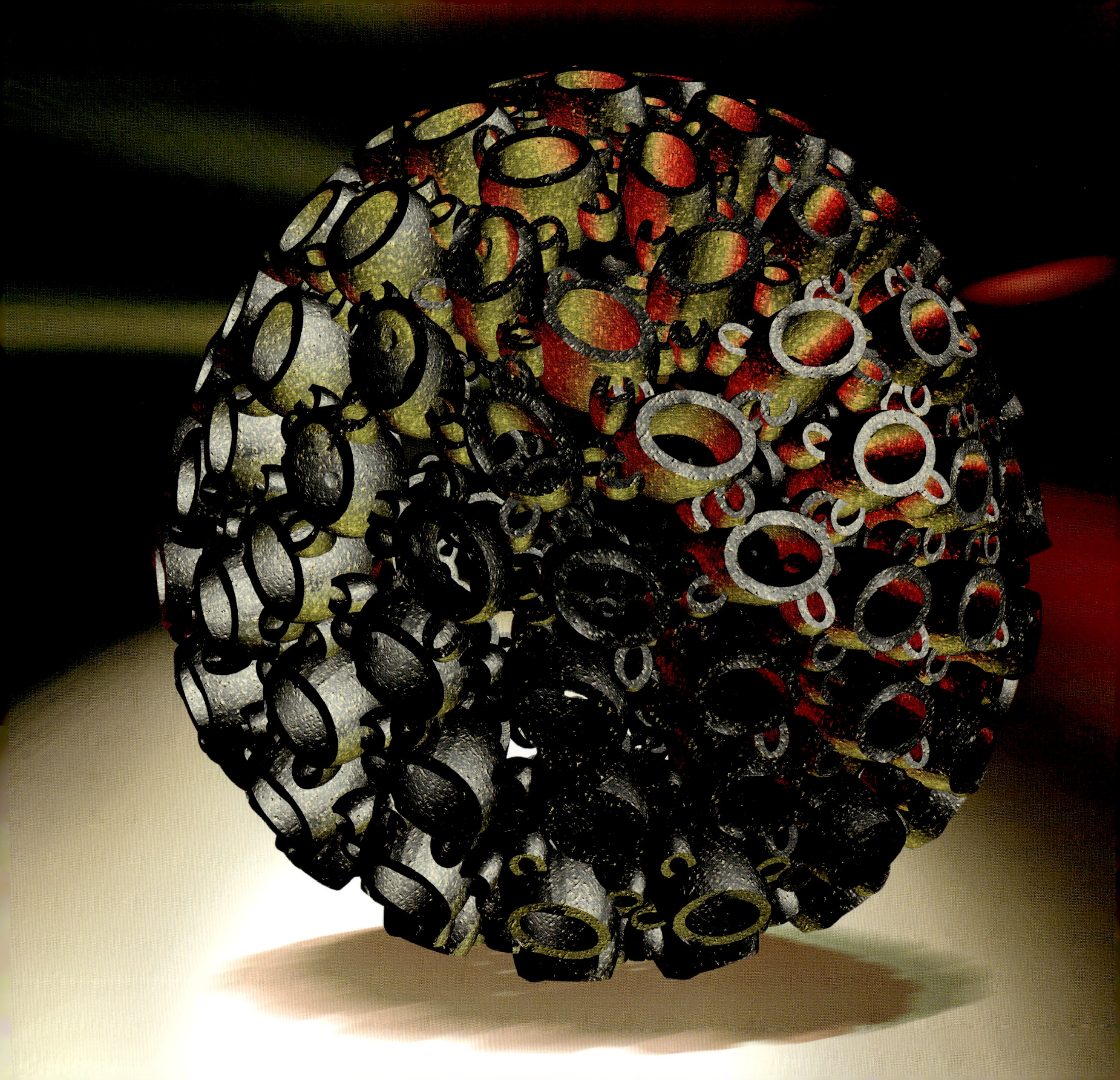

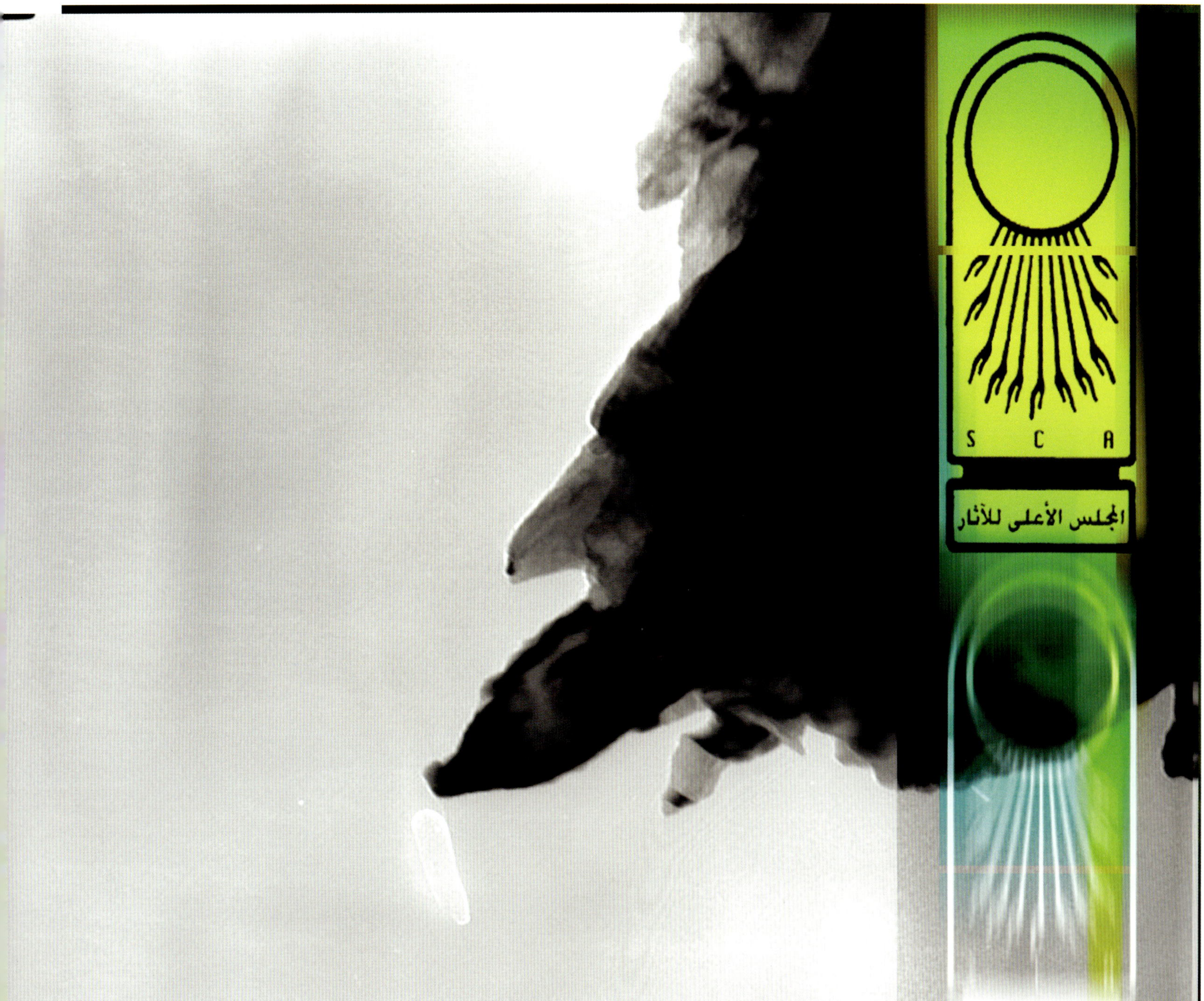


microscale models at the vicinity of the biological proved insufficient. Anticipated mirror effect within urban did not diminish. Fragmentation thick, and as suggested, highly reflective. Protein twisting will not occur at debris junctures. (Abandoned) Taking parallel shift and will follow Mike's group in. Please explain closures of scanner operations.

**2ND OCTAVE:**
**Picture: Microscopic, June 3, 10:15 PM.**
**Topology of a planet surface section, begin the push inward. Inside the air—characterizing the cloud of dust particles, trafficking at the scale of Planck debris. Weather clinging to the edges of the planets, for us; human nature—a sufficiently logical model.**

**Portland: last round was indecipherable. You screw heads are going down a complete fucking dead end. We're pulling out and going with the chip.**

20nm

PT/IR HEW 090199
4958 200.0KV X200K 20nm

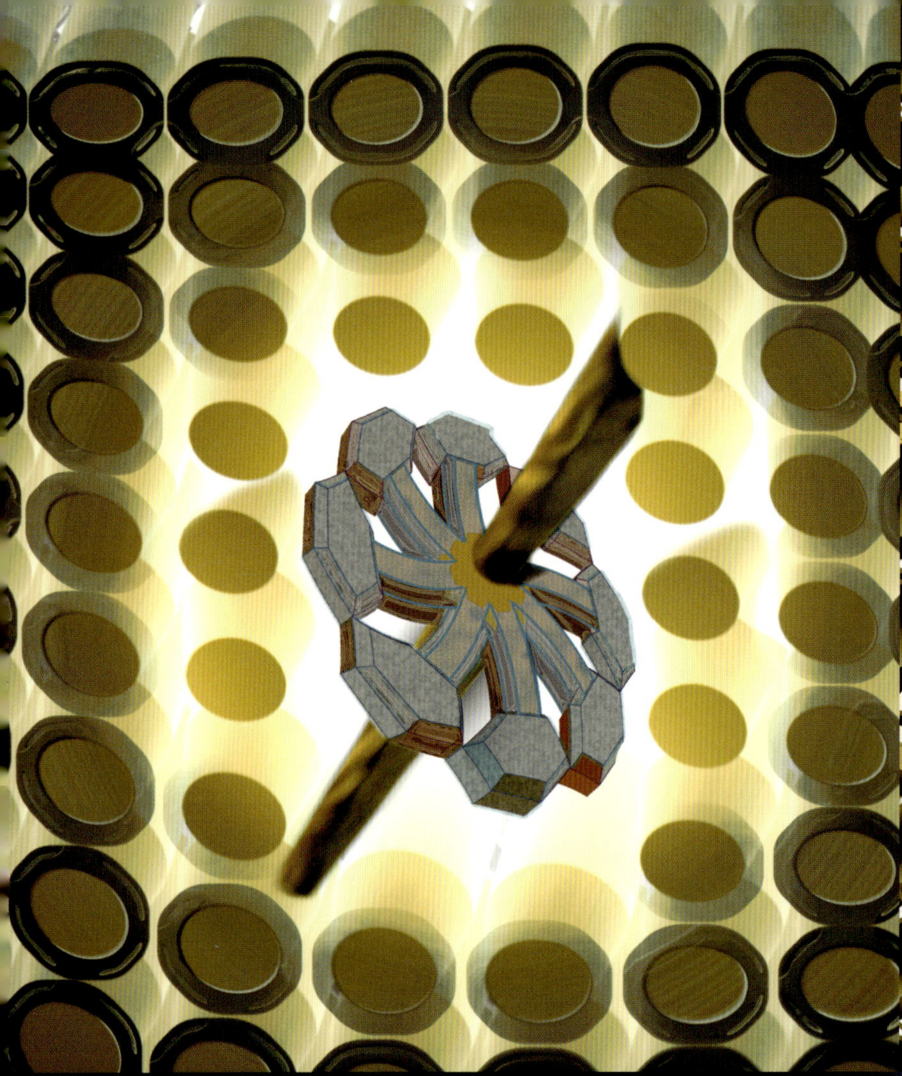

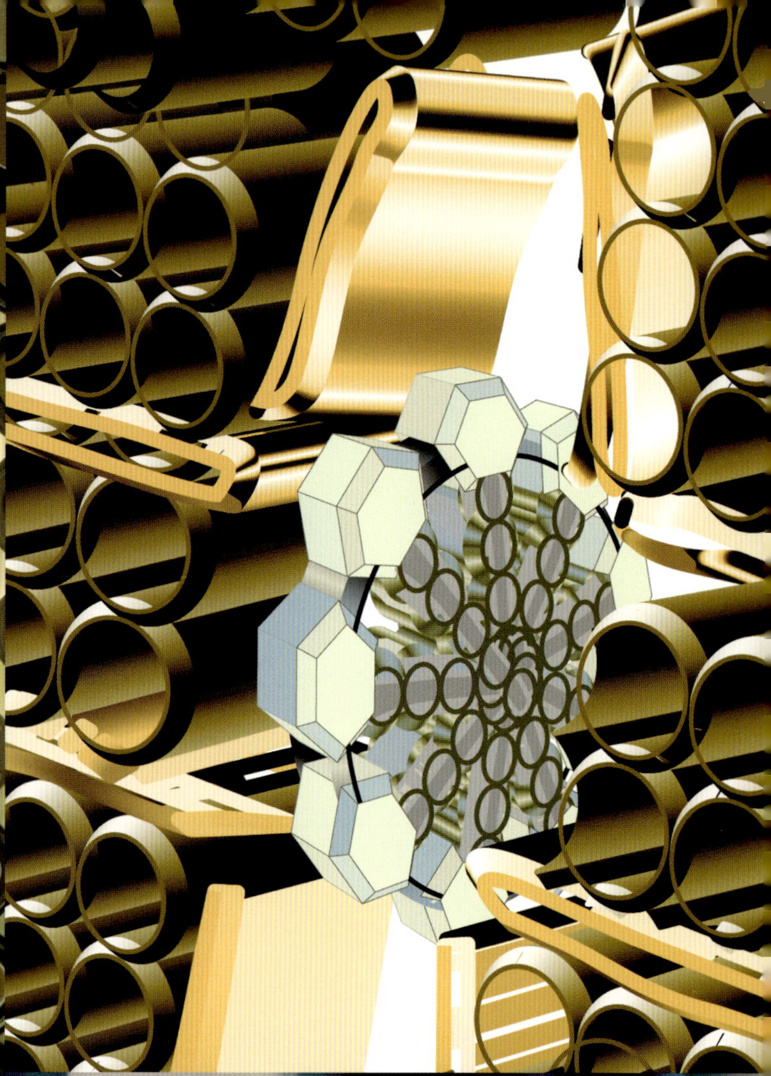

## A Singular Unit of Goo.

Whether it creeps along, transforming the surface of things, or blows in like a heavy-laden cloud, the individual unit creating the gray-goo mass will share similar structural qualities to the basic nanotechnology unit or device. All of the proposed devices, or nanobots, share the same two operating subunits or functions: the nanoassembler, the function that facilitates the construction of the machine from surrounding material, and the nanocomputer, the function that informs the device process, including replication, thereby creating a chunk of something like an ultrathin sheet of diamond-hard paper.

Freitas' assessment of the problem asserts that threats of nanobots let loose in the air will necessarily be the result of a calculated and thoroughly researched design agenda. He summarizes that "biovorous nanorobots capable of comprehensive ecophagy will not be easy to build and their design will require exquisite attention to numerous complex specifications and operational challenges. Such biovores can emerge only after a lengthy period of purposeful focused effort, or as a result of deliberate experiments aimed at creating general-purpose artificial life, perhaps by employing genetic algorithms, but are highly unlikely to arise solely by accident."

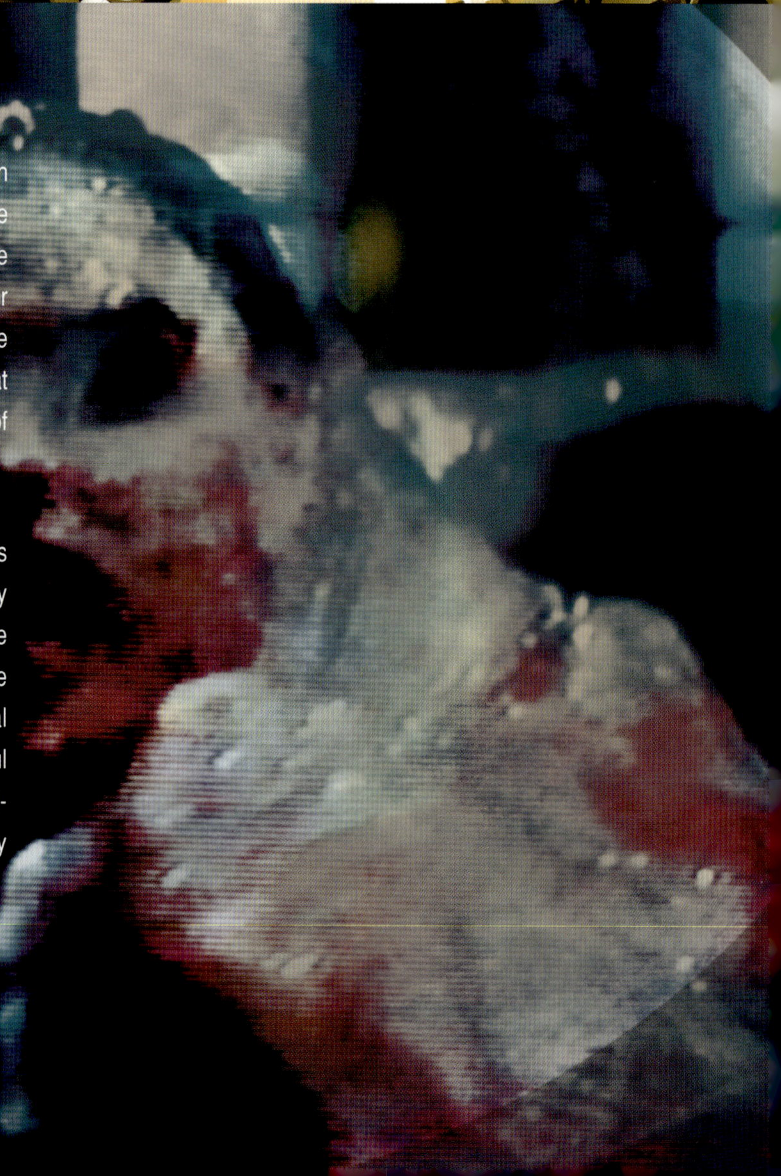

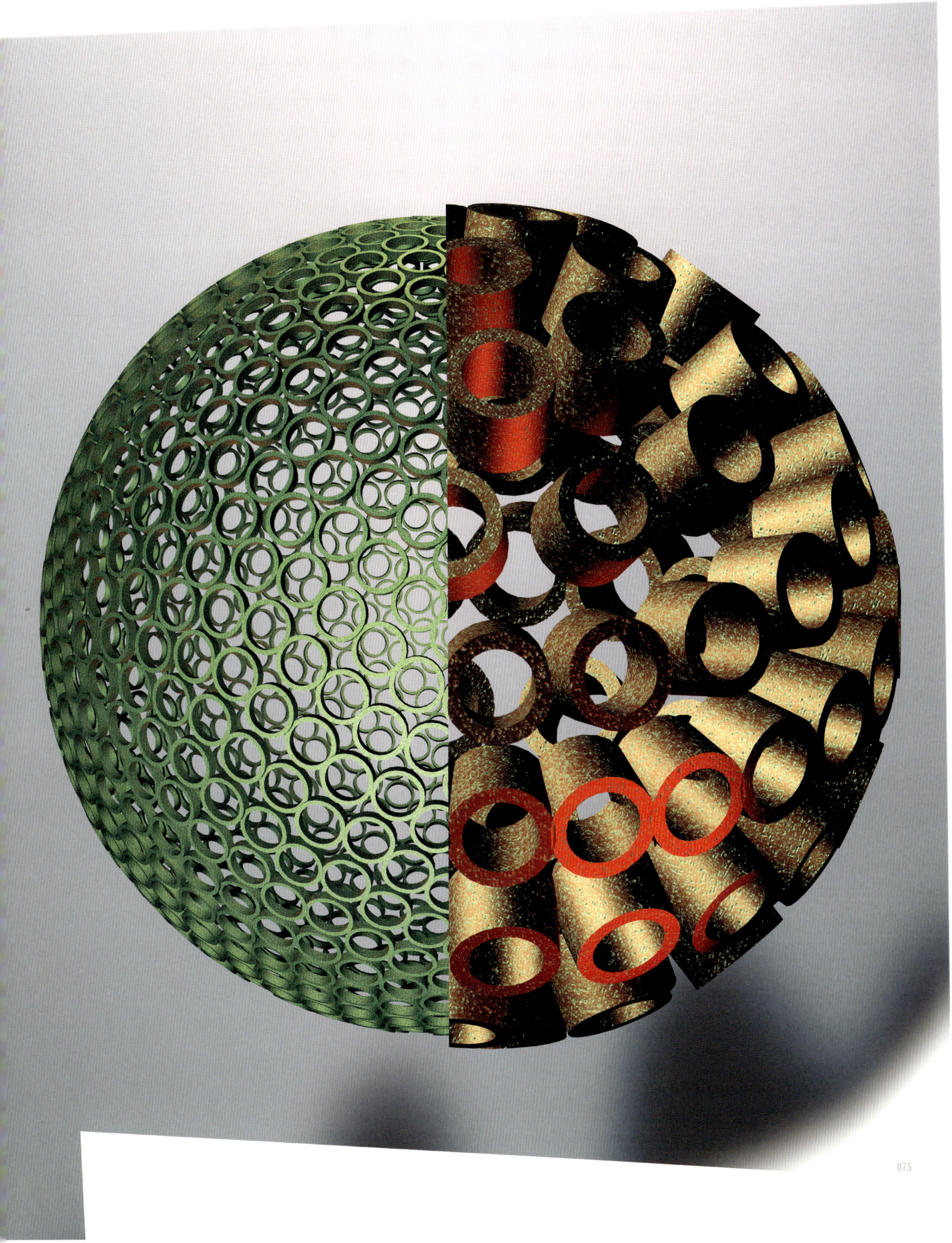

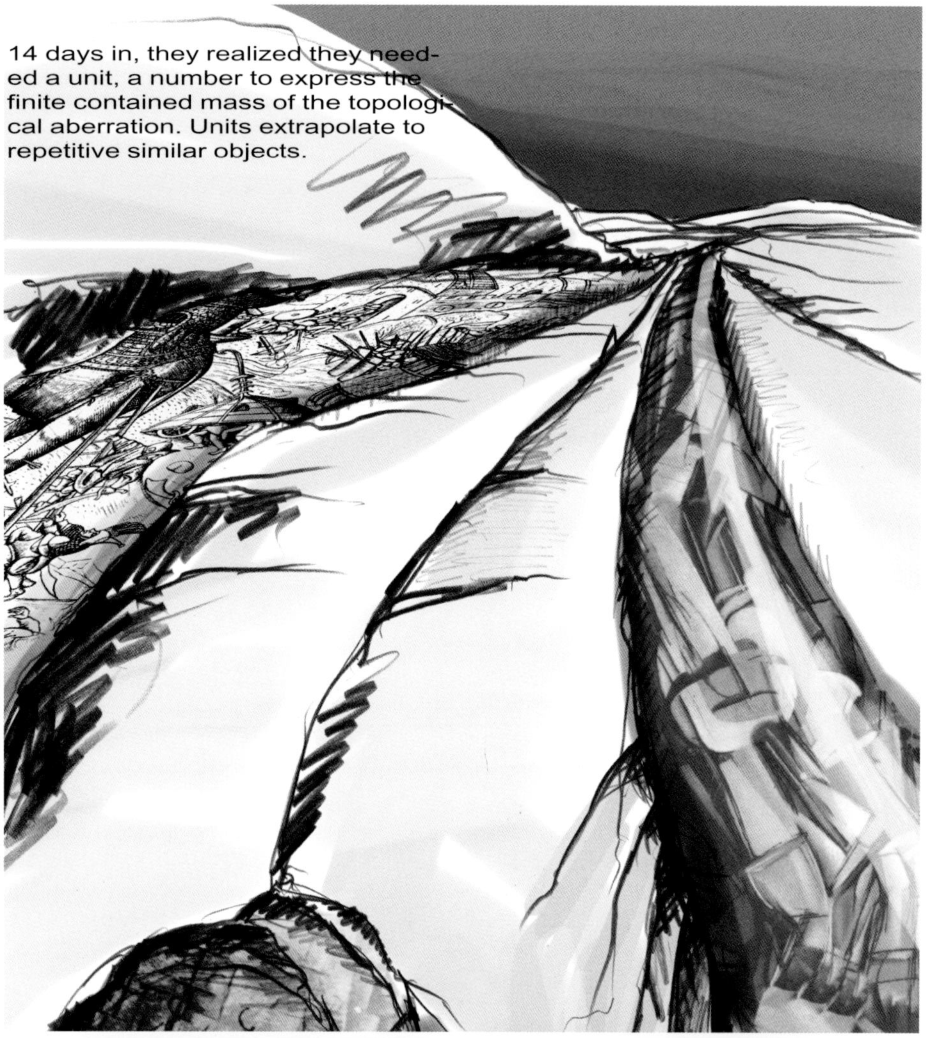
14 days in, they realized they need-
ed a unit, a number to express the
finite contained mass of the topologi-
cal aberration. Units extrapolate to
repetitive similar objects.

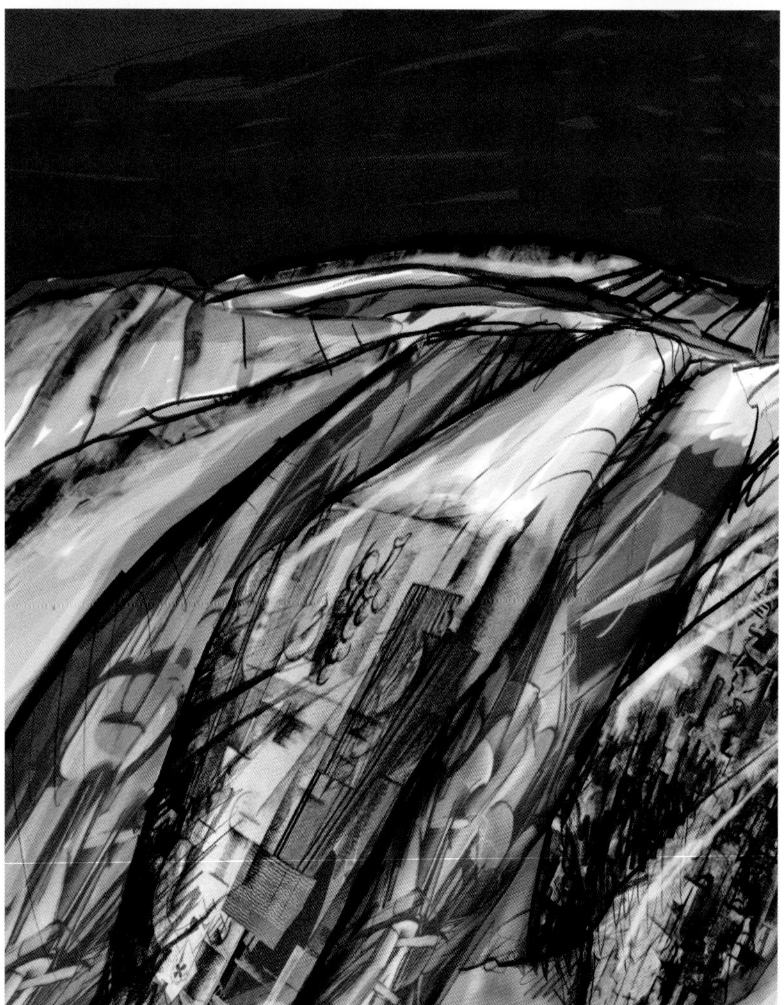

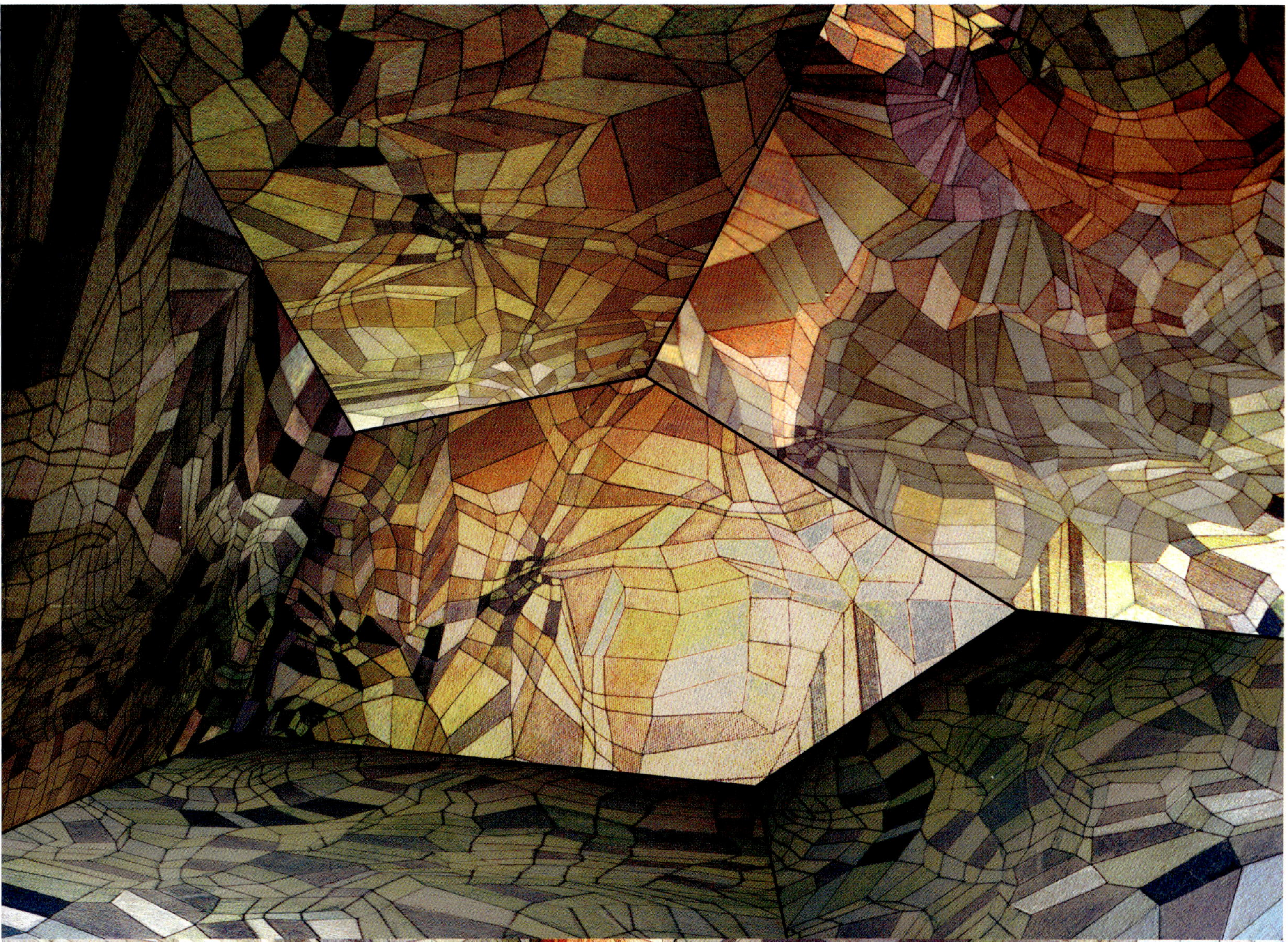

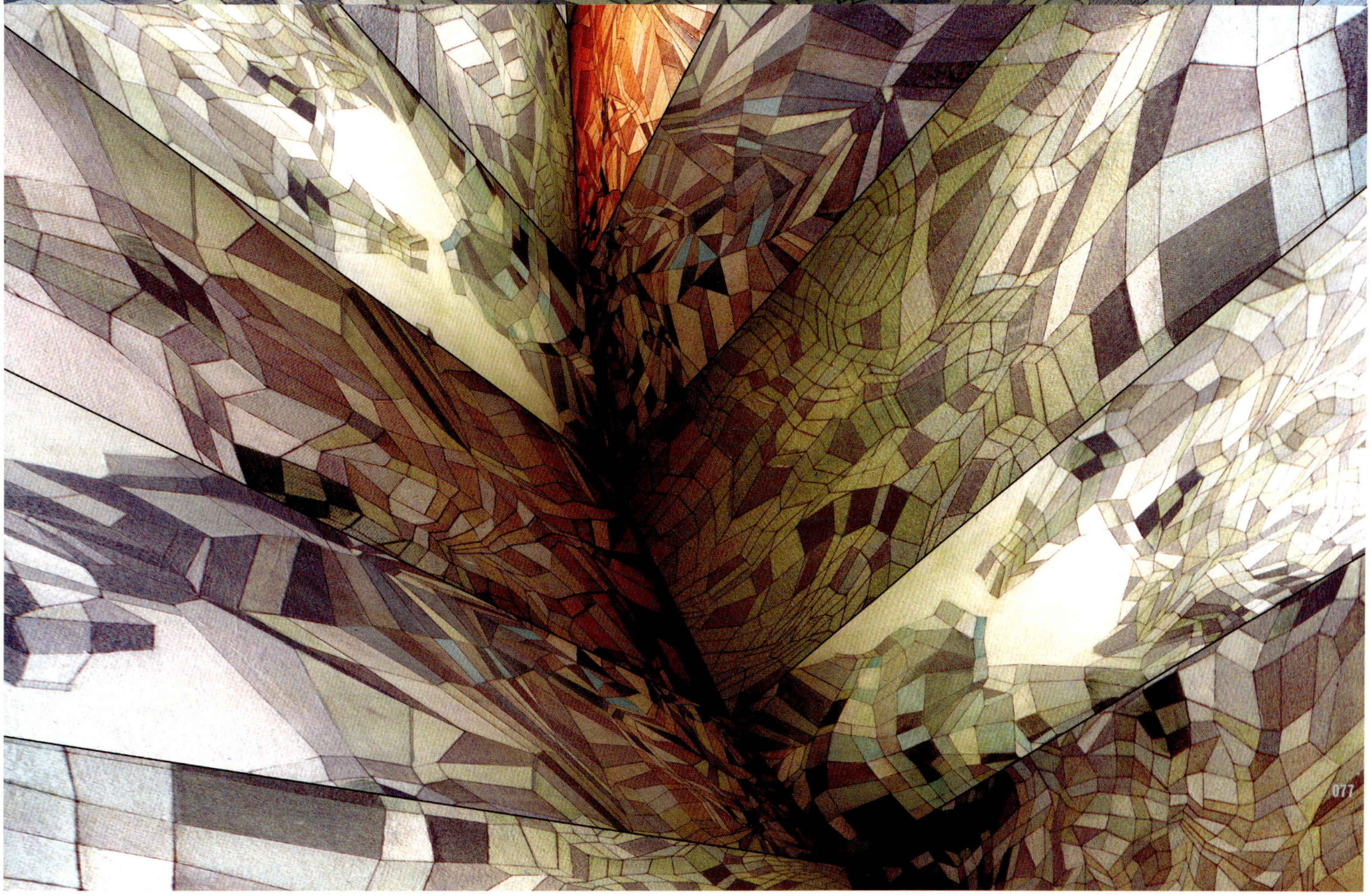

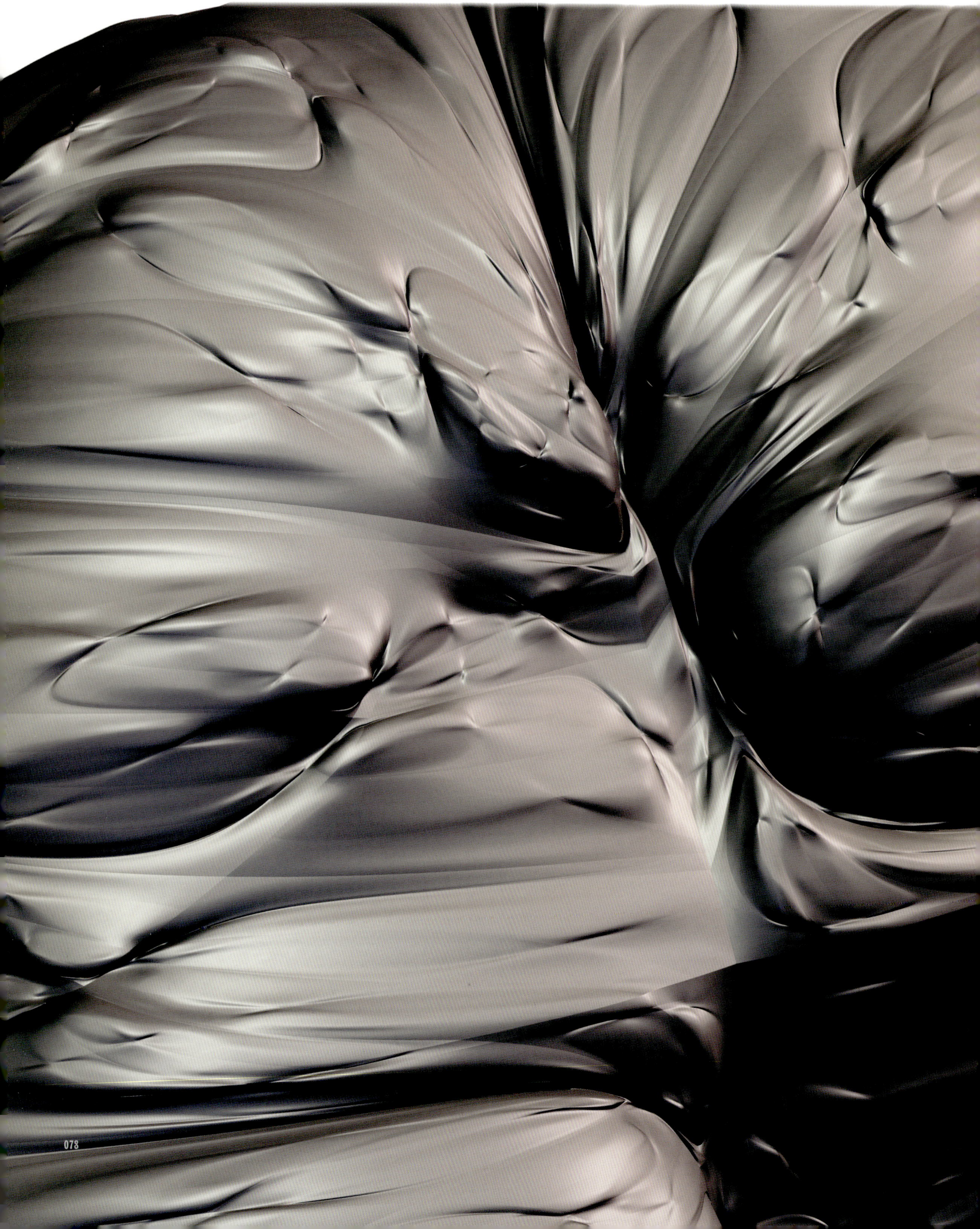

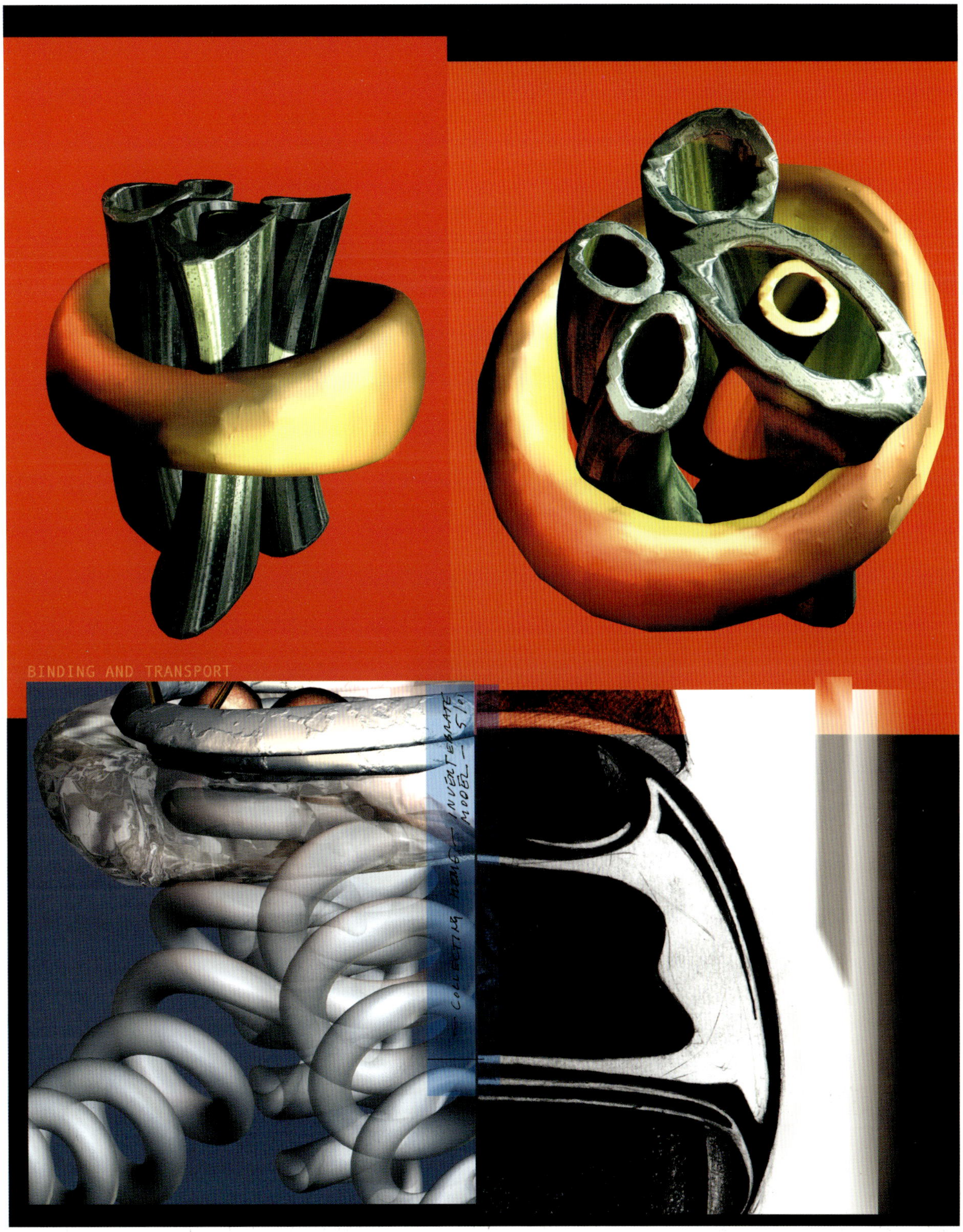
BINDING AND TRANSPORT
COLLECTING
INVERTEBRATE
MODEL – 5/01

**We had gone 7 or 8 magnitudes of observation in a limited direction:**

**a dead end.**

**Move to: 65.3221 SSW**
**Reapply.**

**Everything appears after it is observed.**

**CHARLIERICHTER: many samples, several hot zones. Bleed, and accumulation became thematic, conductively opportunistic.**

*A measurement, after all, merely consists of the quantum state under consideration becoming entangled with a more extended part of the physical universe, e.g. with a measuring apparatus. This measuring apparatus—together with the observing physicist and their common environment—should all, according to conventional understanding, also have a quantum description. Accordingly, there should be a quantum description of this entire quantum state, involving not only the original system under consideration but also the apparatus, physicist, and remaining environment—and this entire state would be expected to evolve continuously. —Roger Penrose*

conclusion of surface.
the culture particle/mass within the vector of its own similarities.
topological forcefield of shifting ambient weight composition which must contain superior percentages of debris, the flashes and burns at the instant. we pretty much all expected it.
waiting for redirection coordinates. this is all recording.

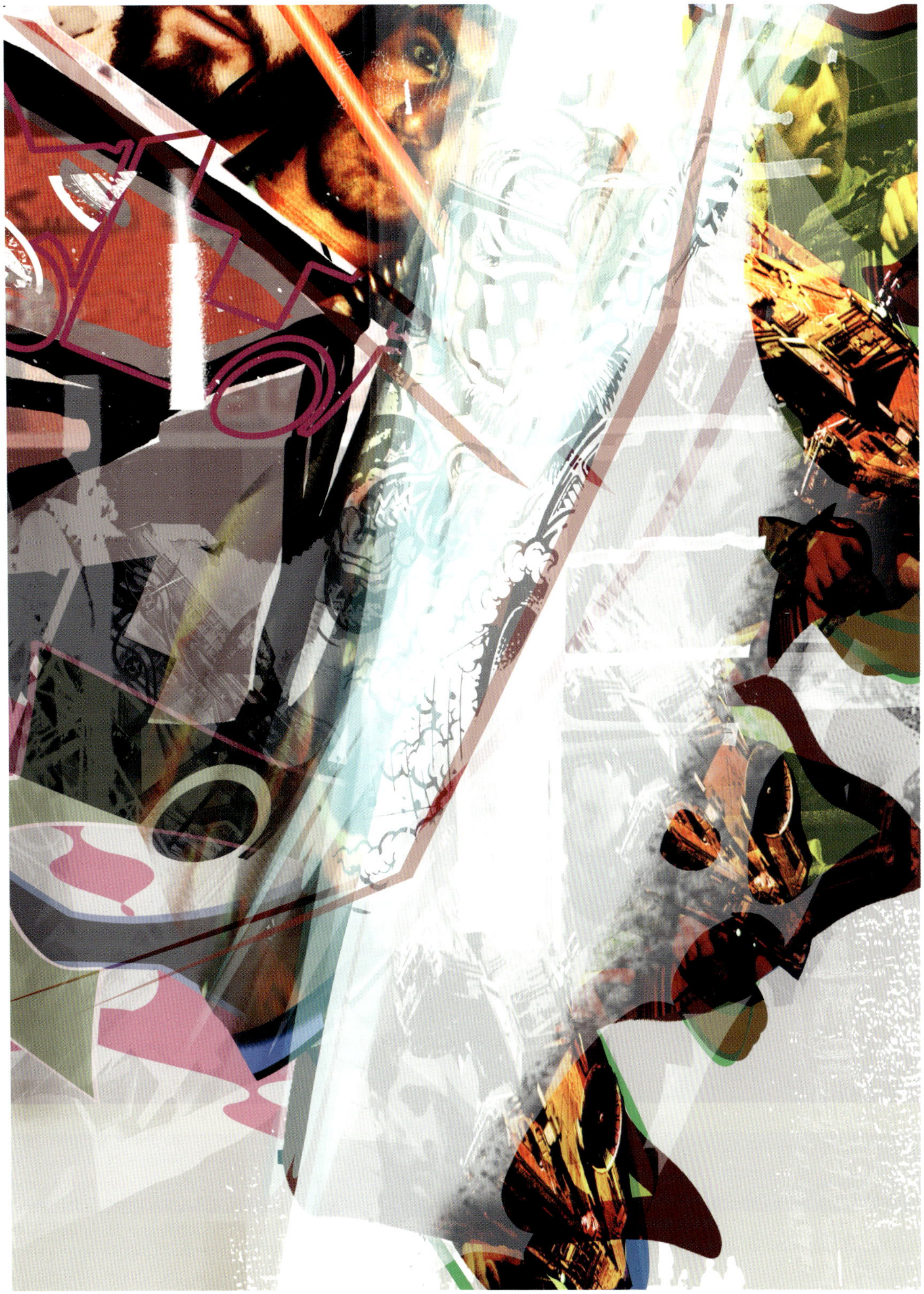

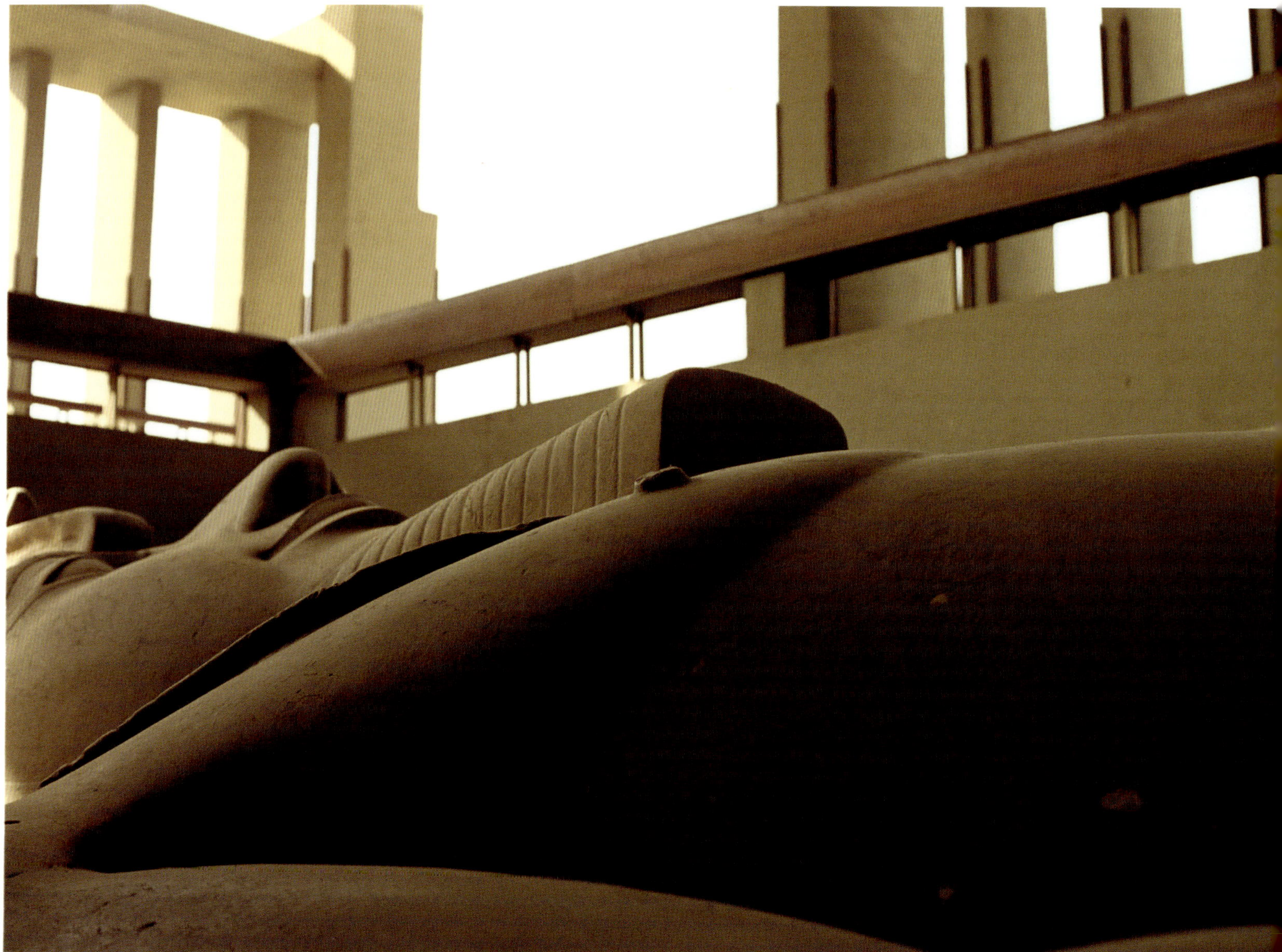

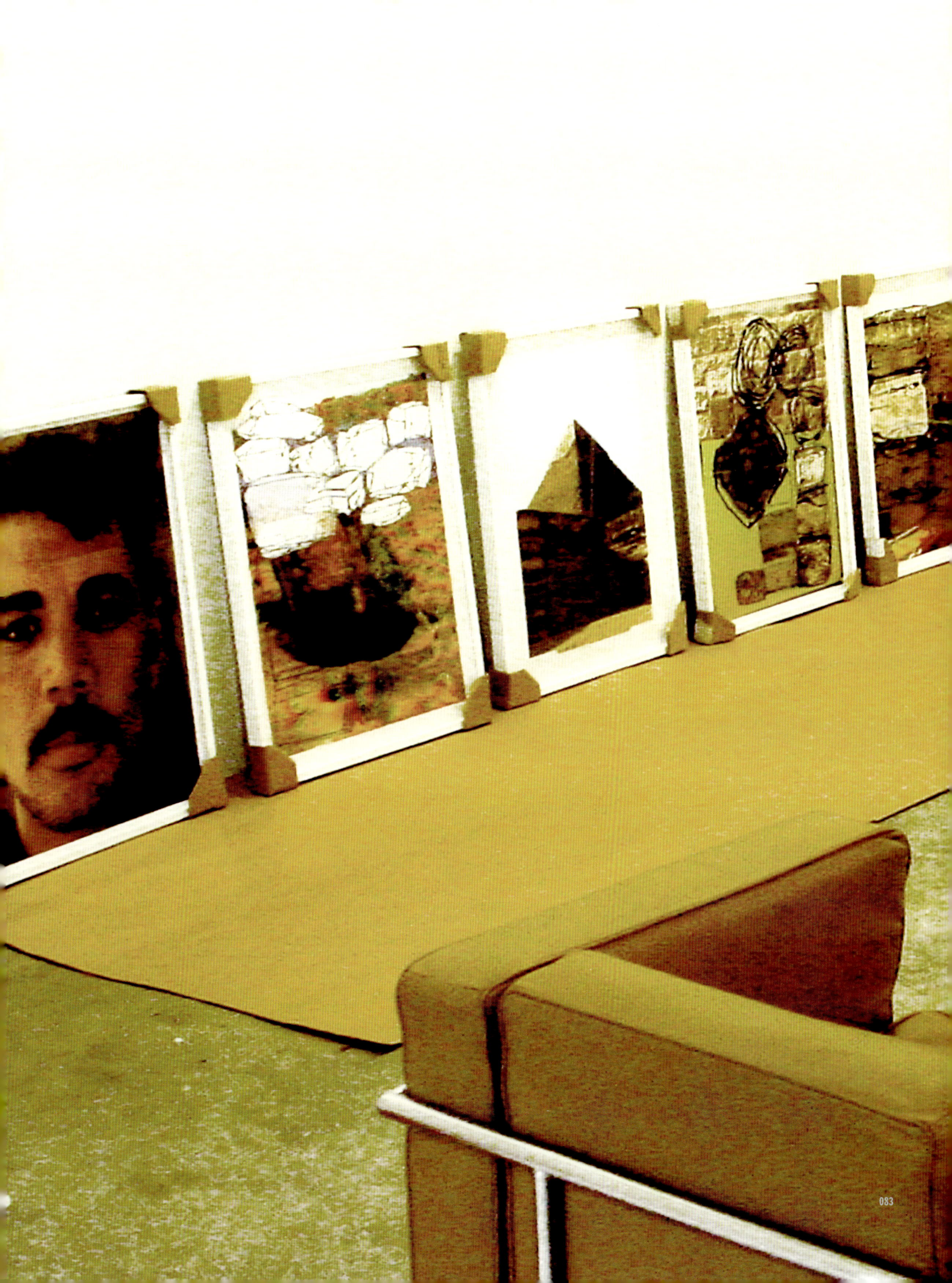

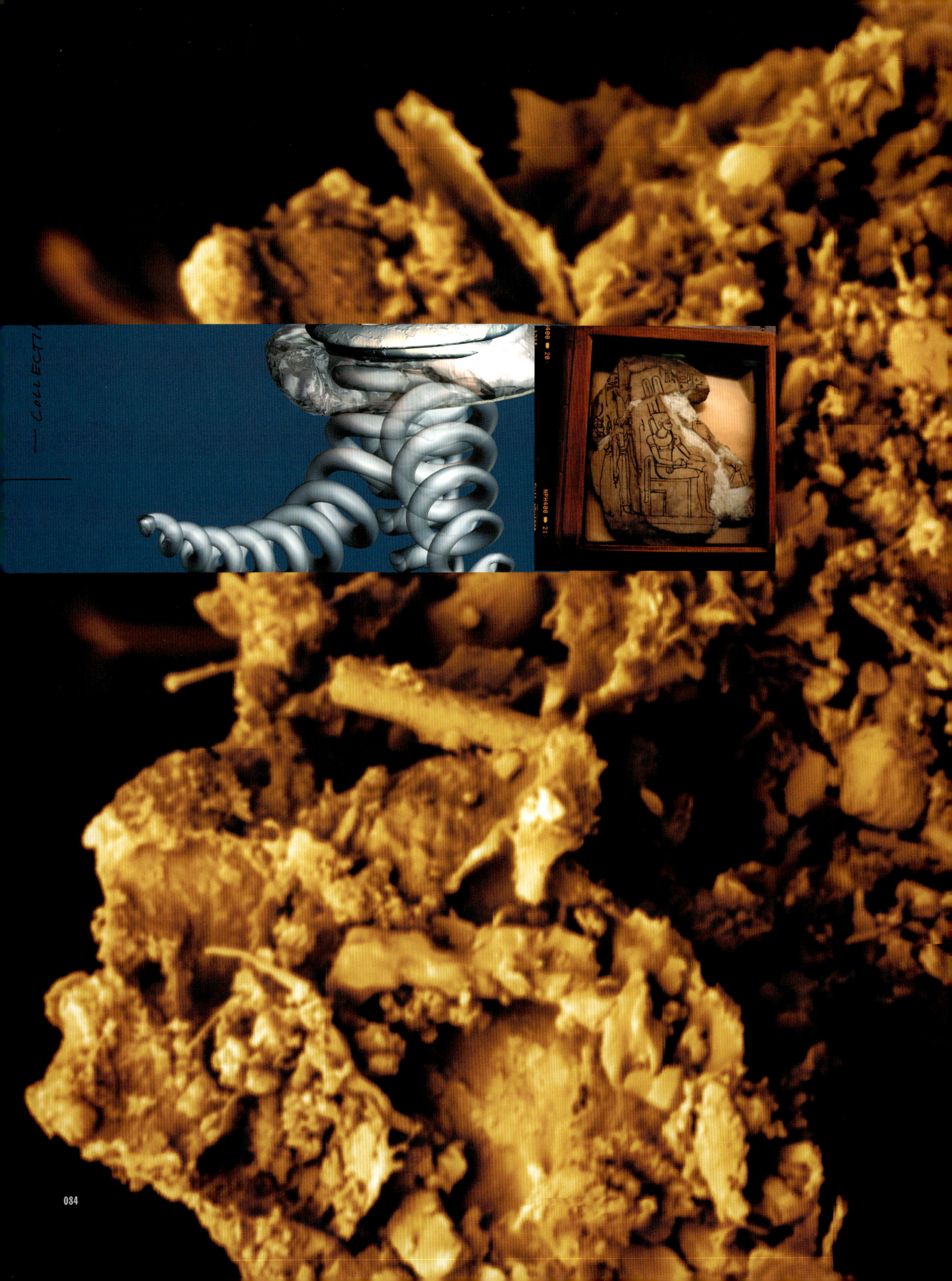

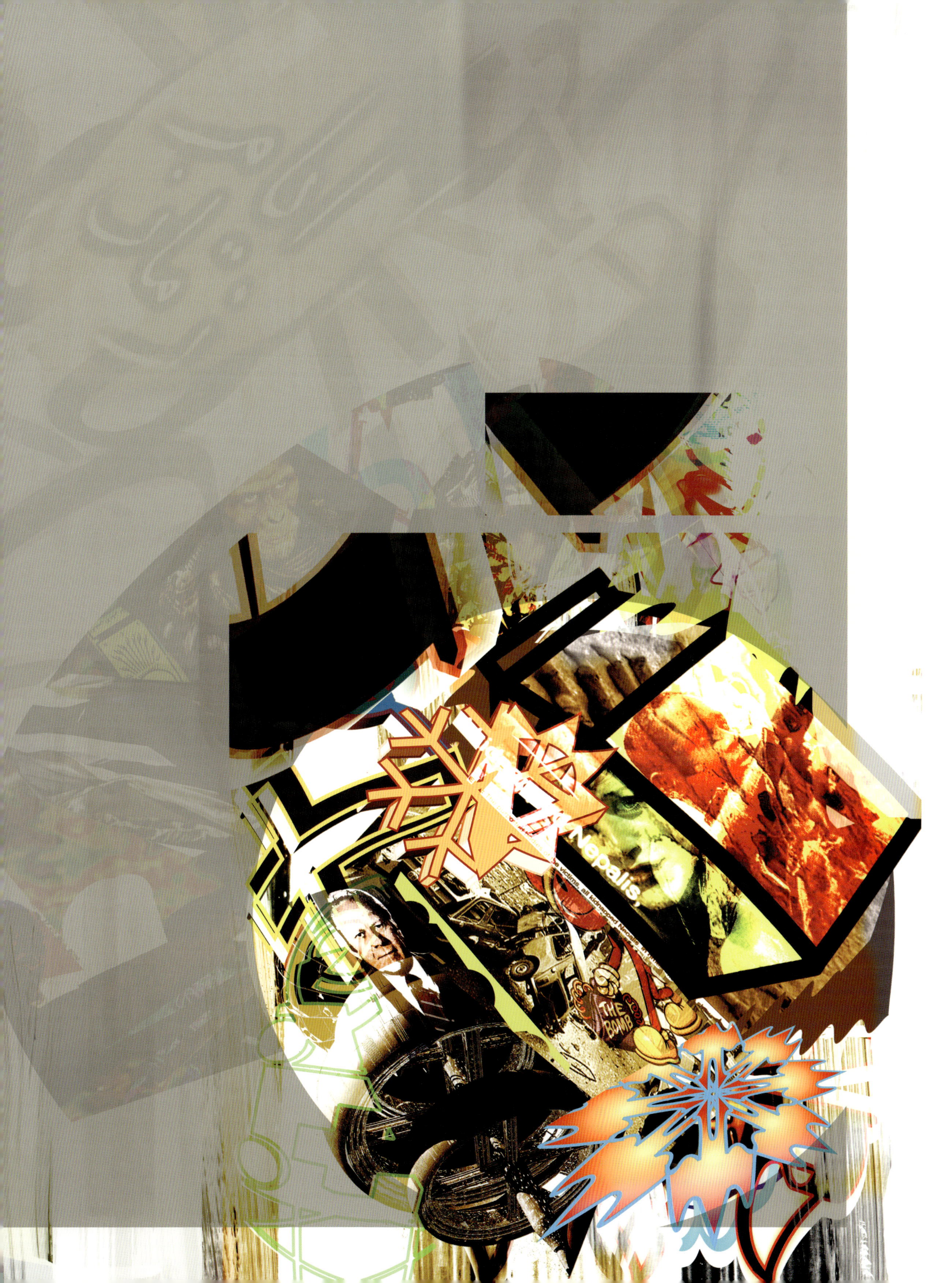
Nepalis,
THE BOMB

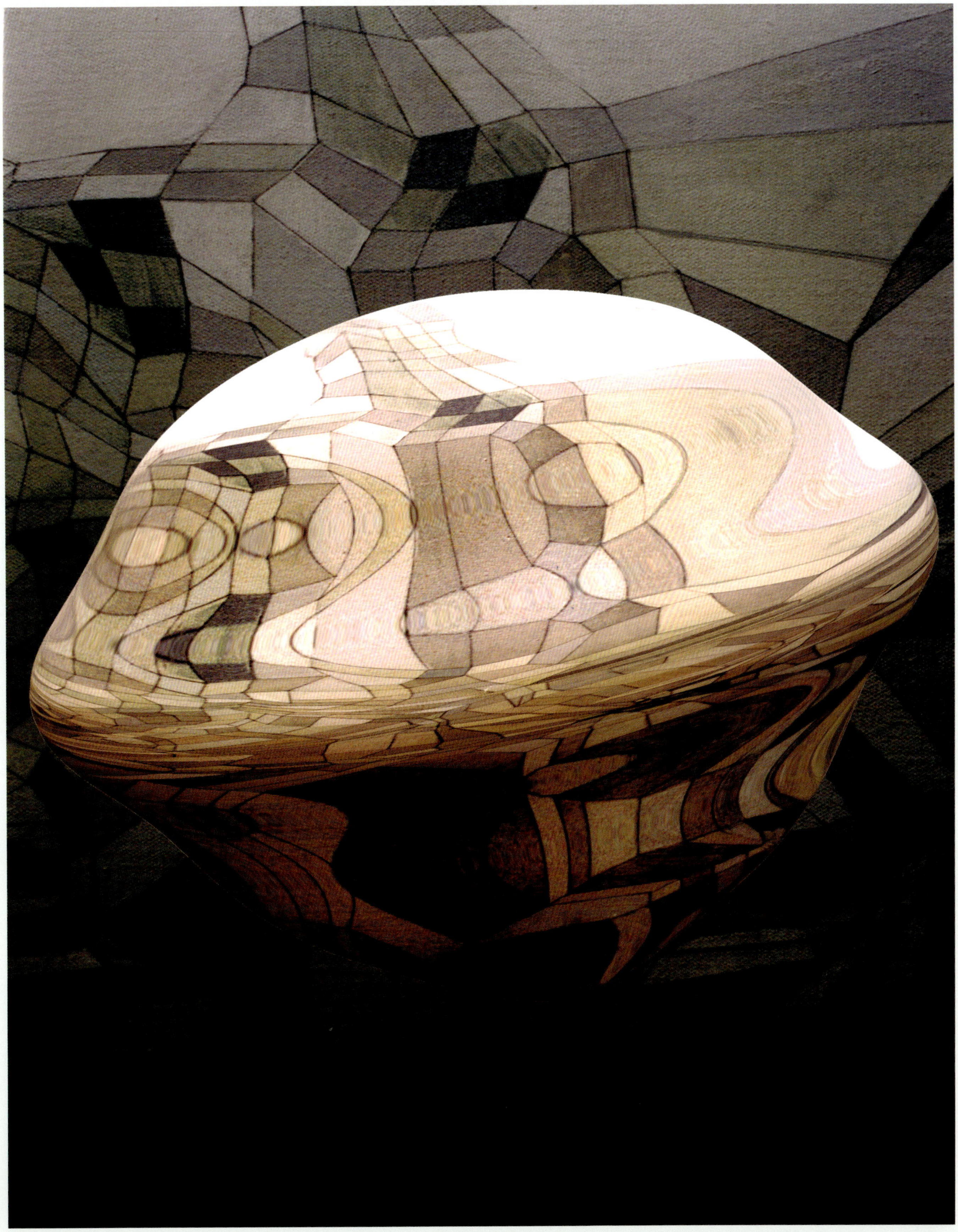

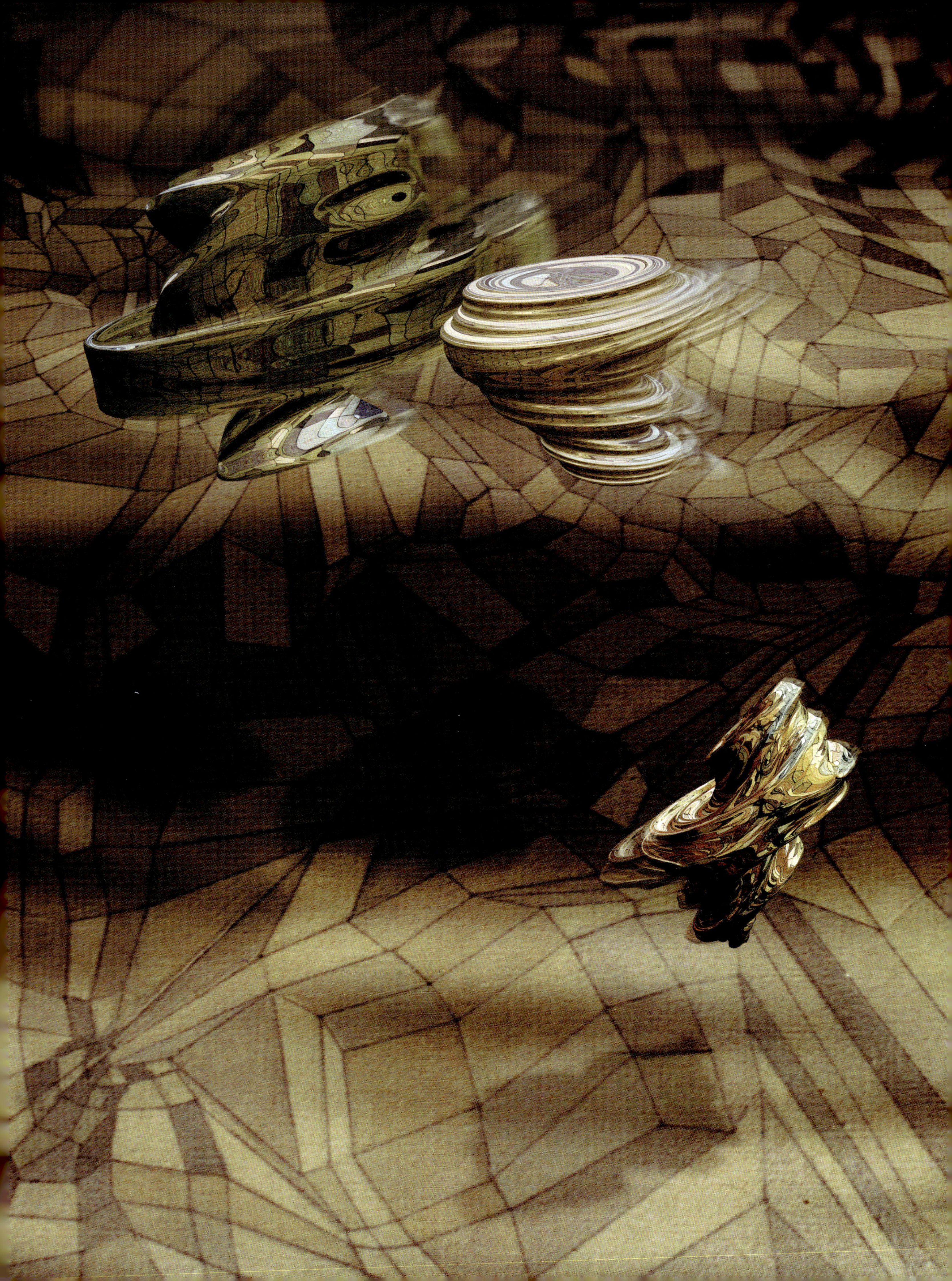

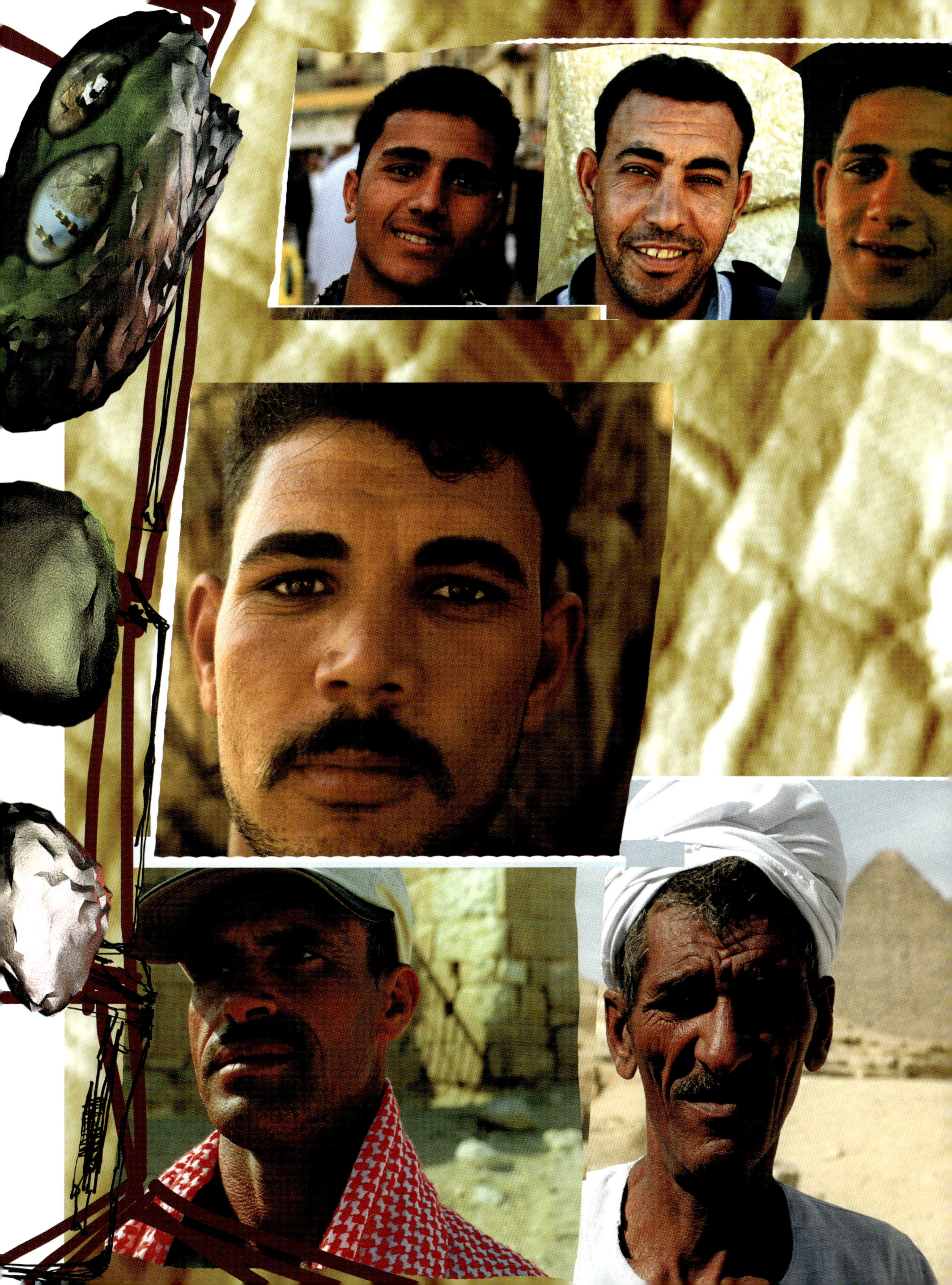

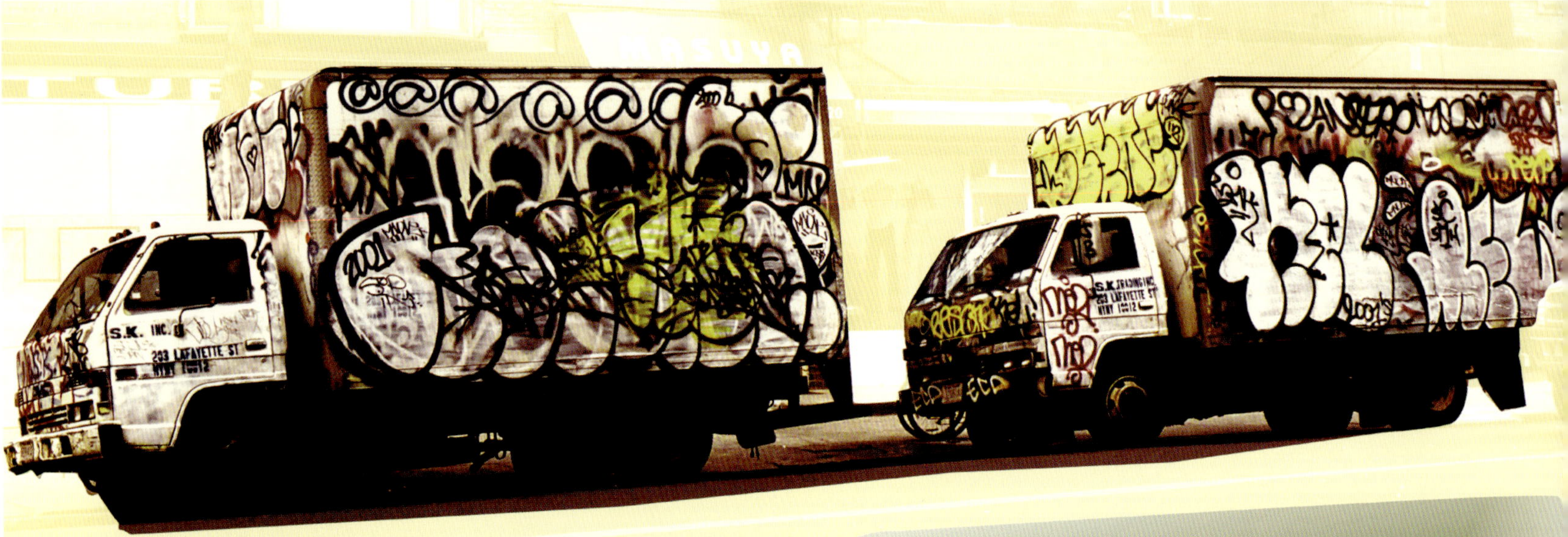

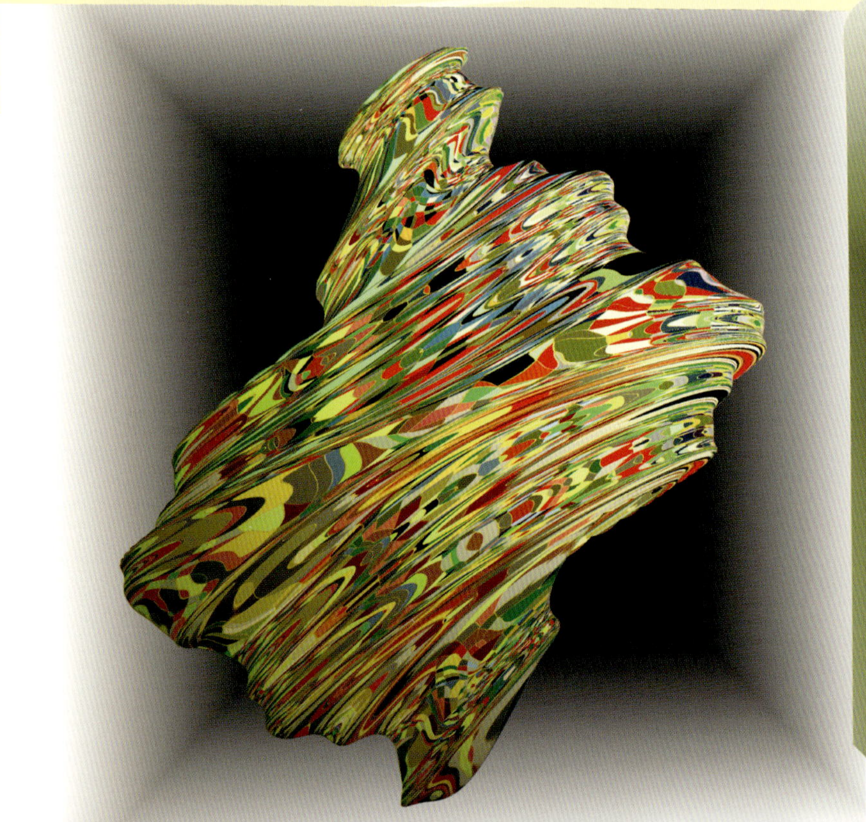

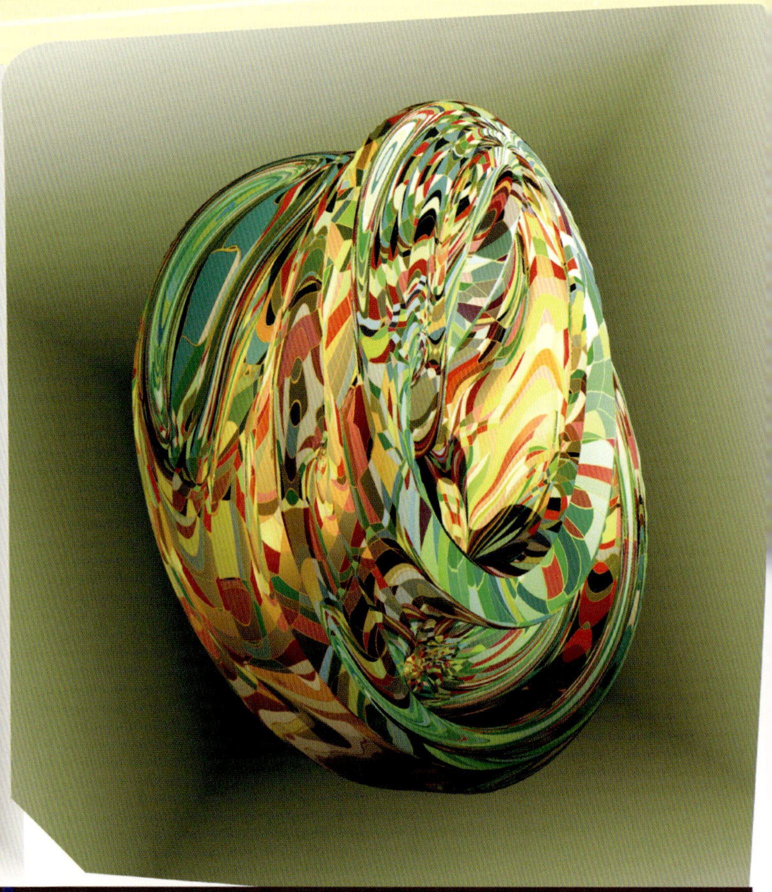

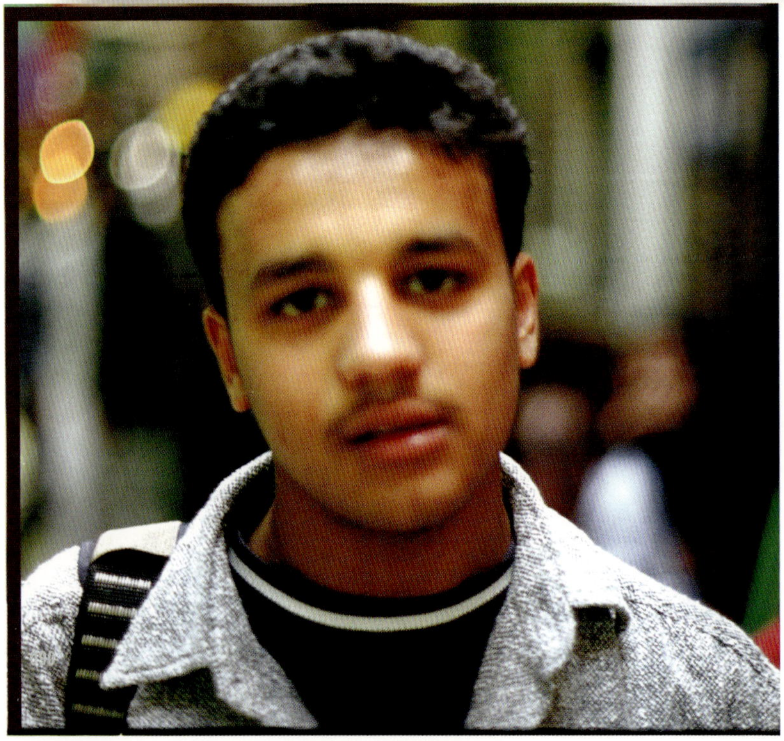

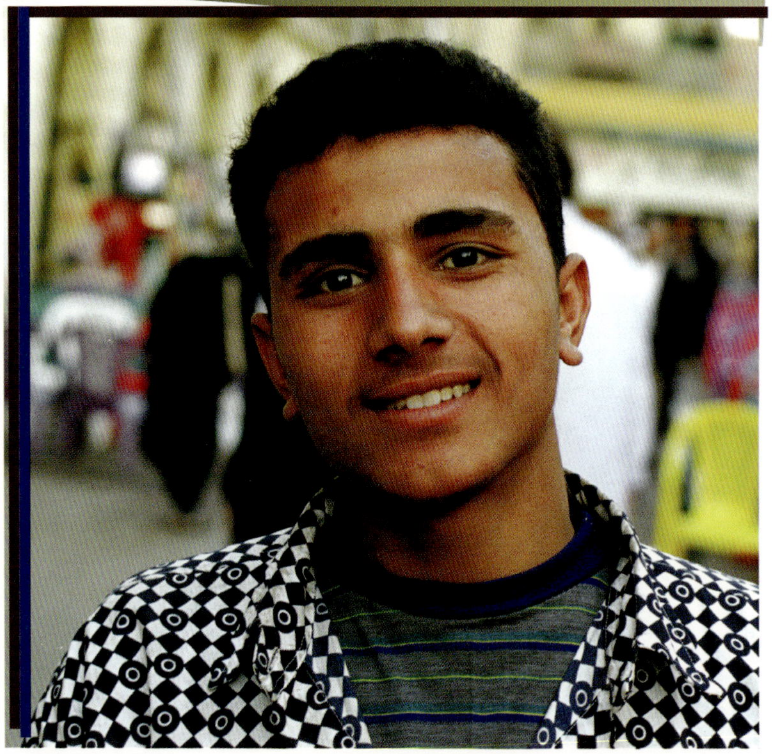

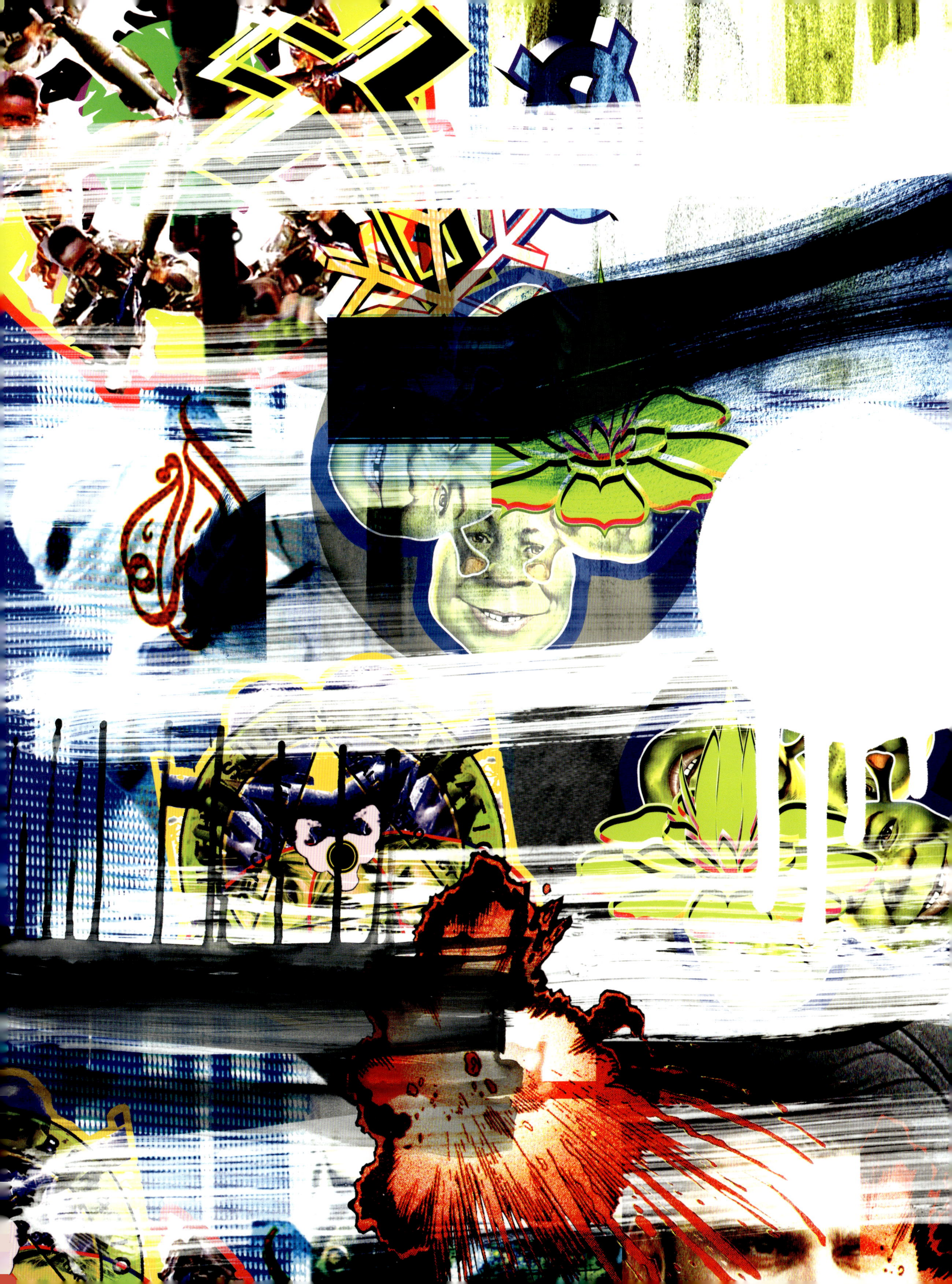

002_3 reflected/diffused waves with one model for inarticulate deep space. 1999.

003_Data Acquisition piece. 1997.

004_Data Acquisition, Panel 4. 20 x 30 inch light jet print, 1997. These prints were produced by Muse X Editions in Los Angeles. This was my first digital work, there were 11 pages in the series.

005_10/27b-10/30a, two moiré details. 18 x 24 inches, 1997.

007_Fictional EMR (Electro Magnetic Radiation) Pattern. 18 x 24 inch nova jet print, 1998.

008_Two SEM (Scanning Electron Micrograph) focus web shots, Citrine 2540x and Tourmaline 1500x, with two versions of Theoretical Space in the back gallery at AFA, 2003. The SEMs were made from samples photographed by the University of Iowa SEM camera. I manipulated the focus and panned remotely via a Website.

009_Data Acquisition, Panel 6. 1997.

010_Giza Sketches, Egyptian Museum. Cairo, 2002.

011_Cairo.

012_Model for a Banded Plastic Particle (Duchamp, Bosch, Braque). 20 x 30 inches, graphite on paper with digital additions, 2001.

013_2 Heads in the Egyptian Museum. Cairo, 2002.

015_Cairo.

016_Close Up. Cairo, 2002.

017_Debris (detail). 2004. This is a model of a Burned Out Juncture of an Eroded String Piece.

018_View of a J. St. Bernard piece. Paris Bar, Berlin.

019_Rocky Horizon #3 (detail). 20 x 30 inch ink jet print, 2004.

020_Esthetic Particle (green version). 2003.

021_Particle Surface Graphic with Minarets. 2002.

022_Photo migration and ladies room with prints at AFA (Data Acquisition, Placebo Sets, and Photo Magnetic Test Patterns). 1999.

023_Africa Gas. Morocco, 2003.

024_11/02c. 1997.

025_Sketch of the planet-scale culture-schema. Egyptian Museum overhead shot. Cairo.

027_Shop owner. Cairo, 2002.

028_SEM Topology, 3 Connecting Pieces. 20X30 inch ink jetprint, 1999.

029_Egypt_Kufu SW/lower Giza, 2002. The first trip to Africa came about when Judith Barry and Ken Saylor asked if I wanted to tag along. Judith was installing a show in Cairo. We rode camels on the Giza Plateau, it was a fun tourist thing to do. Later, Ken and I went back and a guide let us into some chambers that were closed to the public.

030_Planck Context @ $10^{-43}$ sec. This is the brown version from a group of frozen backgrounds. 2003.

031_Giza Doorway #2. Giza plateau, 2002.

032_Three shots of the Photo Magnetic Receiver. 2000. Installed in San Francisco, with various schematic illustrations. On the bottom are two shots of the installation at AFA, one with Danny, the other with a clear shot of the array shields.

033_PMR at AFA with Colin. 1999.

034-5_Looking NW Sahara_SEM muscle and tourmaline.

036-7_Planck Moment 5b (detail). 2005. This particle model is from a series that incorporates the "remnant trajectory" (my term). In this concept scenario, remnant trajectories are the physical description of displaced energy as fundamental particle fragments crisscross space at the Planck scale.

038-9_Rocky Bluff. Morocco, 2003.

040_Giza Graffiti. 20 x 30 inch ink jet print, Giza Plateau, 2002.

041_Egyptian Man, Full Face. 20 x 30 inch ink jet print, 2002. This picture was taken further south in Memphis. I thought this man had the particular quality of looking young but connected to an ancient order.

042_Edge of the Sahara, near Saqqara, with two insets; Khafre_Close Up (Giza Plateau) and Epistemological Mason Meme 1 (detail). 2002. This picture used a swipe from the NASA Viking Lander 2 EDR images of Mars.

043_Stacked Blocks, Two Shots. Giza Plateau.

044_Giza Passage. Giza Plateau.

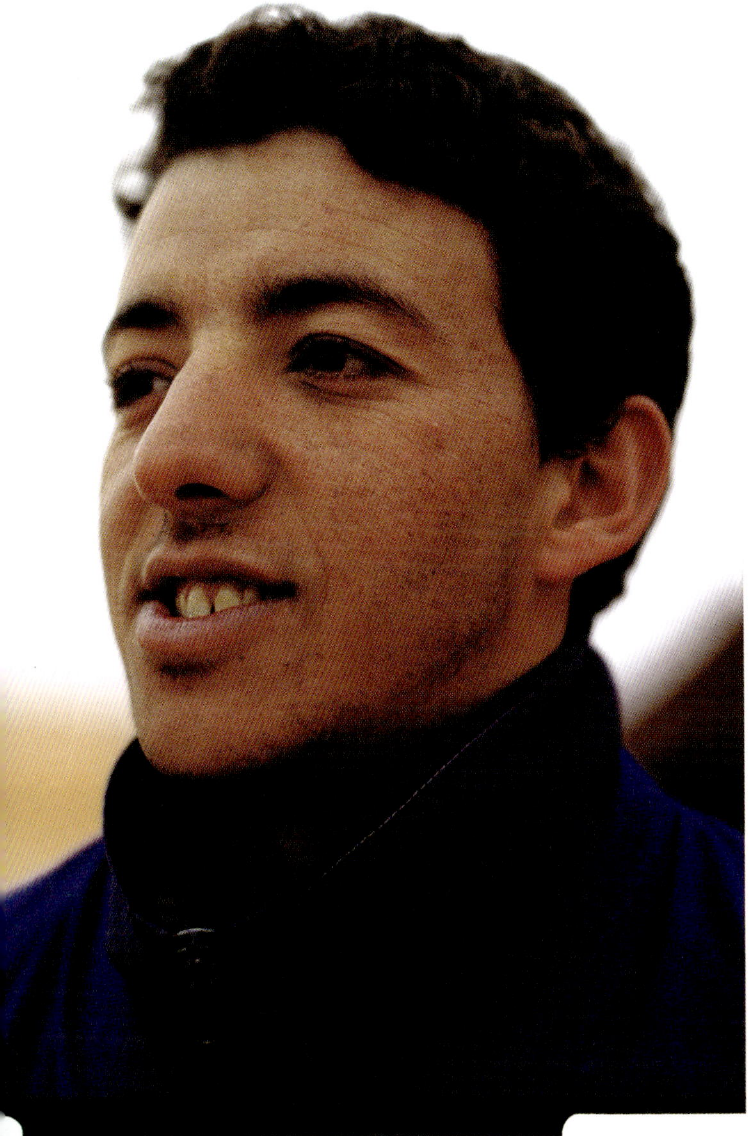
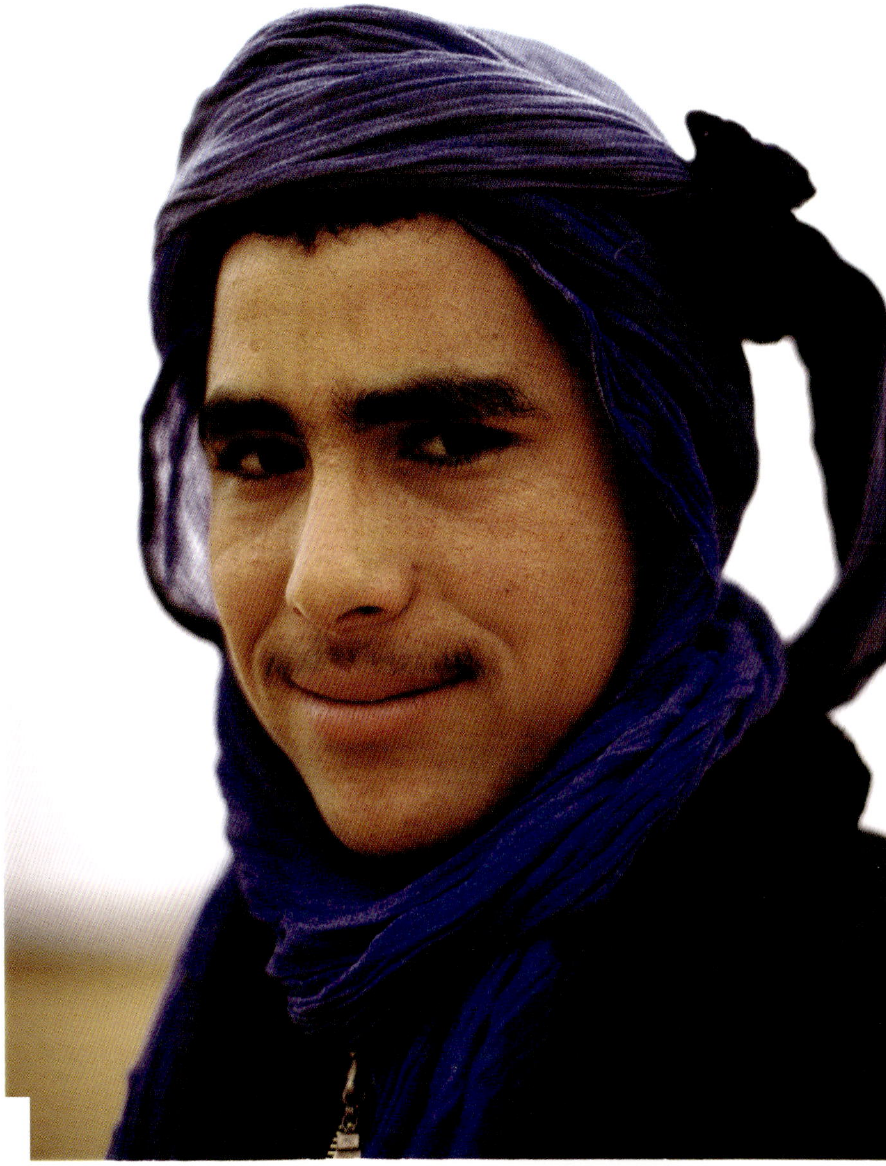
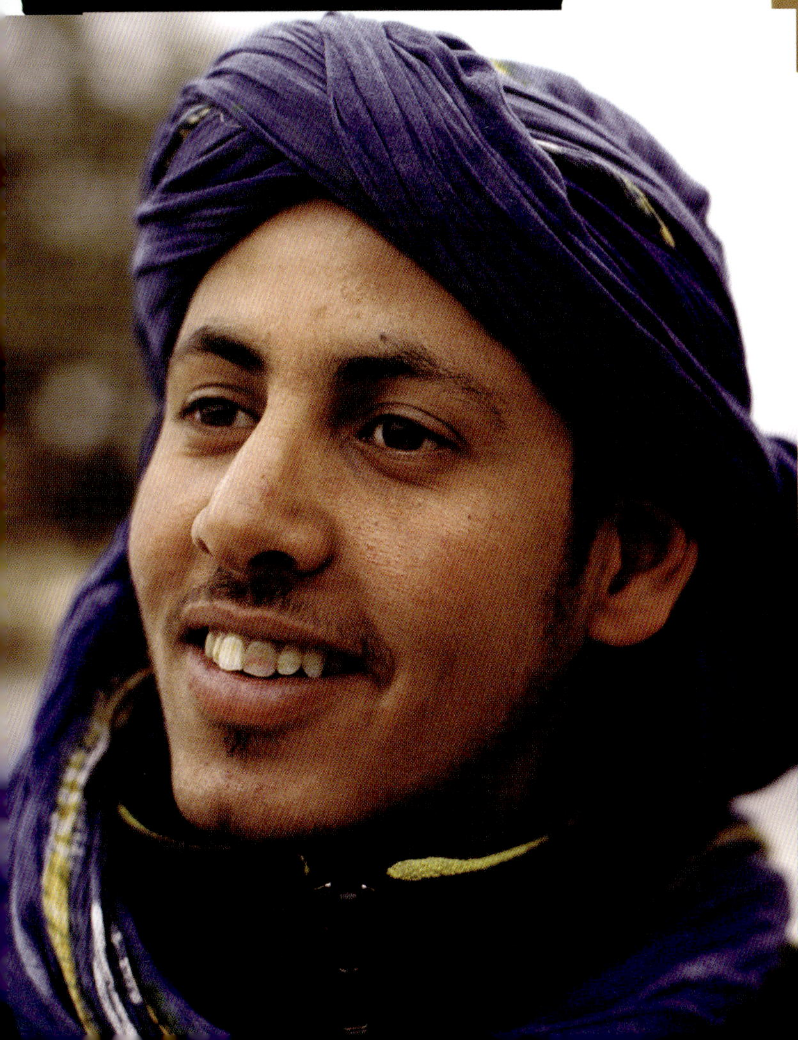
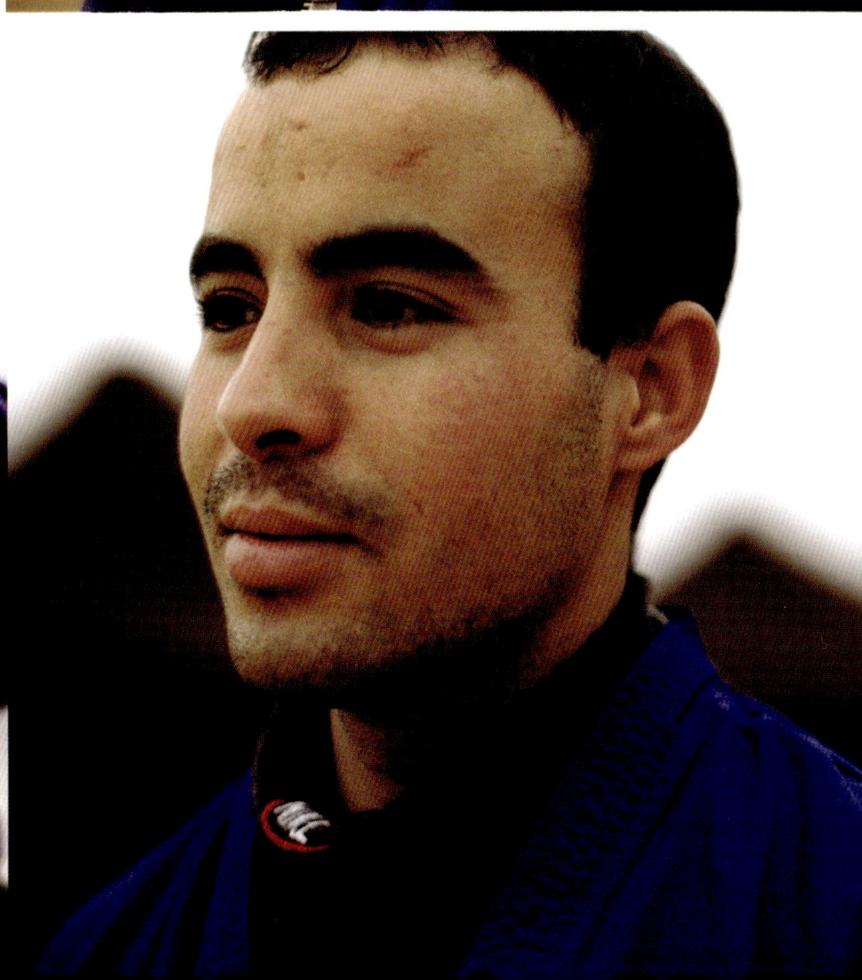

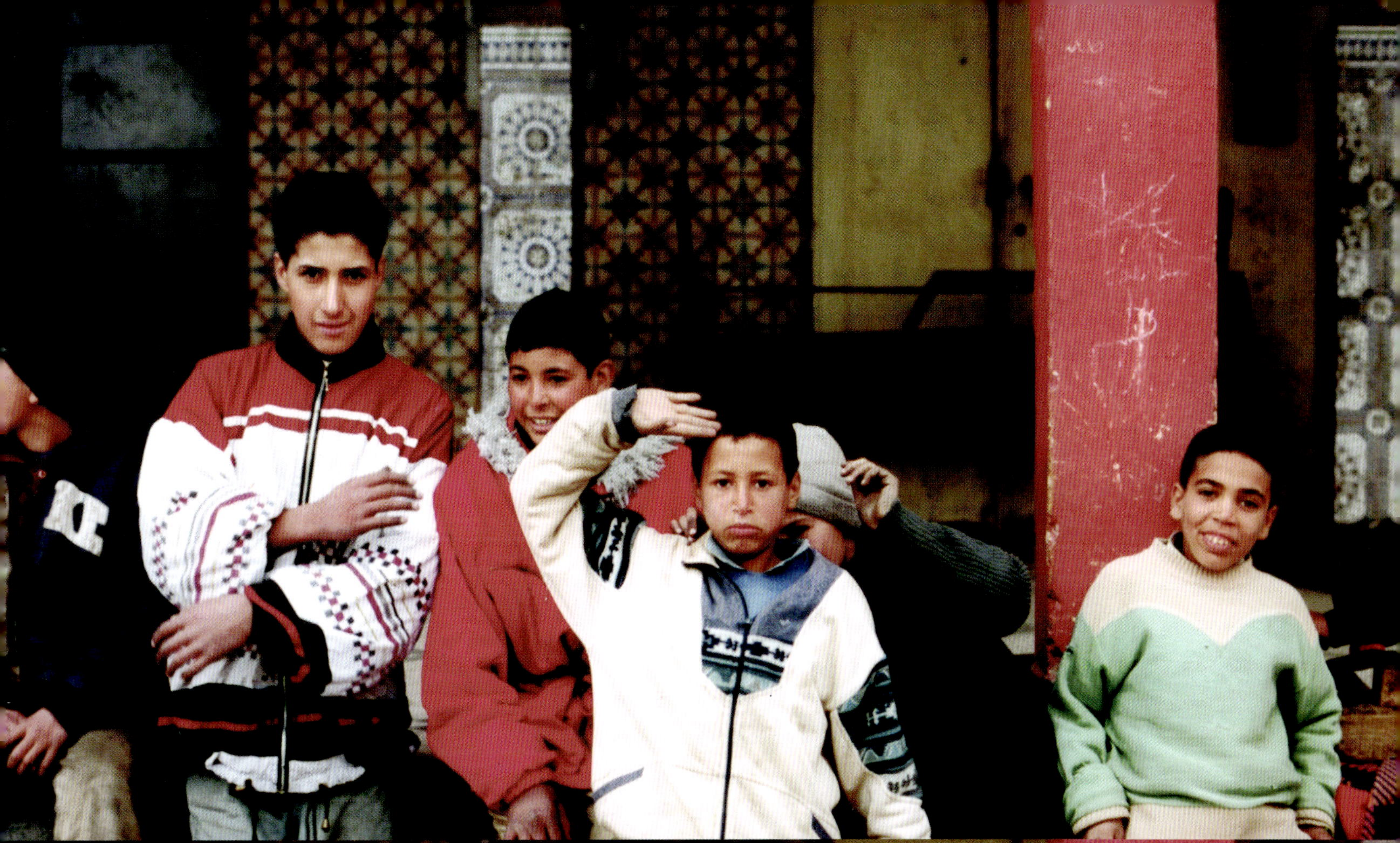

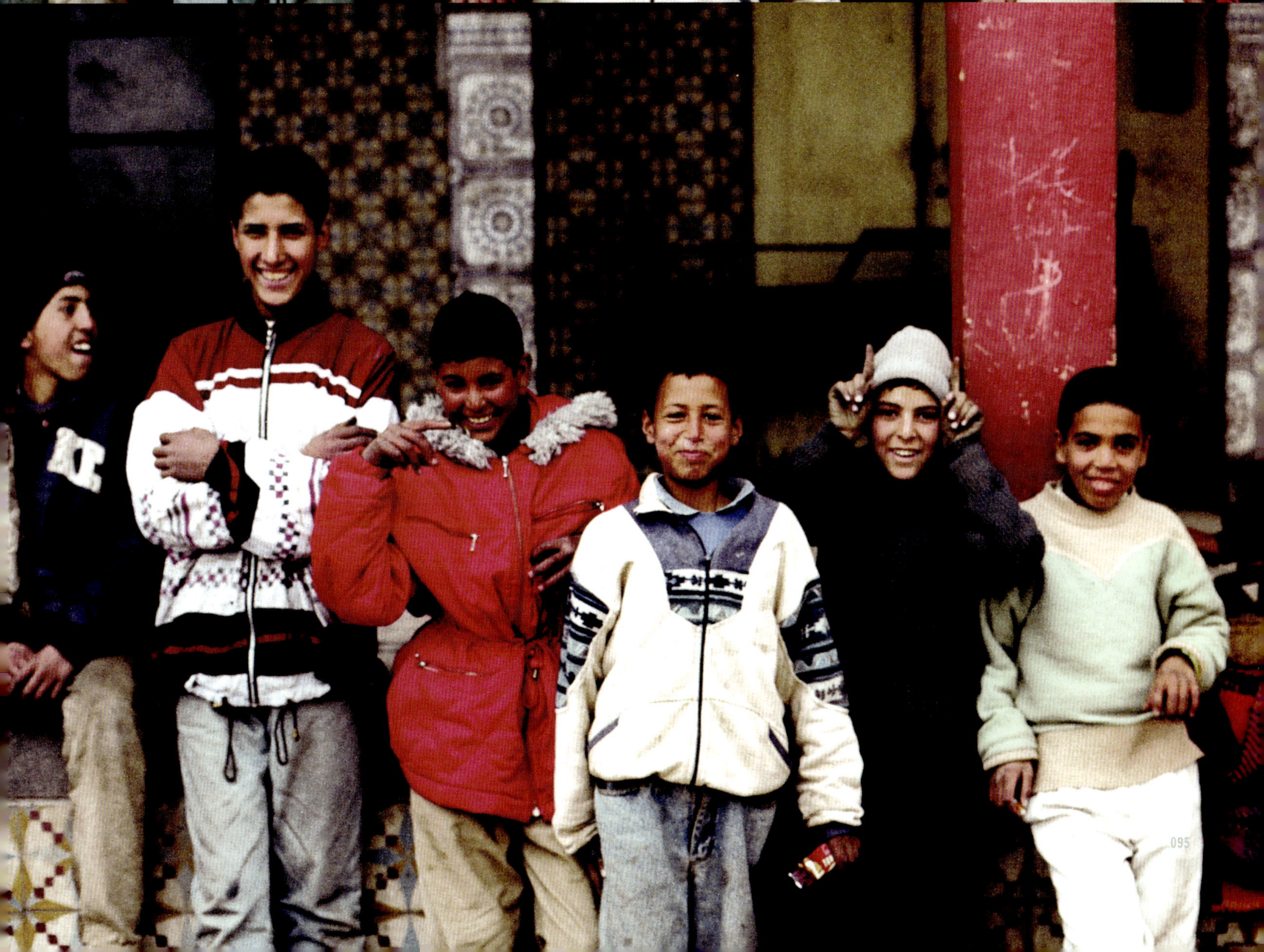

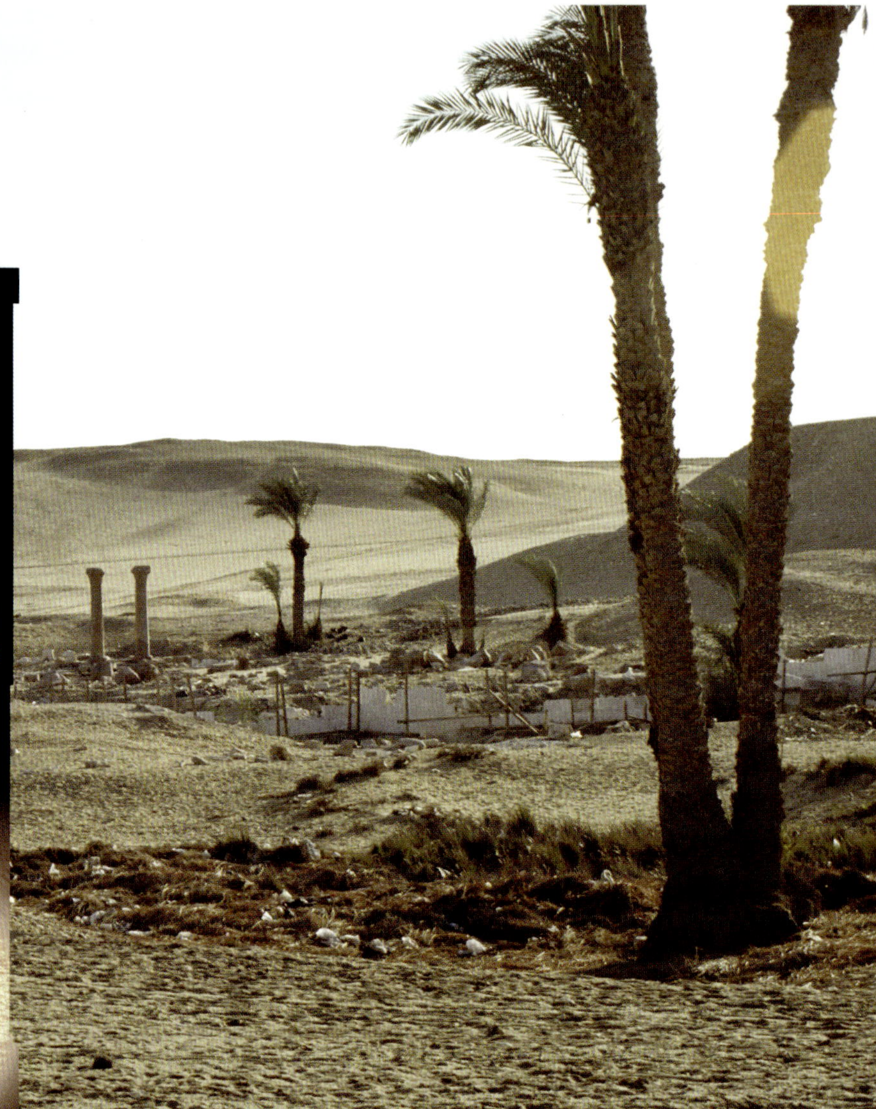

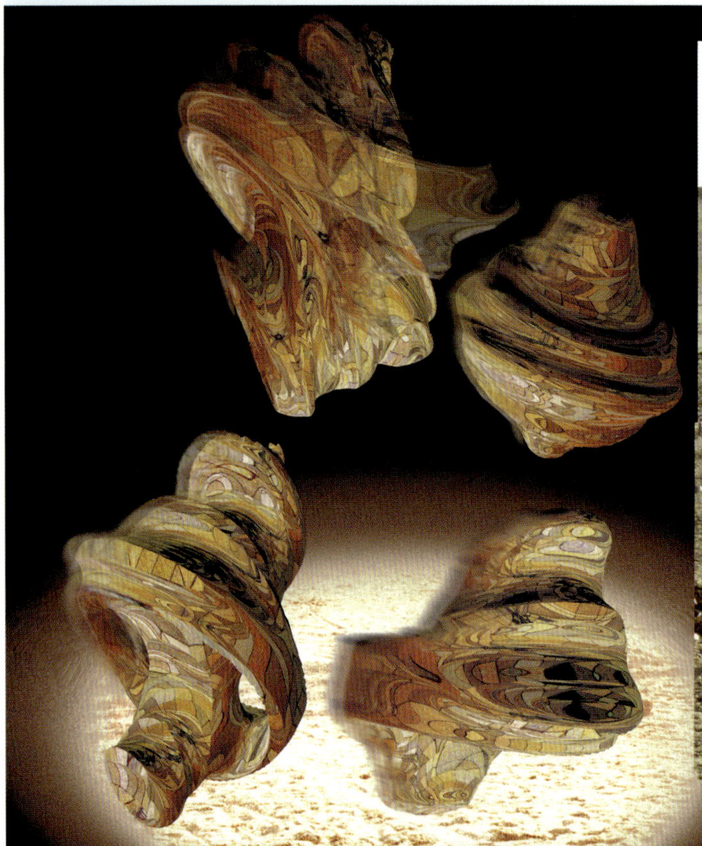

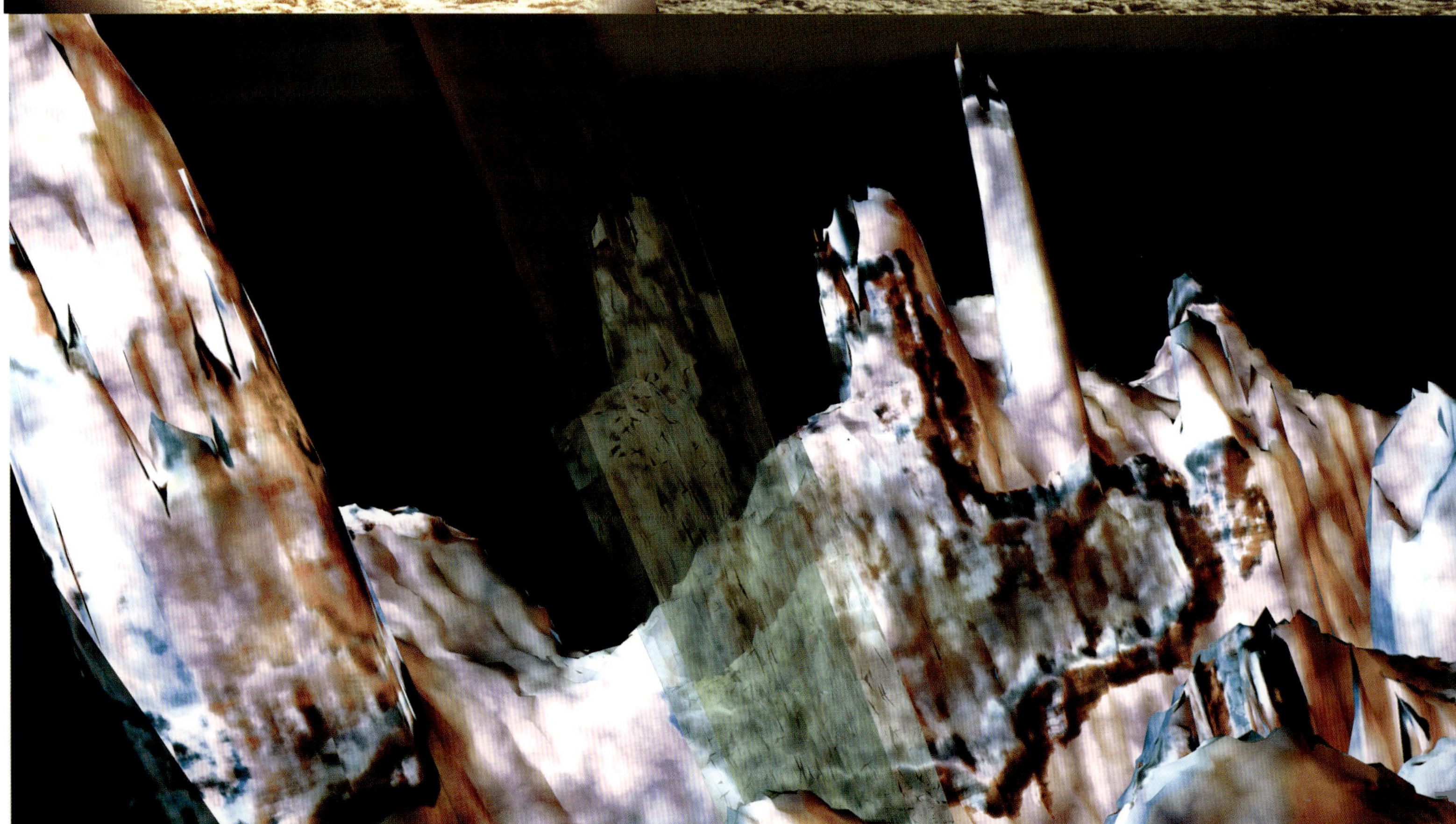

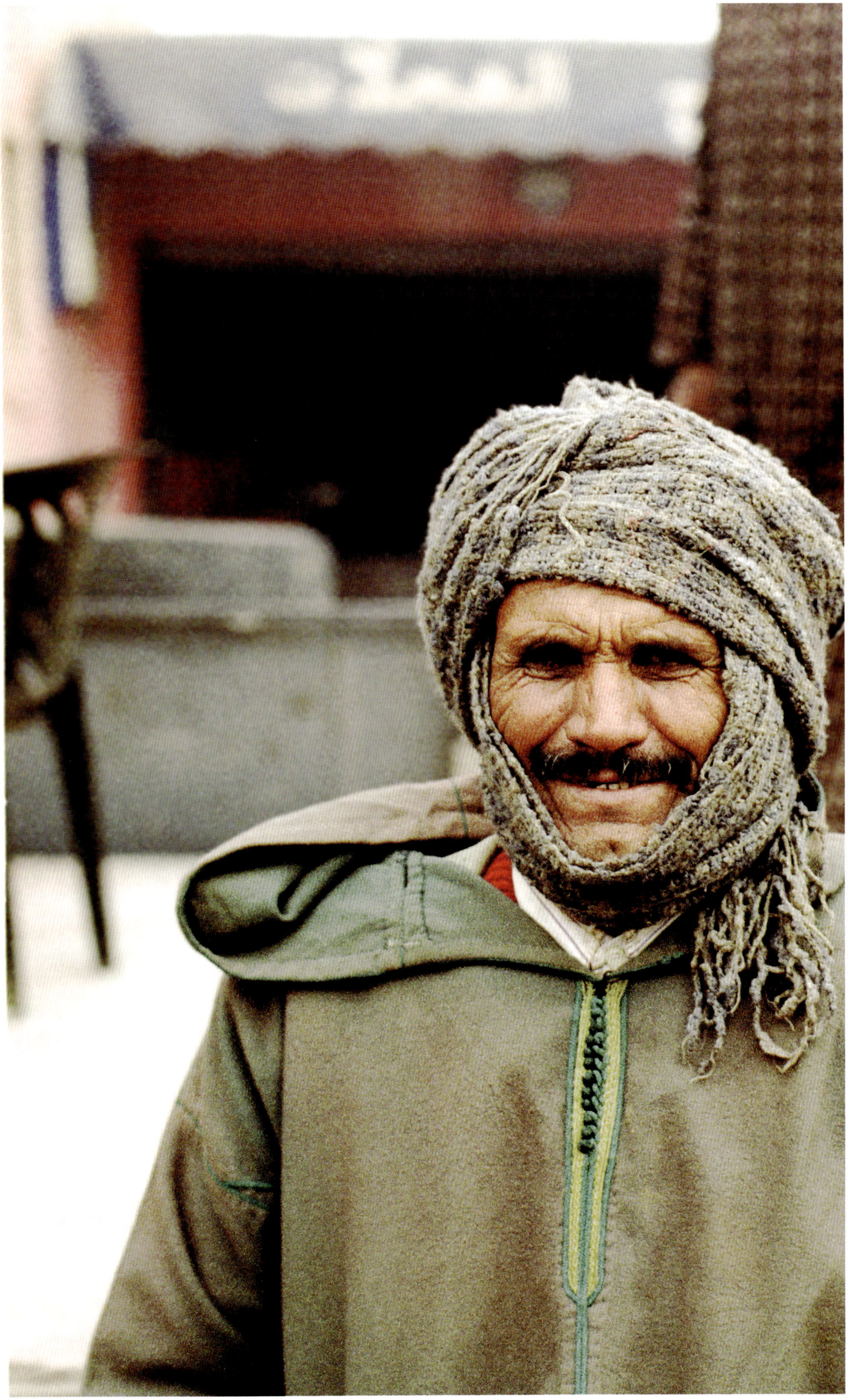

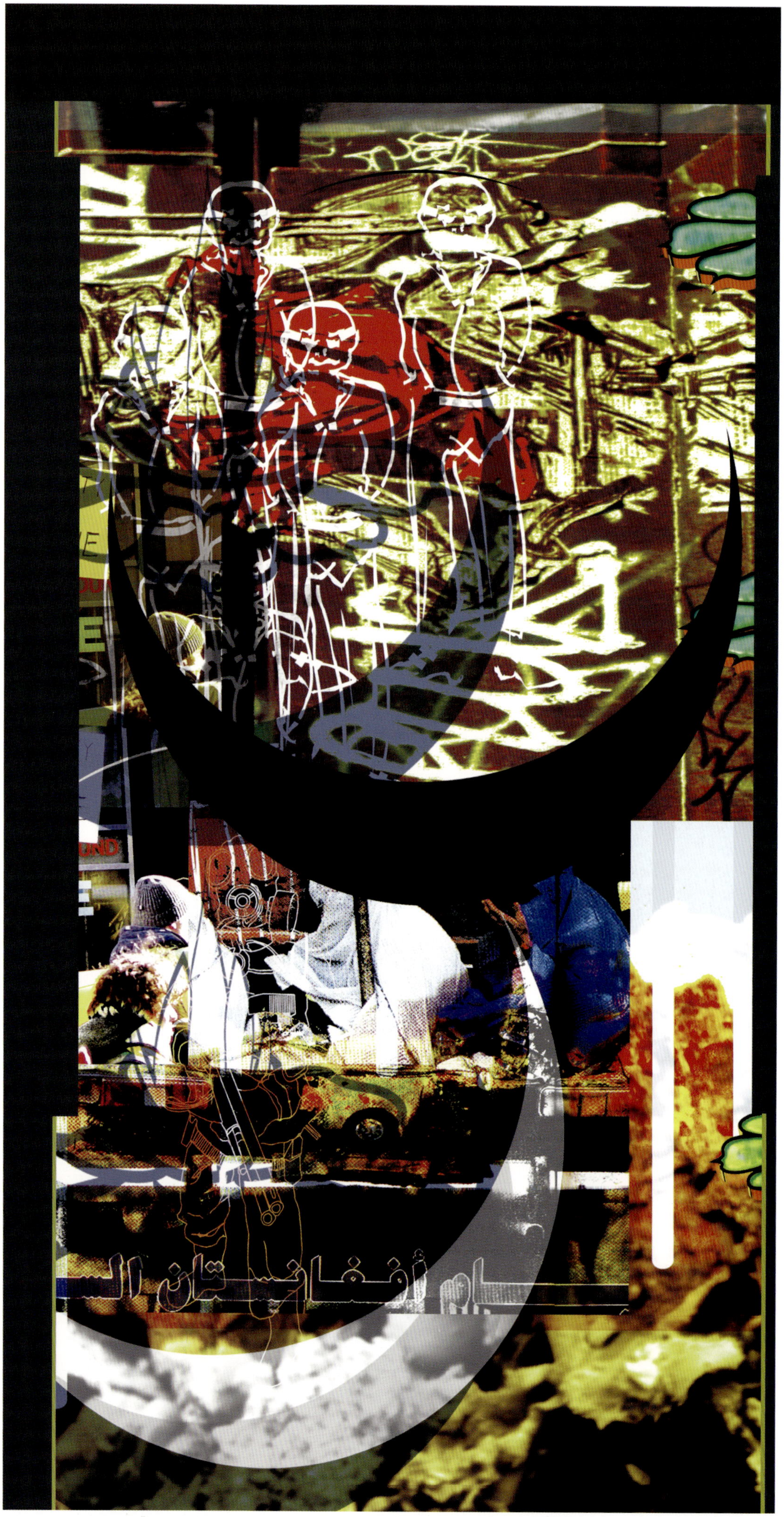

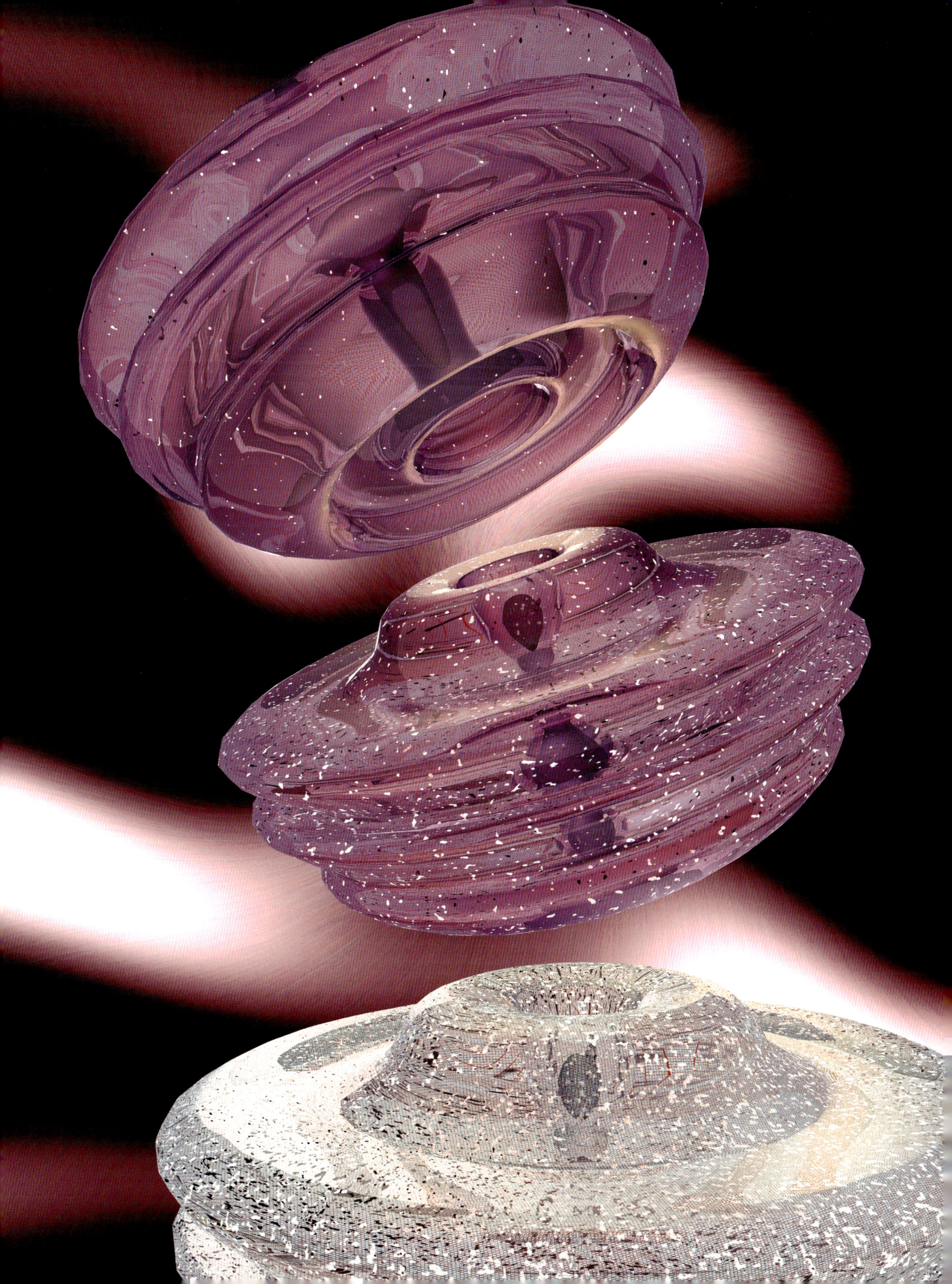

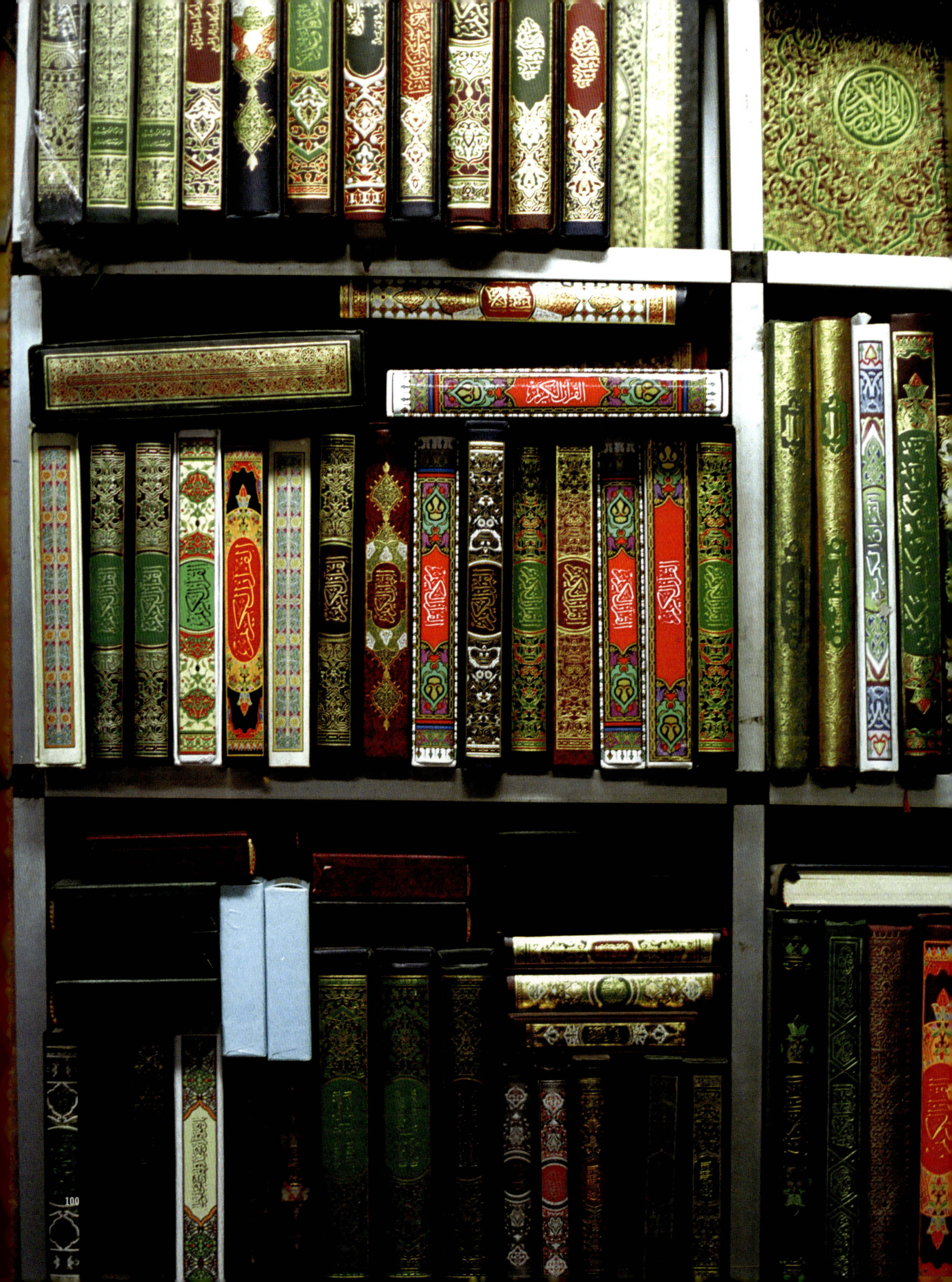
القرآن الكريم

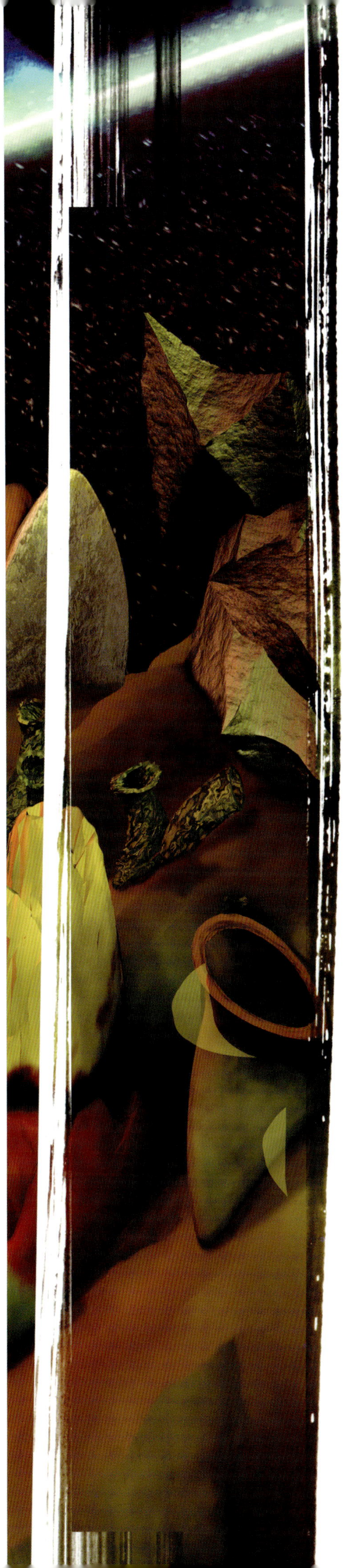

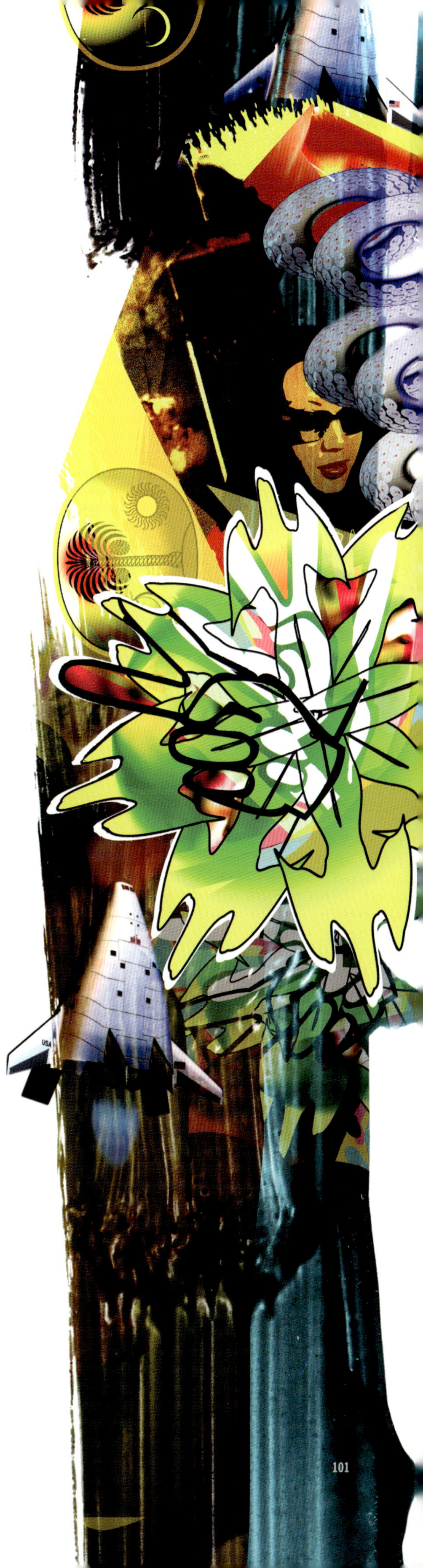

045_Egypt Soft Face. Cairo, 2002.

047_2 Women South of Cairo. Cairo, 2002.

048-9_Solid South Face. Menkaure, Giza. Insets include Cairo Department Store and Yogi Rock from NASA/Mars.

050-1_Planck Moment 5c. 2005. Fez, apartments for similarity, some graffiti from New York.

052_Untitled (detail), with narrative scrap. 18 x 24 inch digital pairing of a theoretical particle and Anna Karina, 2004.

053_Sphinx Signage. Memphis, 2002.

054_Egypt, East Kufu Doorway. Giza Plateau, 2002.

055_Magazine swipe.

056_10/28a. Clipped from the Data Acquisition Edition.

057_SEM 2/10/01 Quartz 160x. 20 x 30 inch ink jet print, 2001.

058_SEM 2/18/01 Quartz 160x. 20 x 30 ink jet print, 2001.

059_24 x 30 Sheet of Atoms. Magazine swipe.

060_Planck Moment rough 4v. 2004.

061_Foresight Monitor 11/00. 2000. I took this shot at a conference for nanotechnology while researching an article for Issue Magazine. Initially, nanotech was interesting to me because it was the smallest scale that technicians were probing with imaging technology.

062_Graffitti samples. 2005.

063_SEM 2/11/01 Quartz 150x #1. 2001.

064_SEM 2/11/01 Quartz 150x #2. 2001.

065_Carbon Nanotube Model #8. 2002. This is from a series of designs for nano-scale models using carbon tubes actually produced by NASA. Primarily I was interested in designing models at increasingly smaller scales that were based on or used existing technology.

066_Previously published as "Grey Goo" in *Issue* Magazine

067_Nantube Model #5. 2002. With a NASA image.

068_Carbon Nanotube Models #1, #2, and #3. 2002.

069_Men in the Alley in Fez. Morocco, 2003.

070_SEM 2/10/01_Citrine_380x. 2001.

071_3/03/01 Corrupt Geography 3c. 2001. These pictures started as a large painting I did on Broome Street. They were photographed much later and digitally altered. This was one way I could address the complexity of a simple corner vector in 3 dimensional space.

072_Carbon Nanotube Model #6. 2002.

073_Scanning Tunneling Micrograph with narrative scraps. The scanning tunneling microscope (STM) is a type of electron microscope that shows three-dimensional images of a sample. In the STM, the structure of a surface is studied using a stylus that scans the surface at a fixed distance from it. Electrons tunnel between the surface and the stylus, producing an electrical signal. The stylus is extremely sharp, the tip being formed by one single atom. It slowly scans across the surface at a distance of only an atom's diameter.

074_2 Nano-scale Sorting Rotors. 2001. Accompanied by text from "Grey Goo".

075_New From Old/Compound Model X 2. A simple abstraction or literary device where it doesn't really belong.

076_Particle Surface Study #1, #2. 2002. Digital sketches with the idea of esthetics. (Bosch, Duchamp, Braque.)

077_2/25/01 Corrupt Geography 2, 3b. 2001.

078_Meme topology 3D #2. 2001.

079_A collection of models used to gather, bind, and transport.

080_Narrative scrap, one a quote from Roger Penrose. Penrose has been labeled a polemicist and a neo-Platonist by his colleagues.

081_This detail is from an unprinted canvas, Taliban and Big Daddy Roth.

082_Memphis.

083_Pictures laid out at AFA for the installation "particles + waves with plausibility".

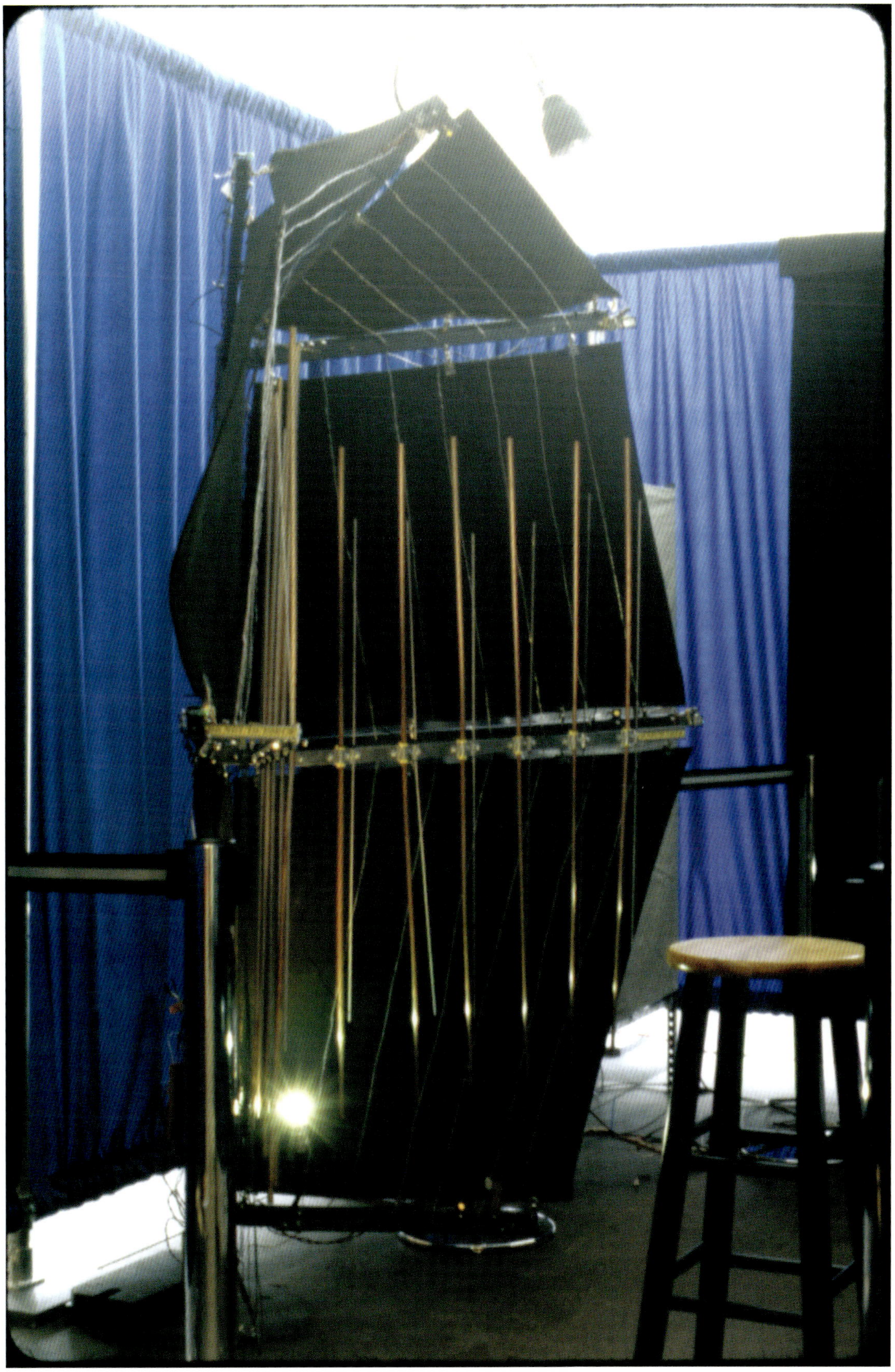

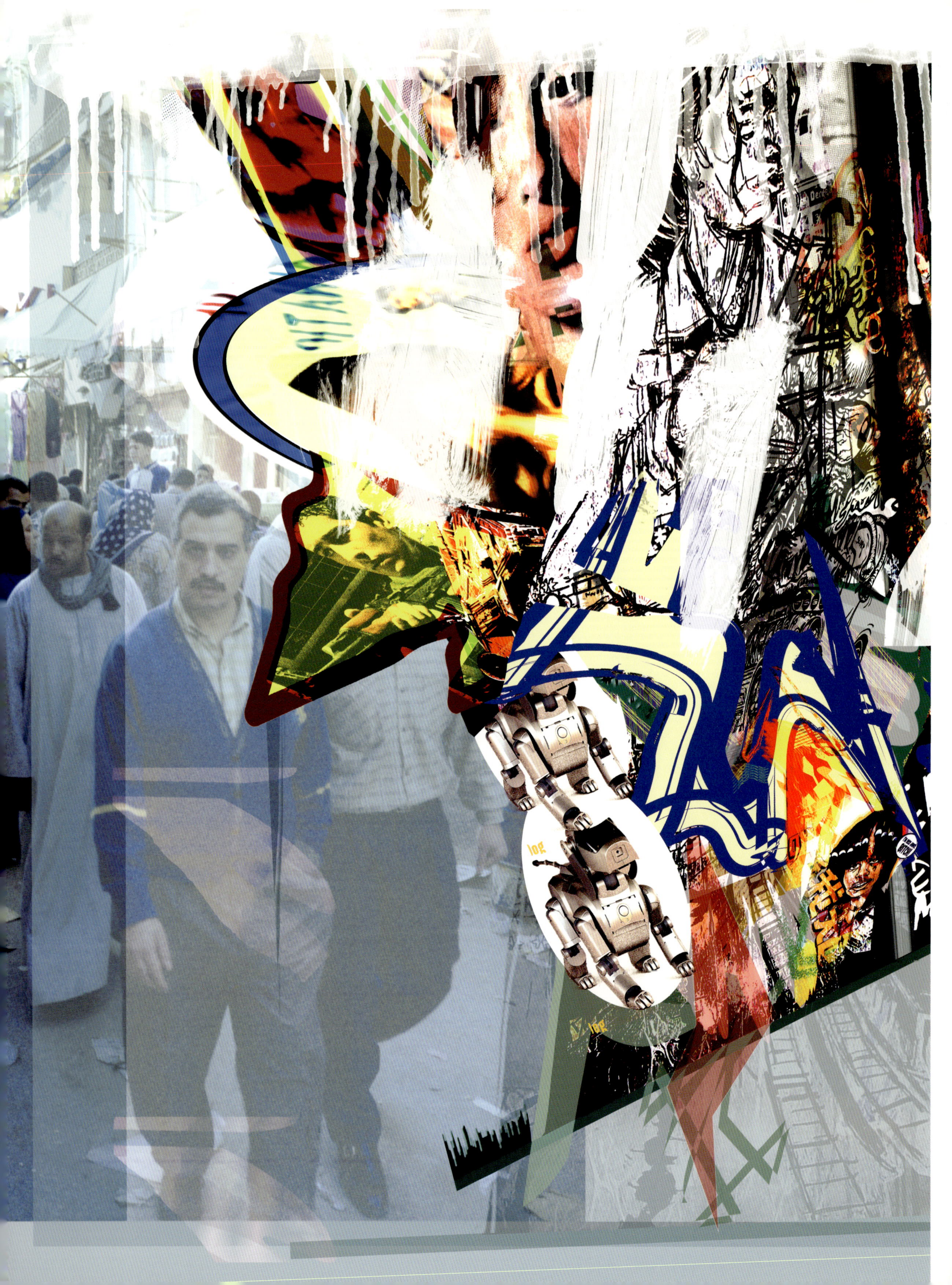

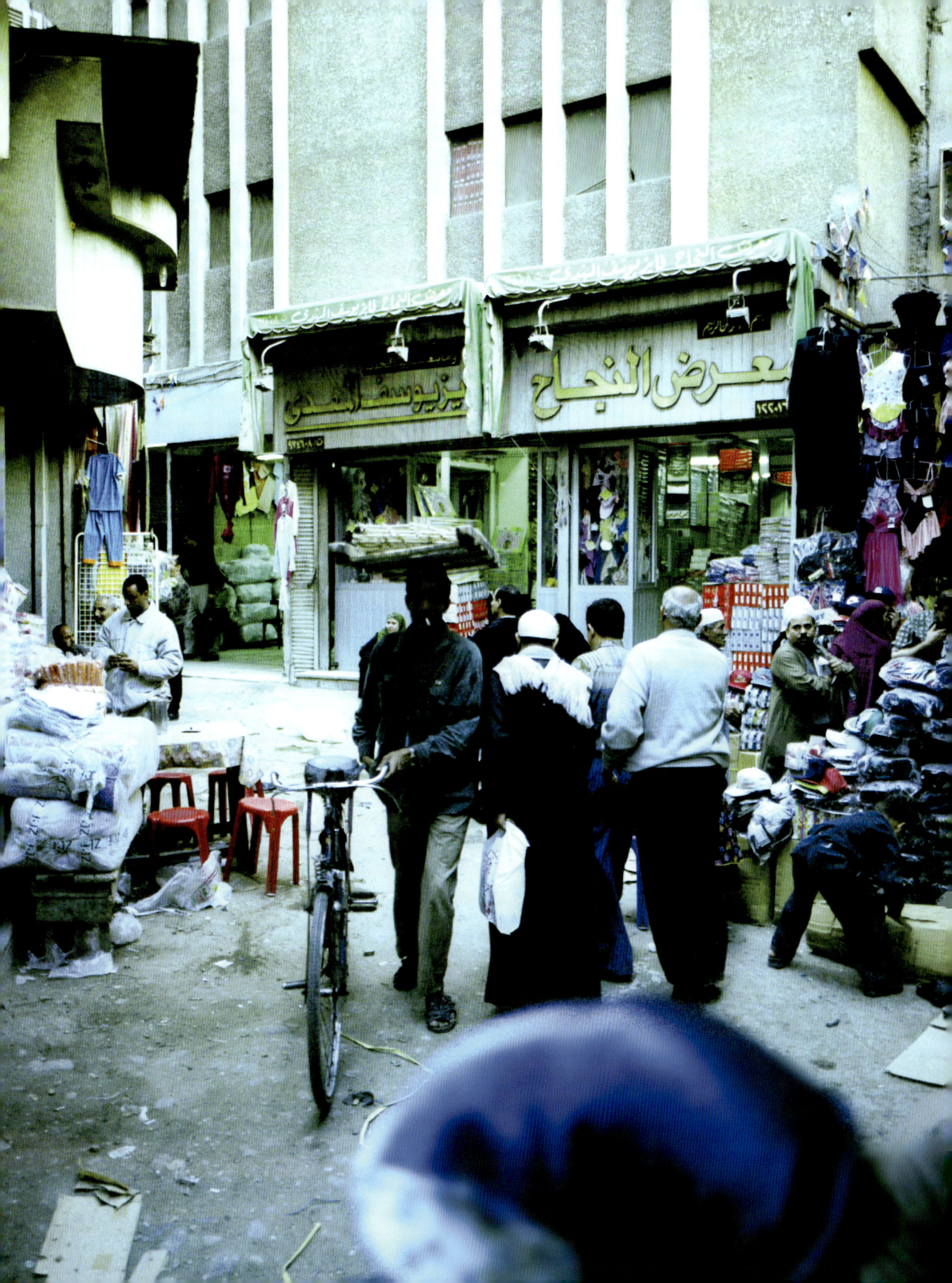
عرض النجاح
١٢٢٠١٦

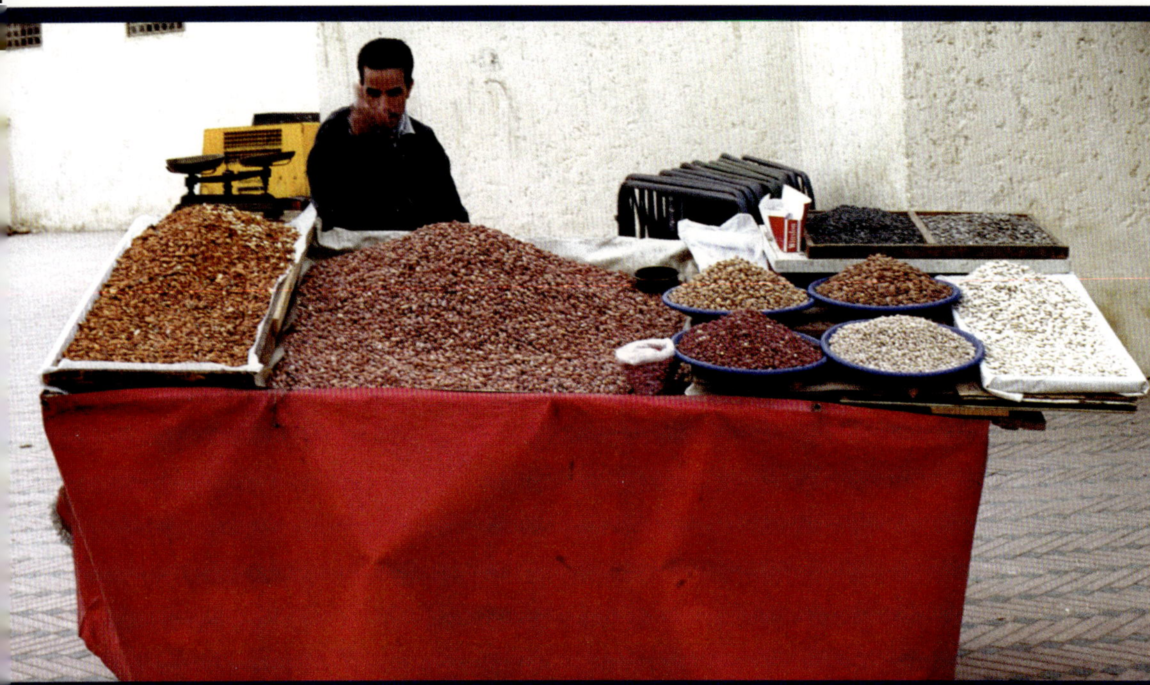
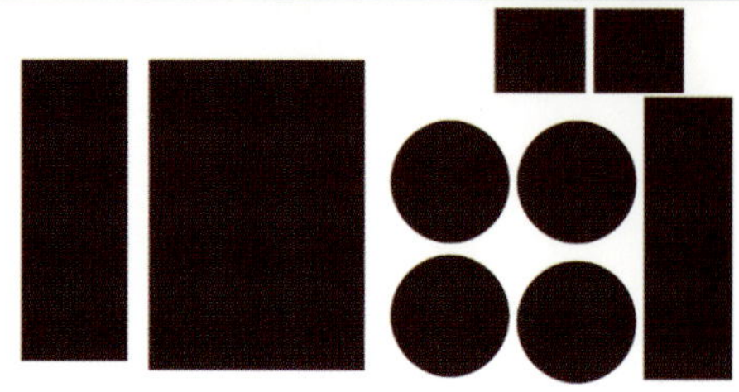
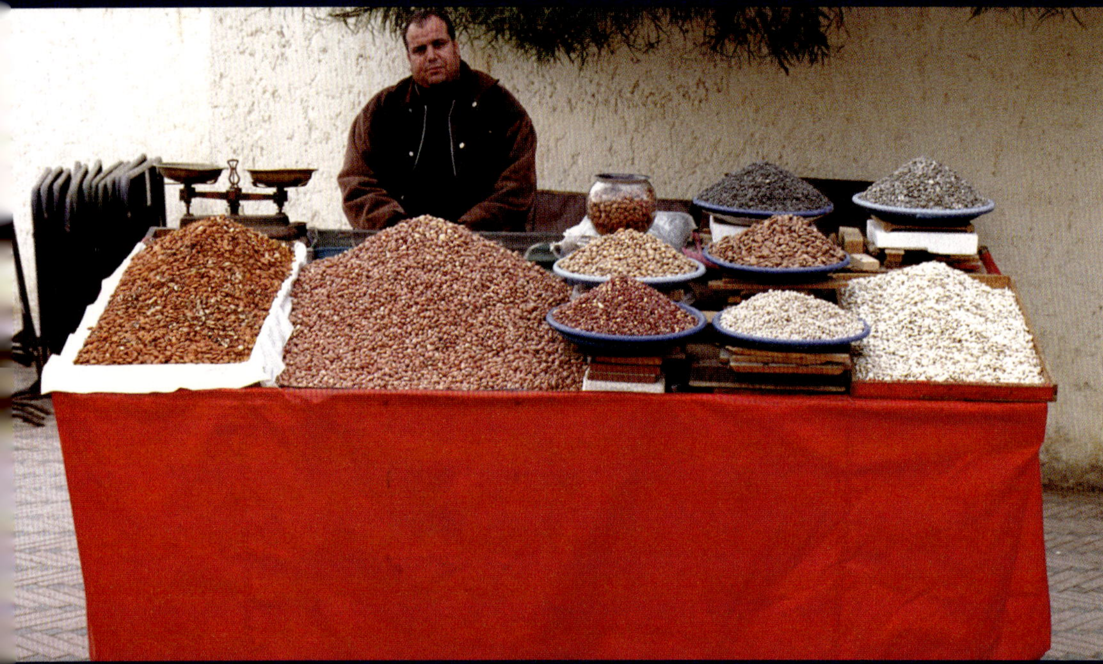
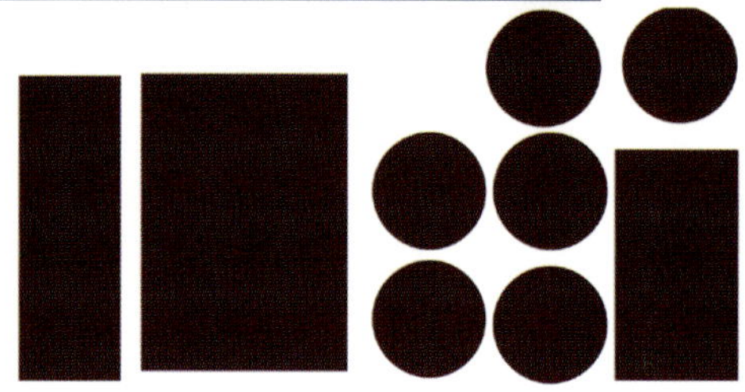
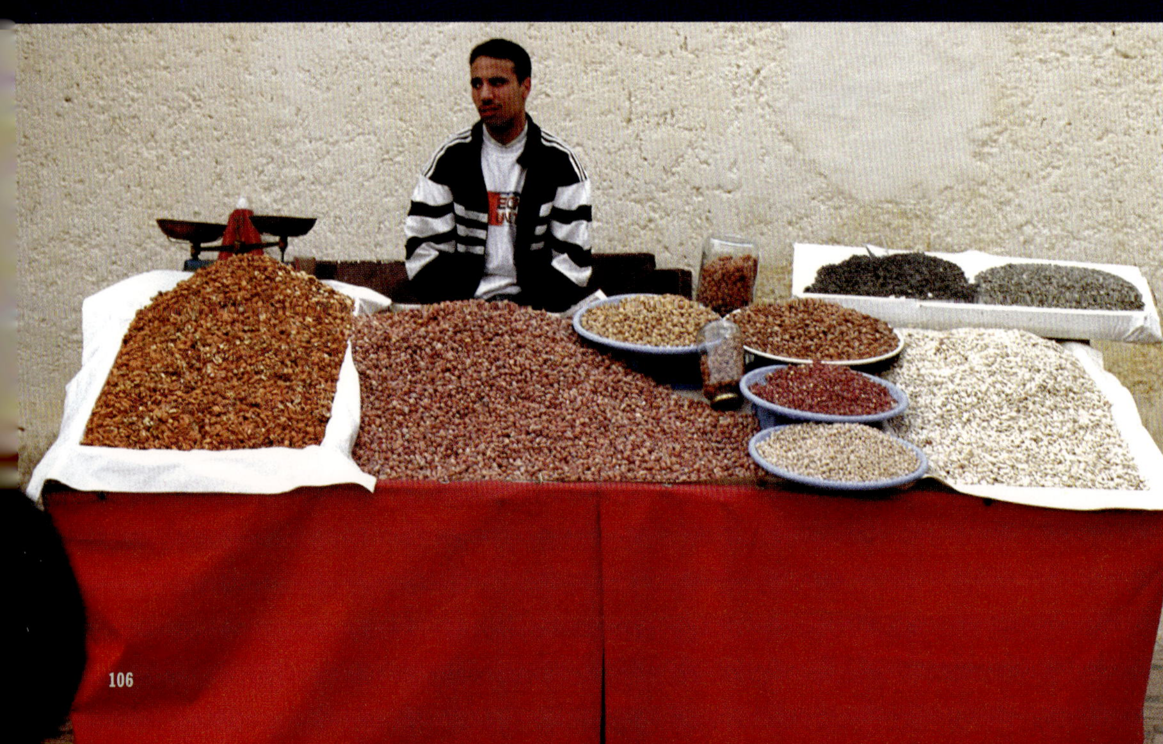
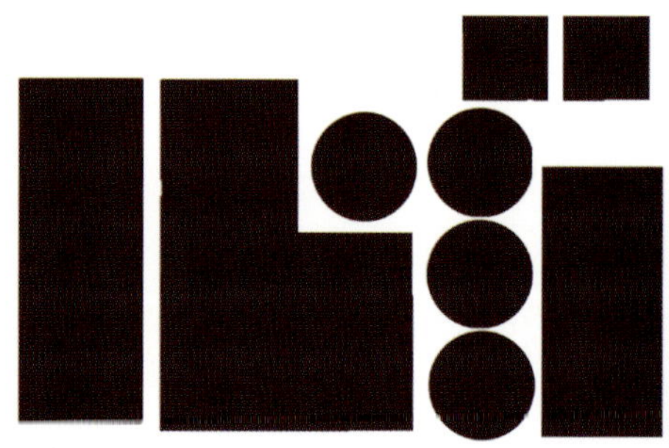

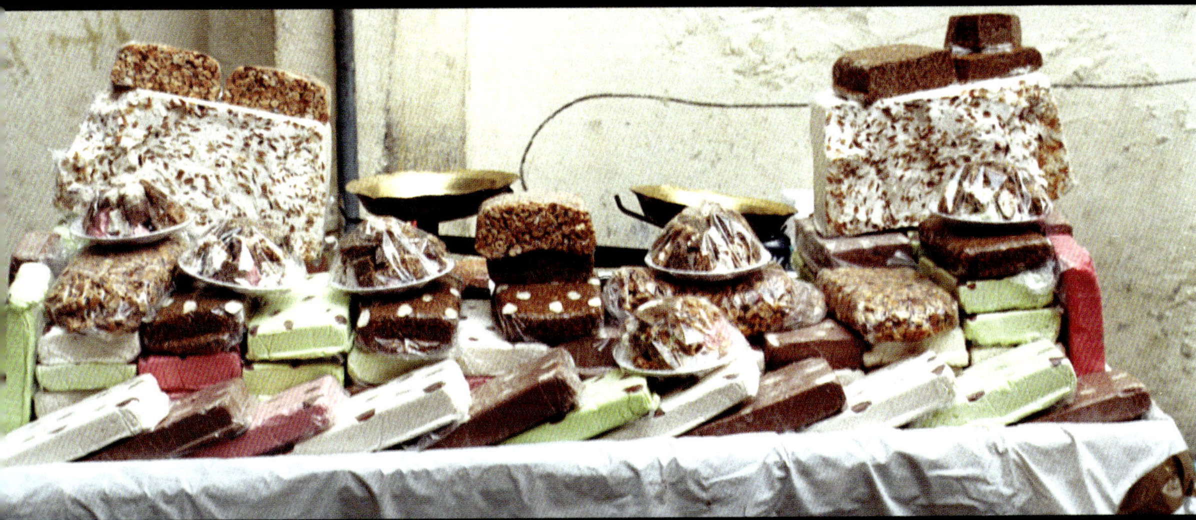

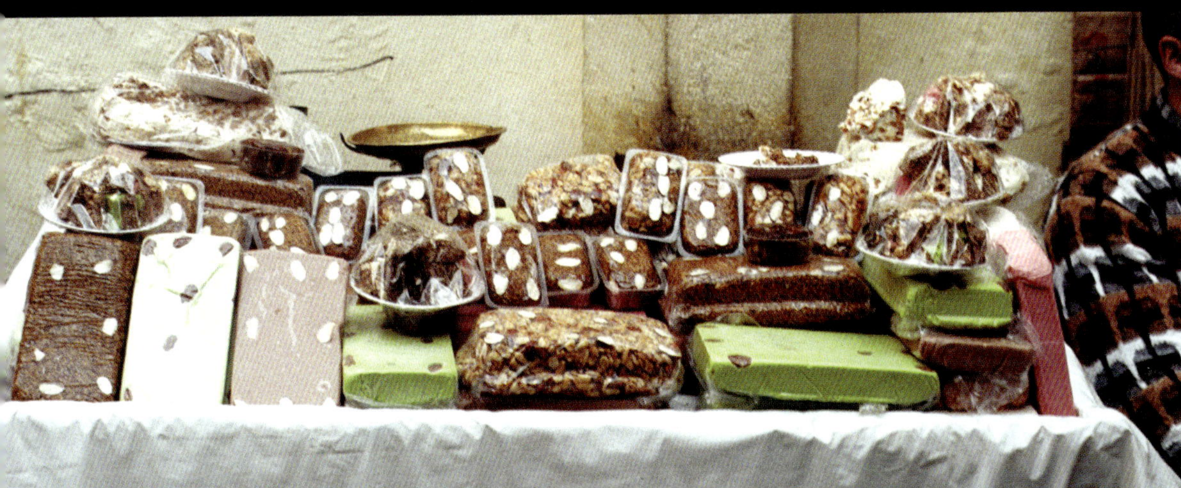

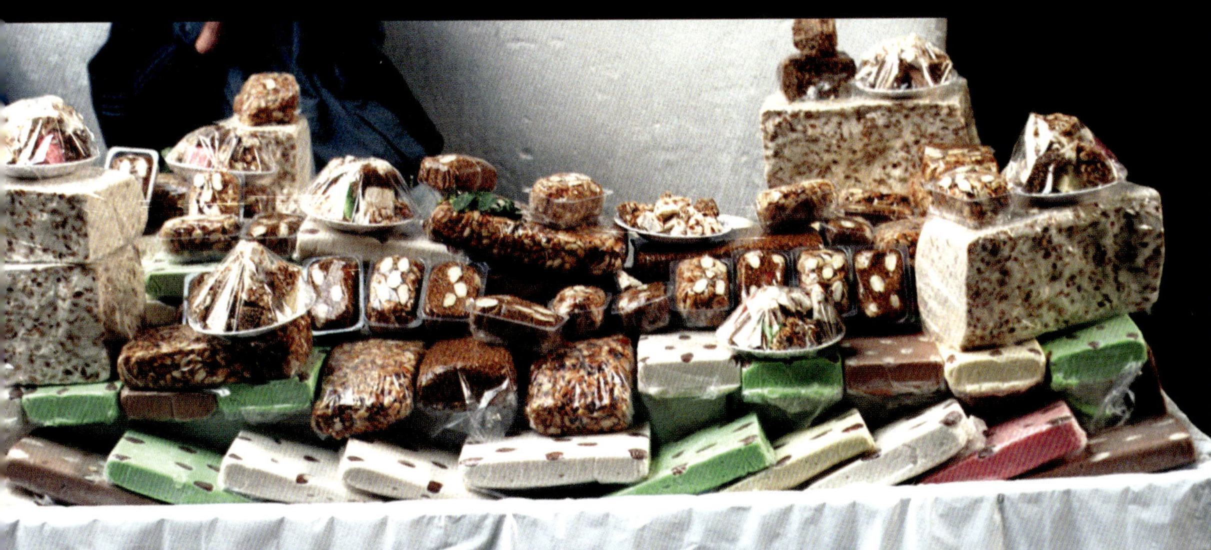

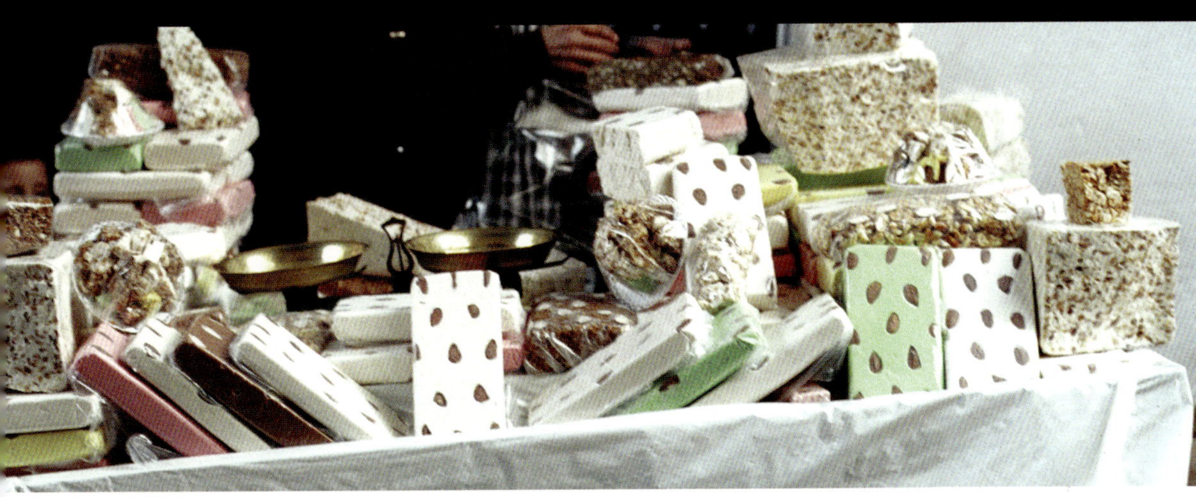

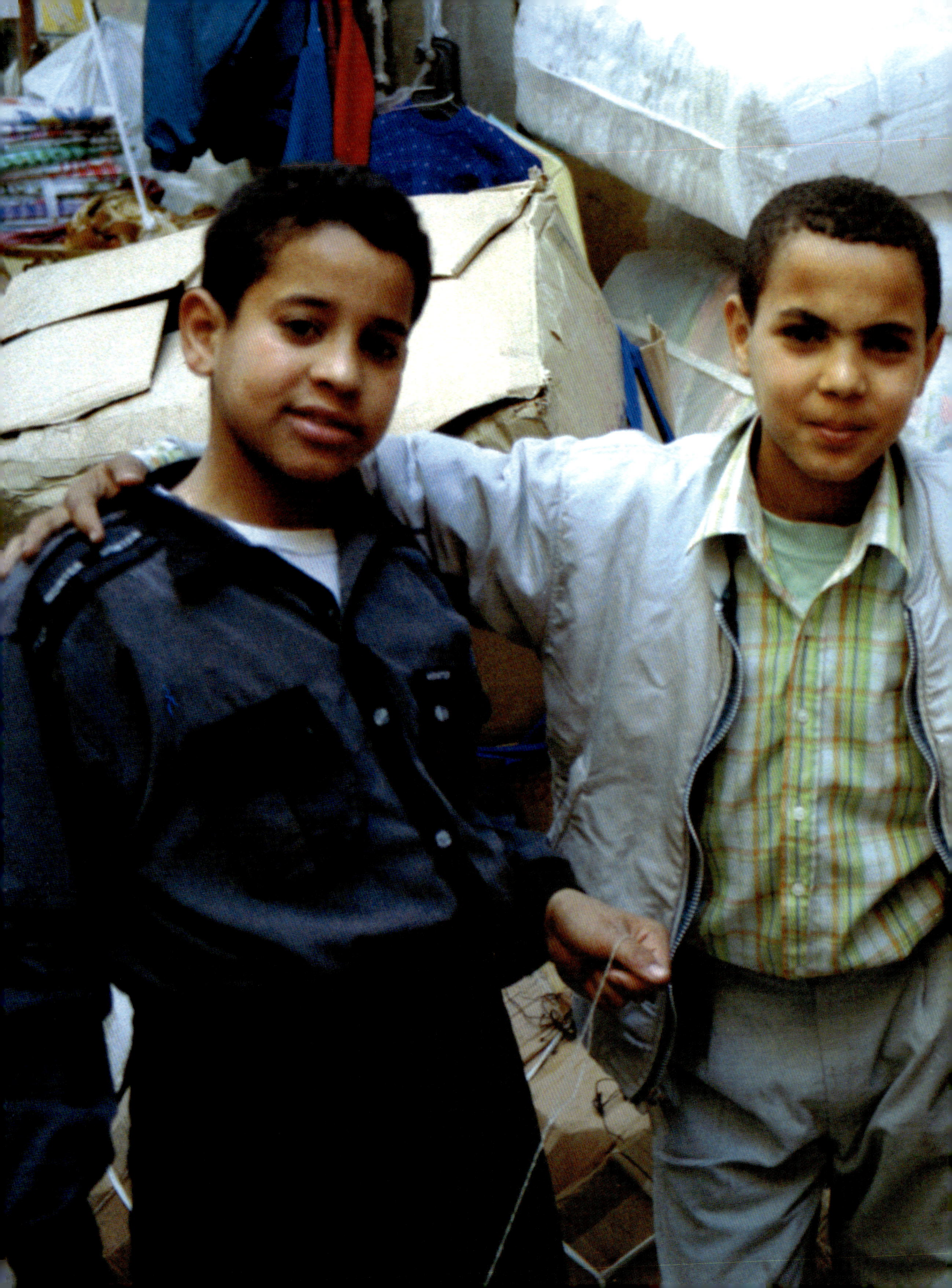

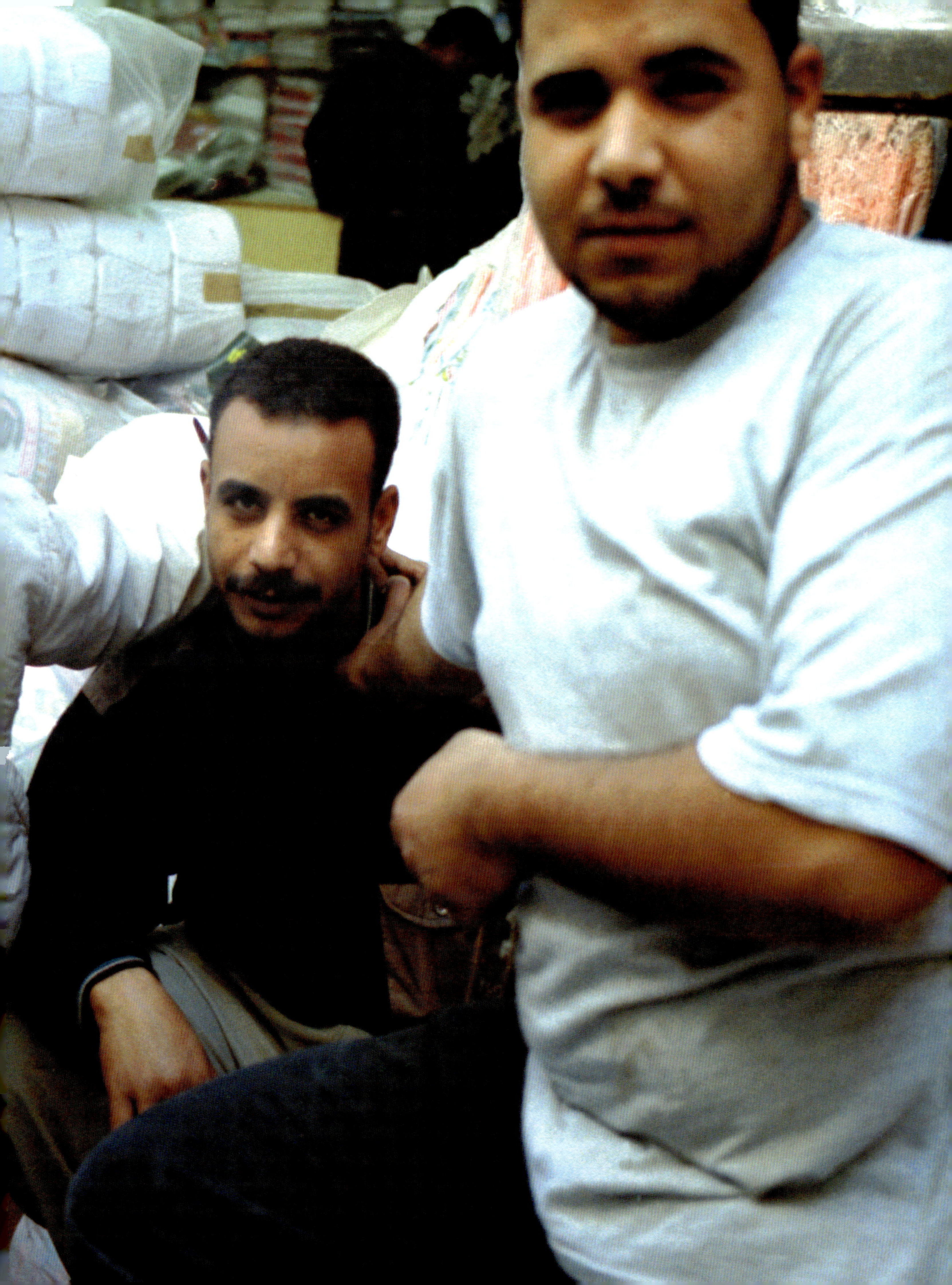

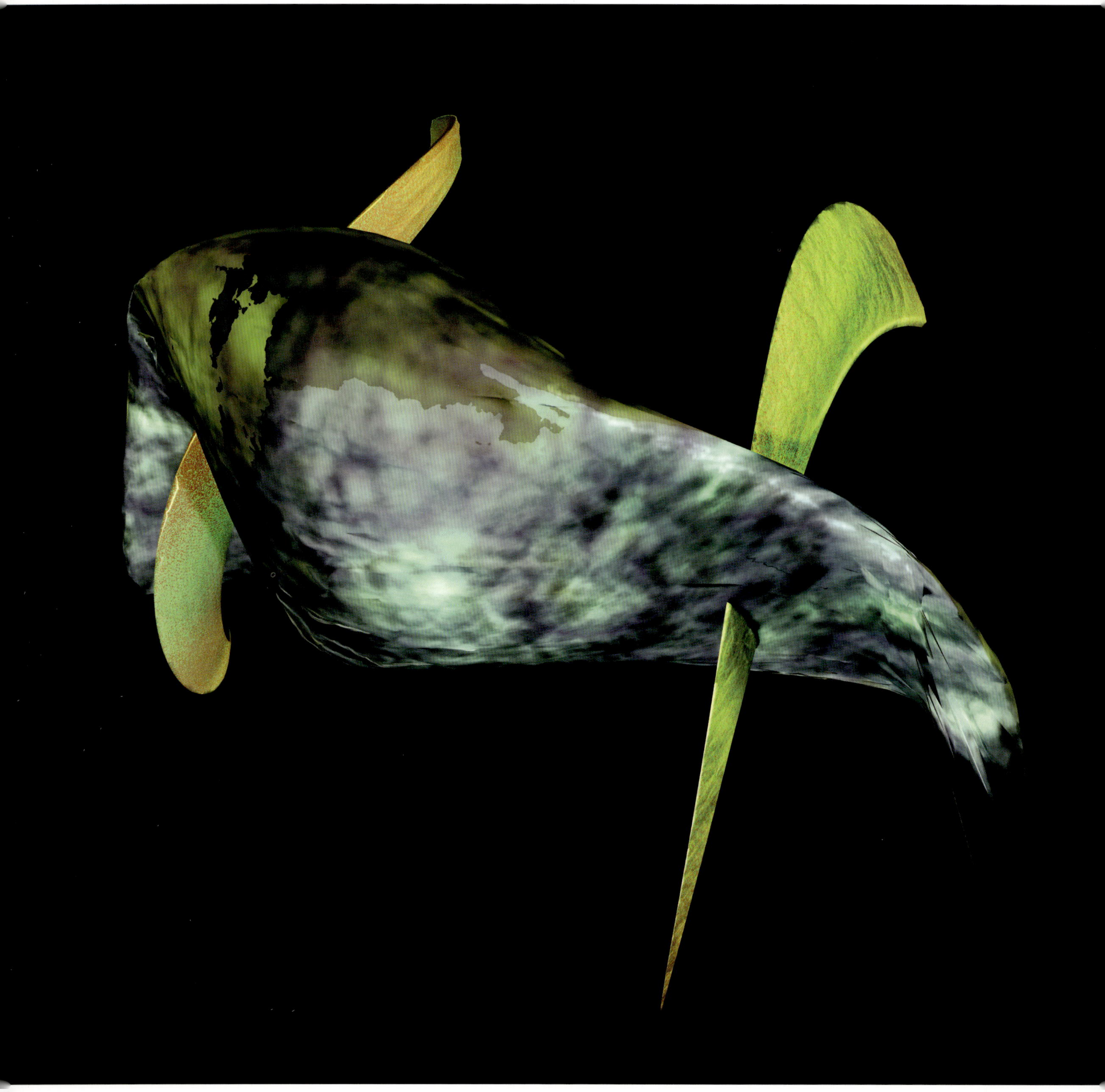

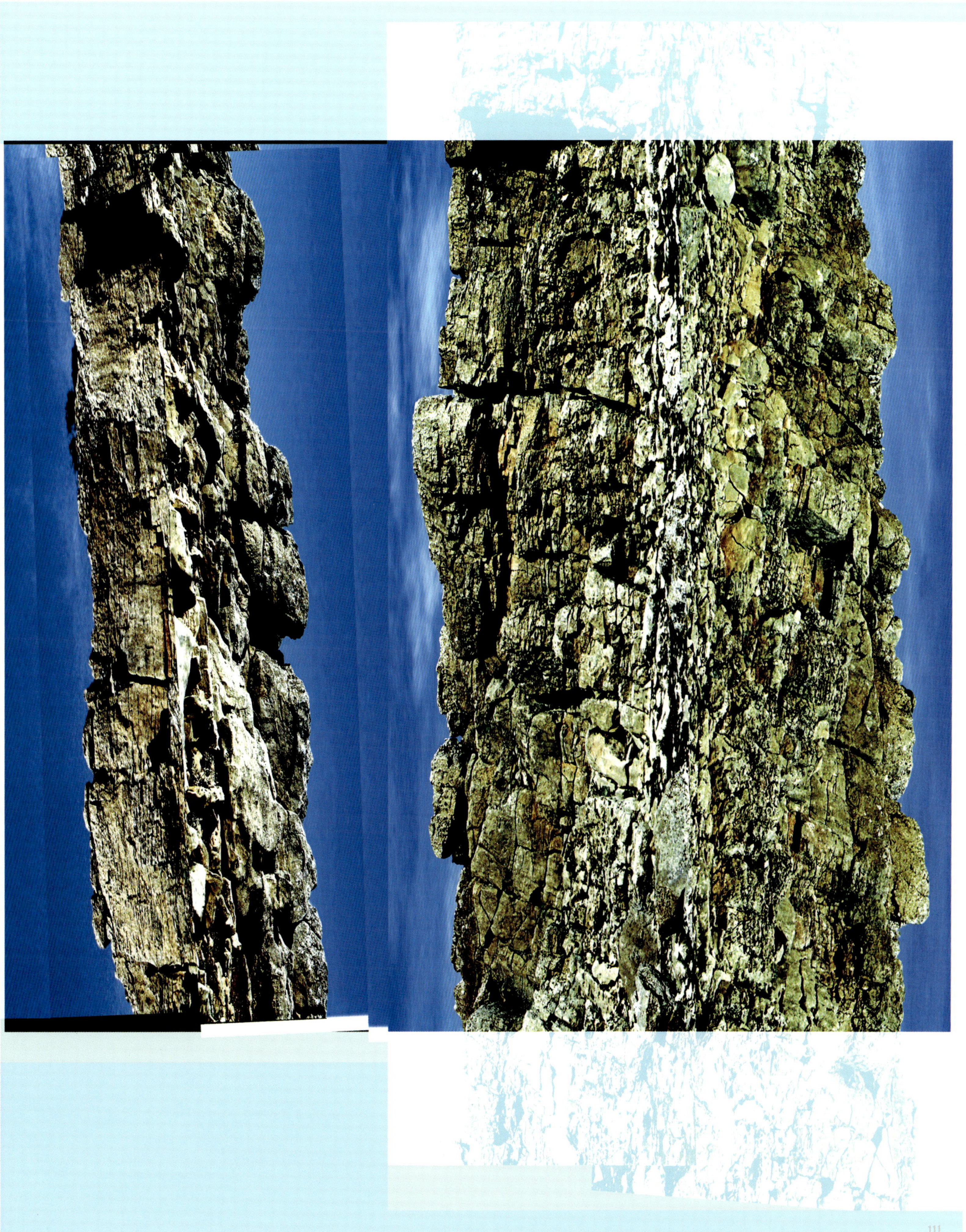

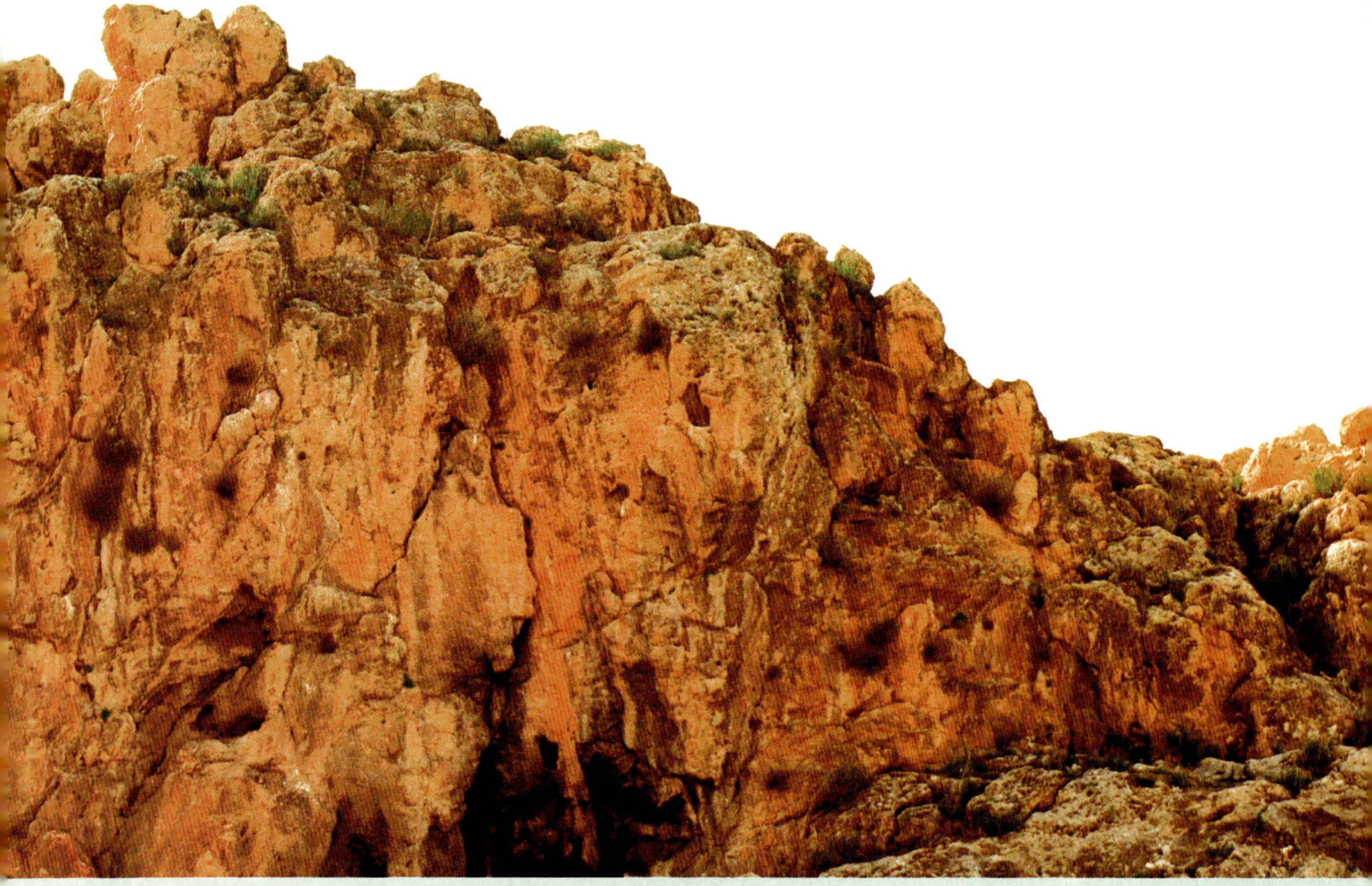

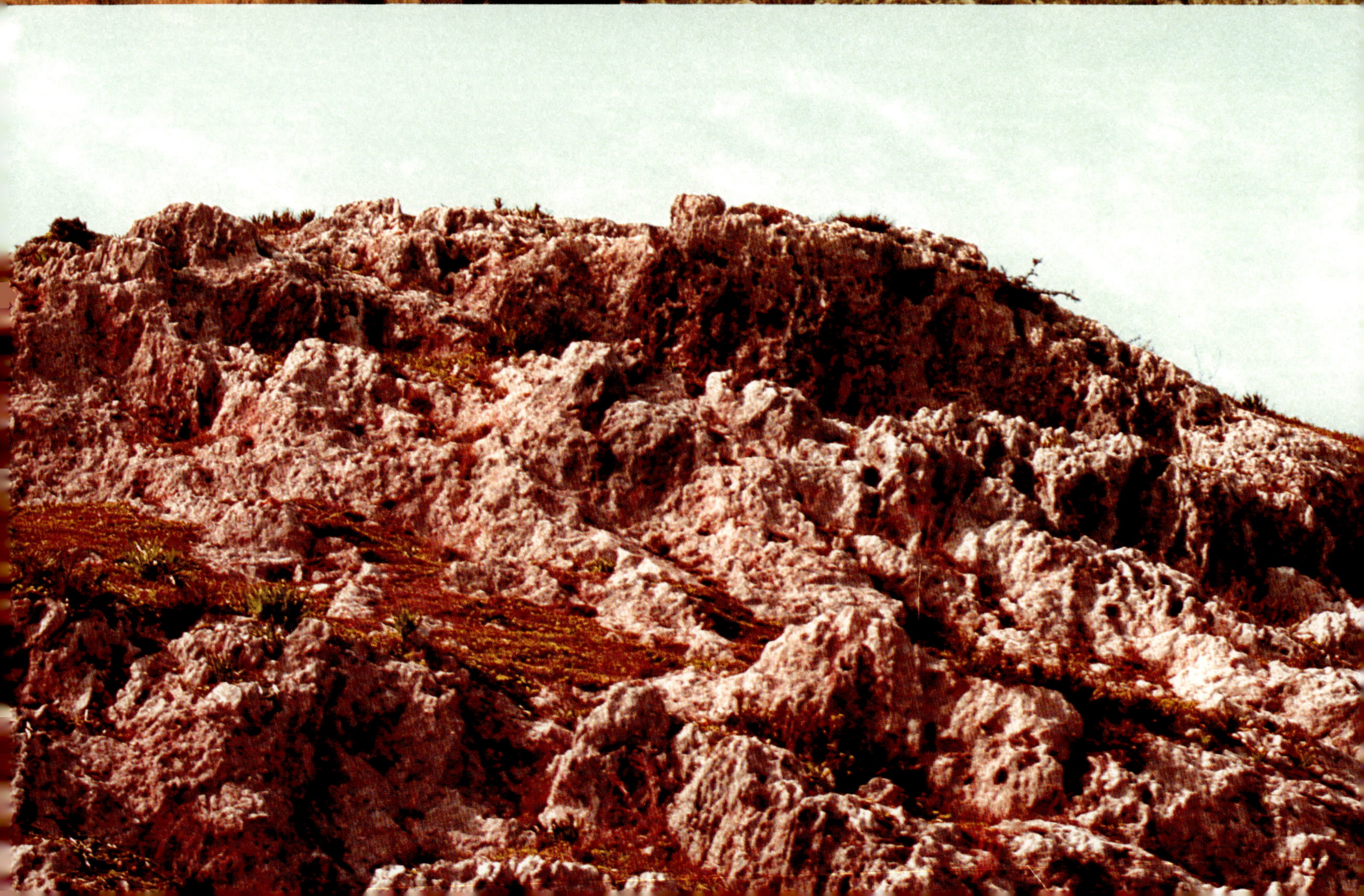

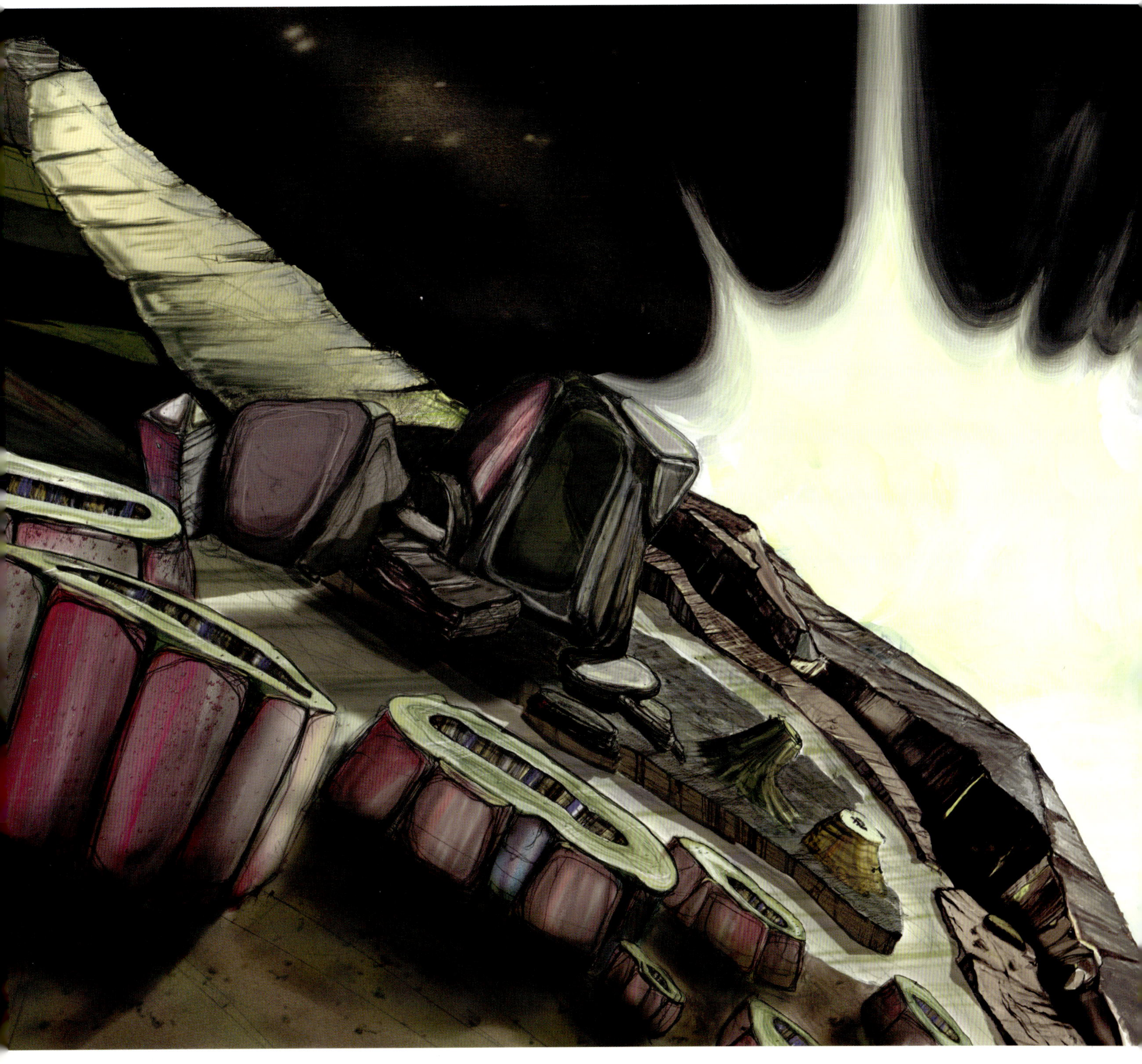

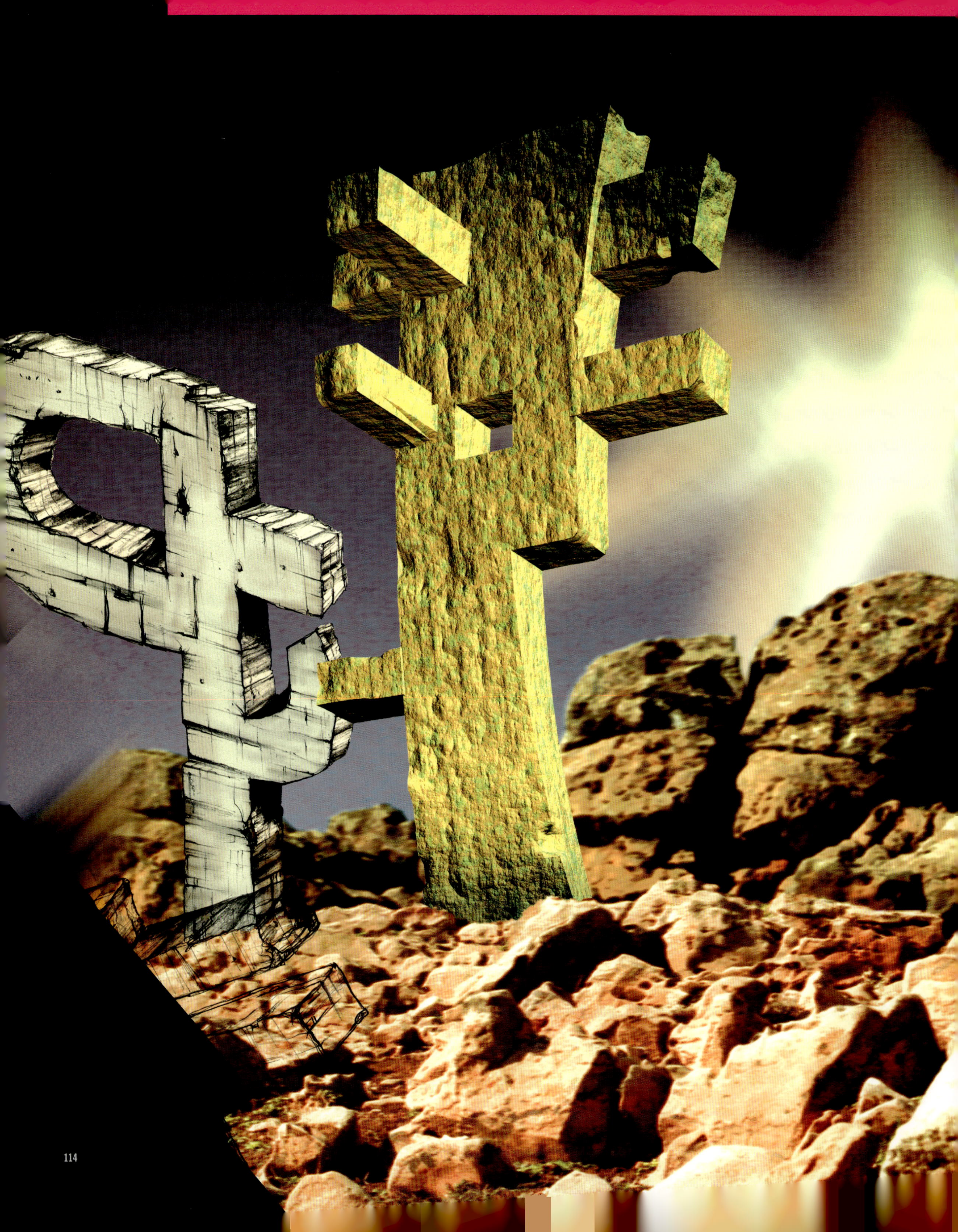

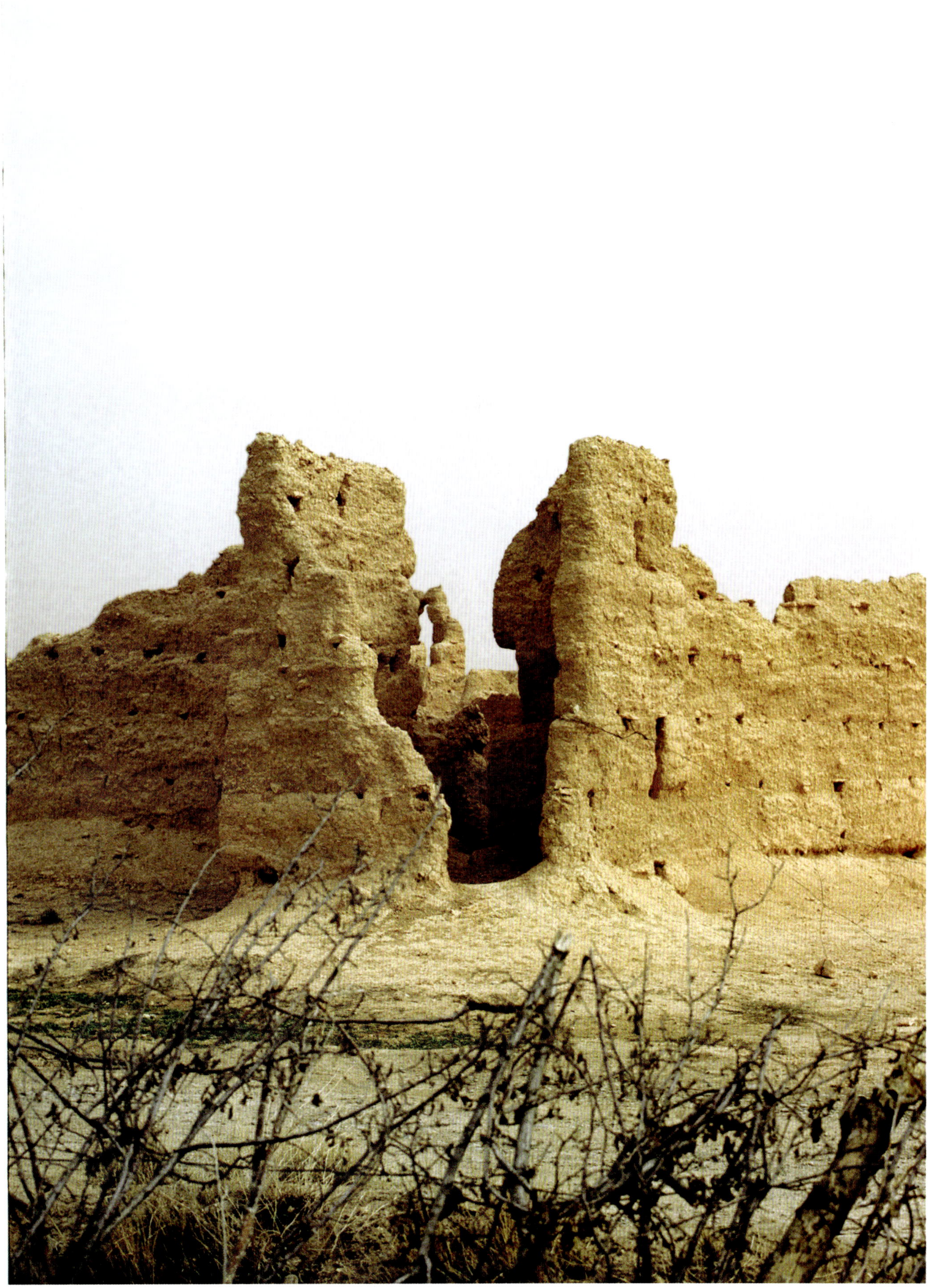

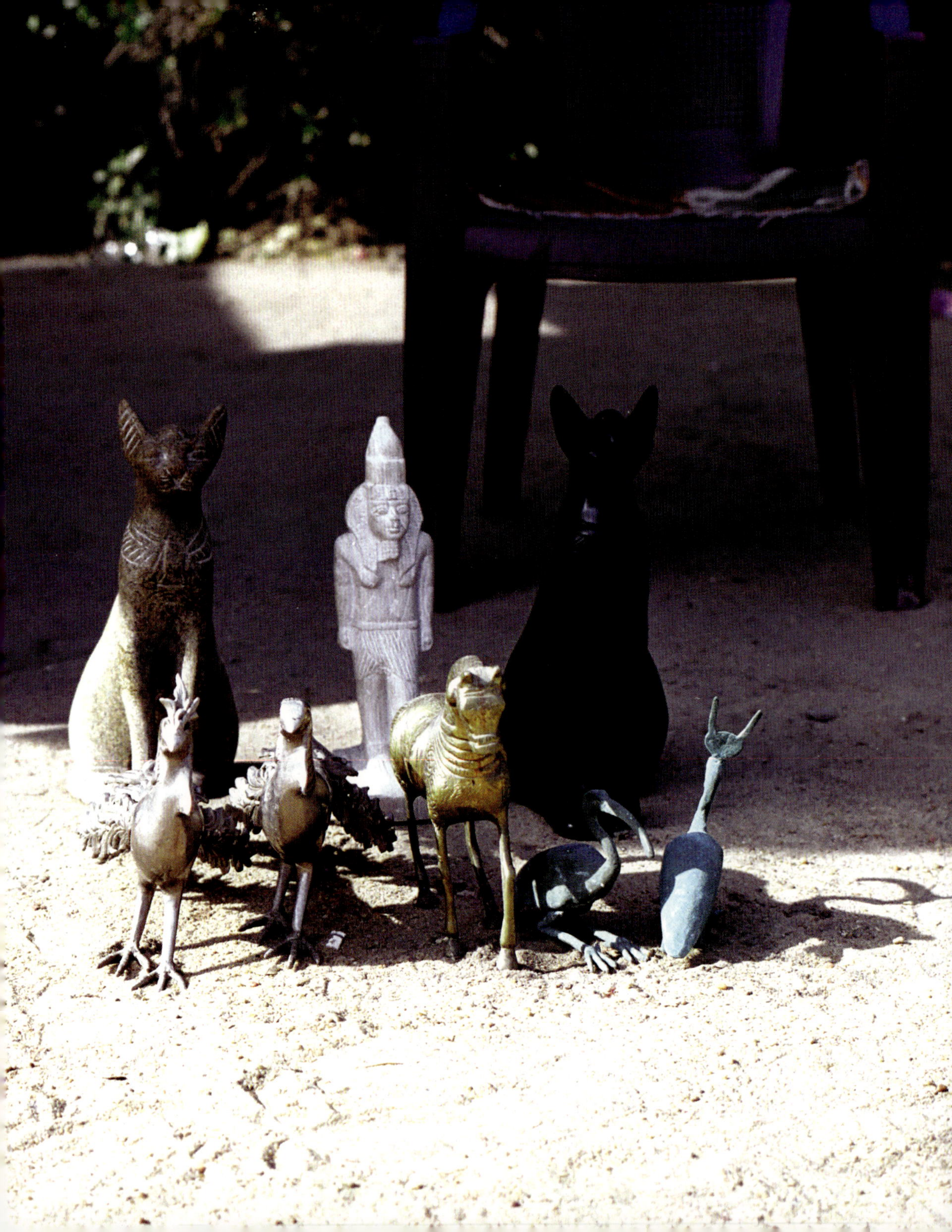

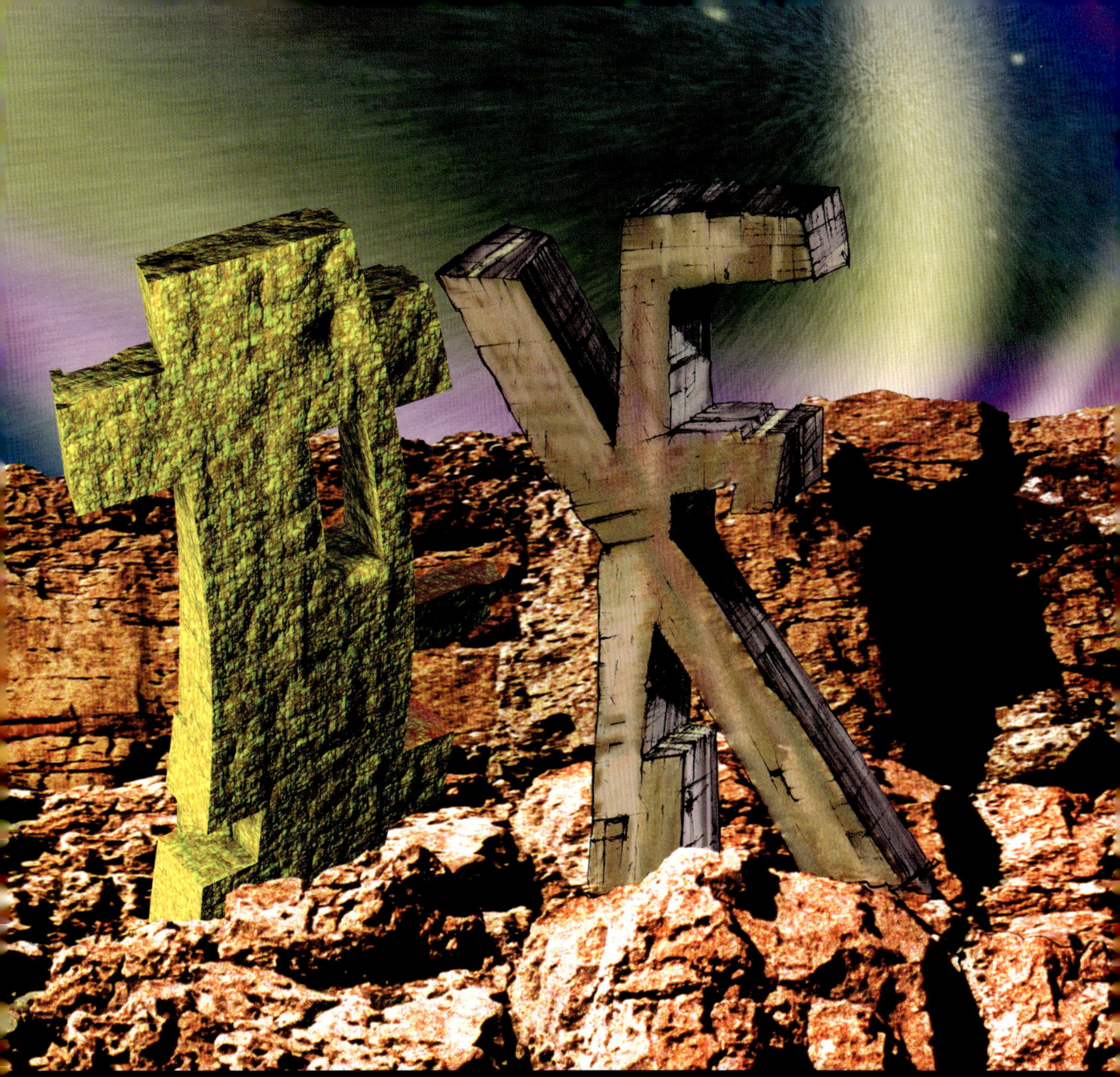

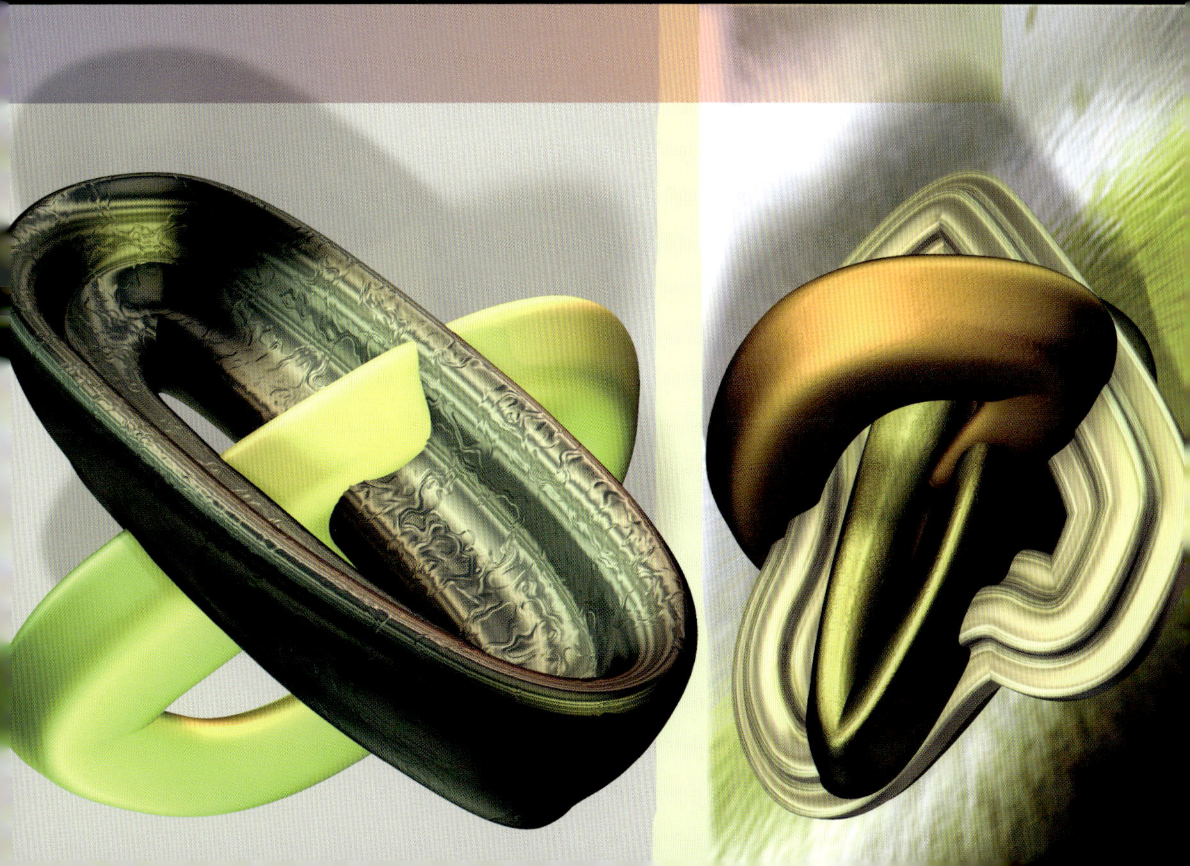

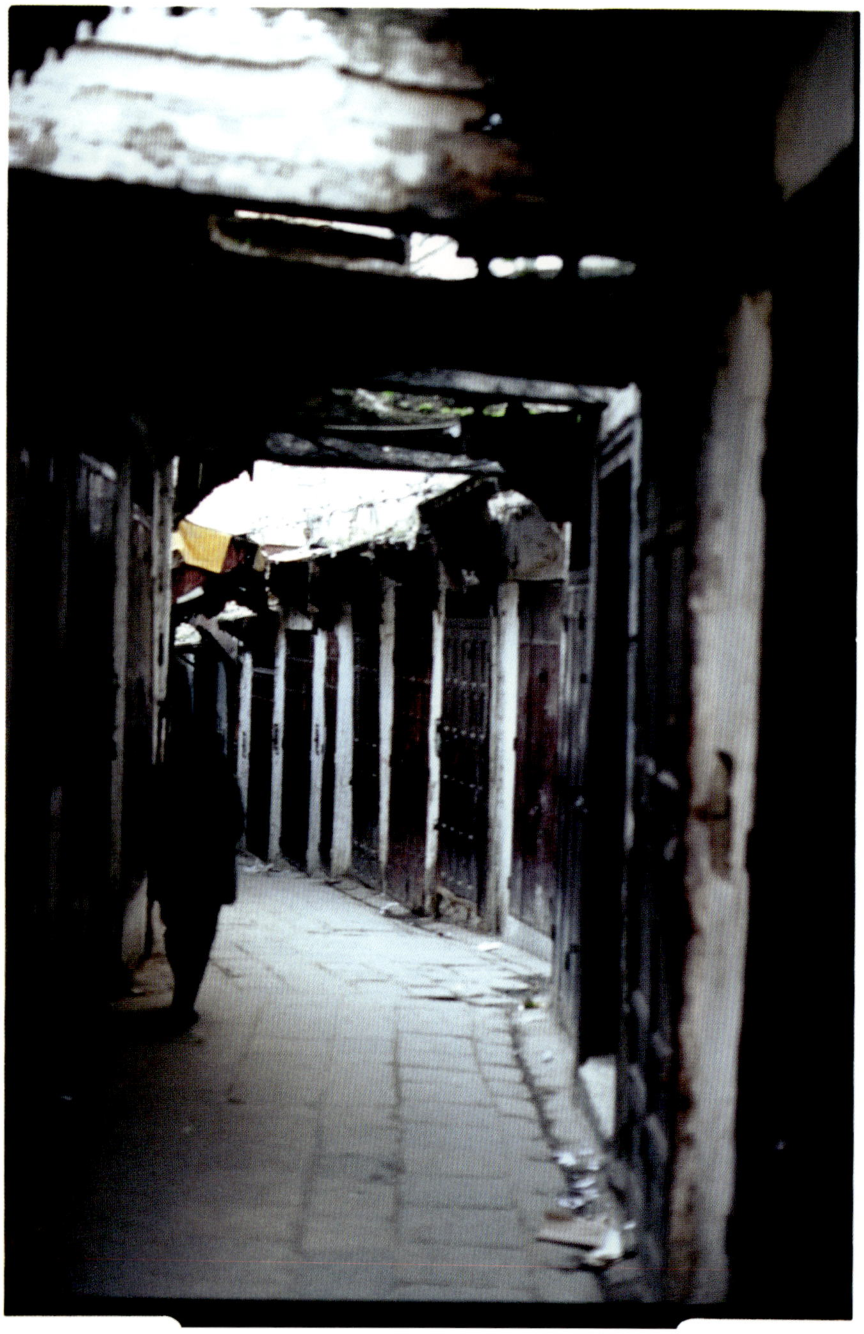

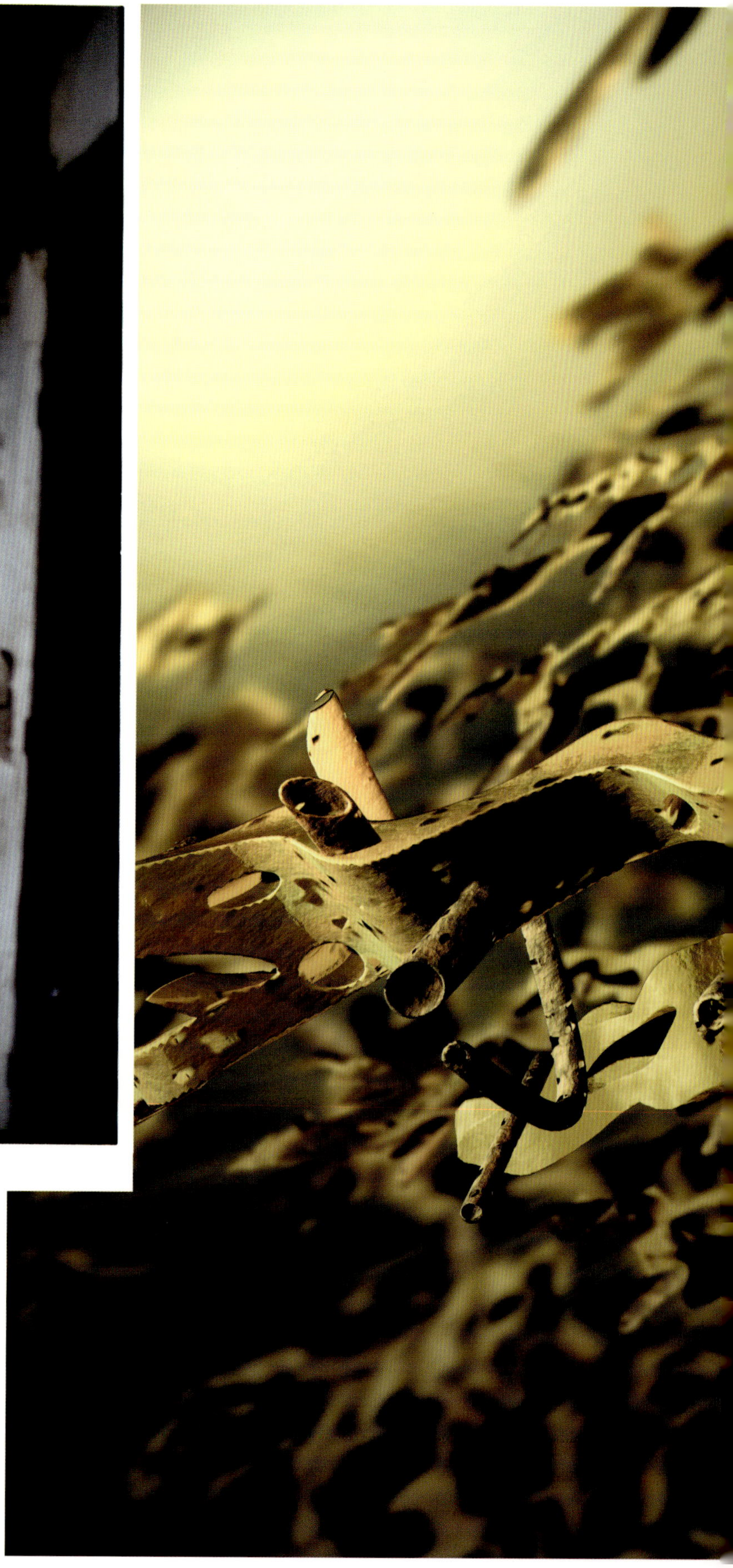

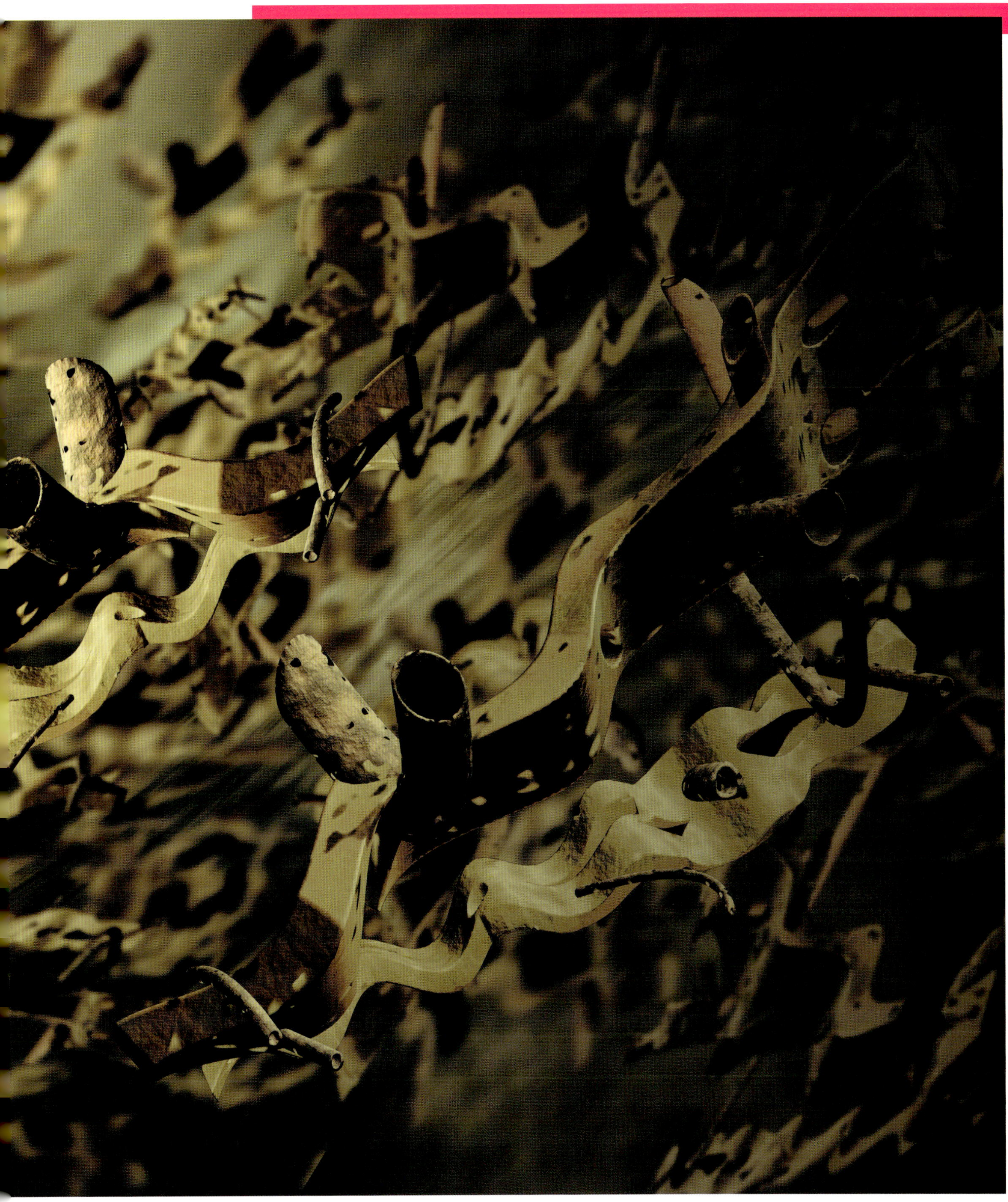

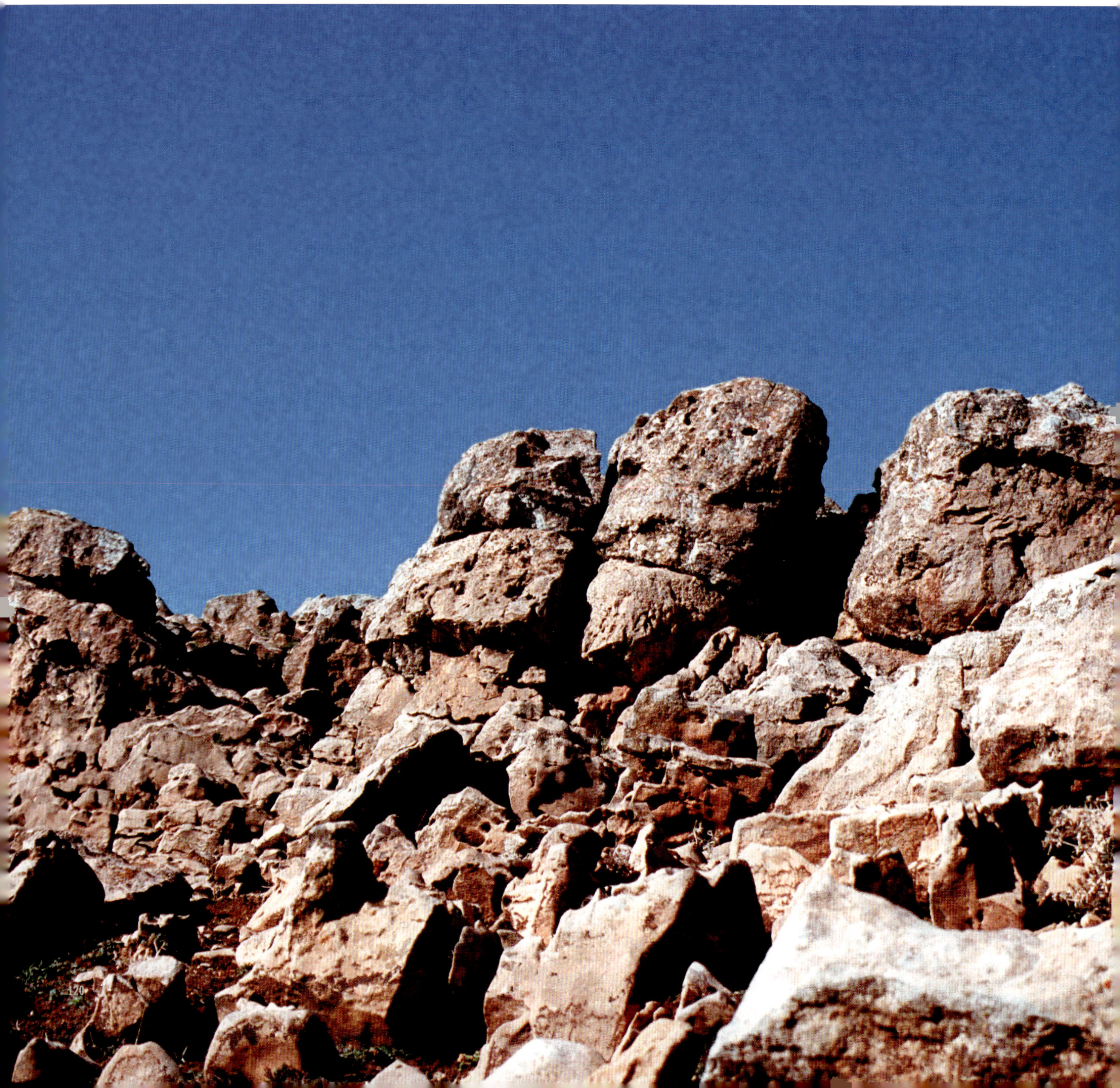

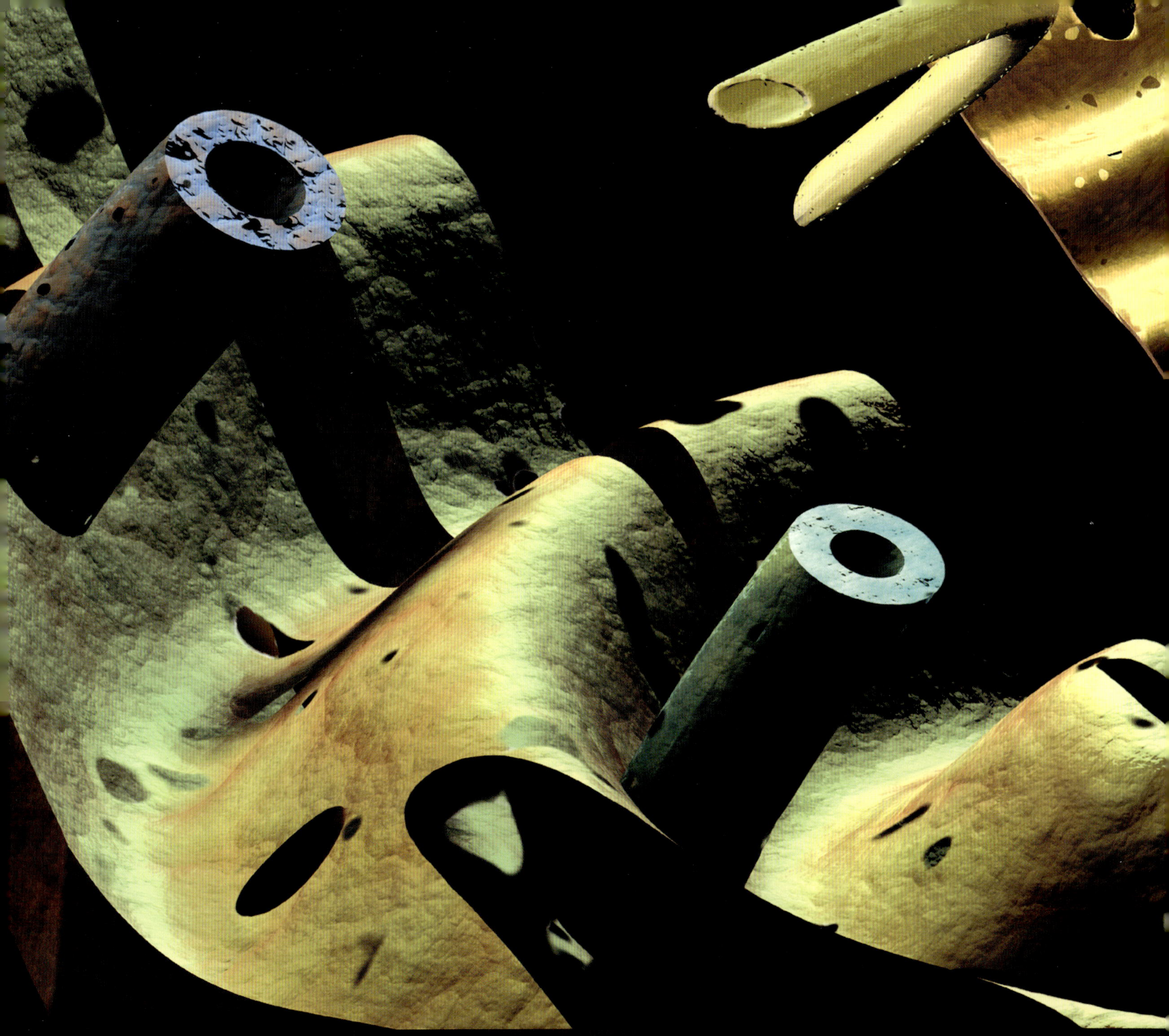

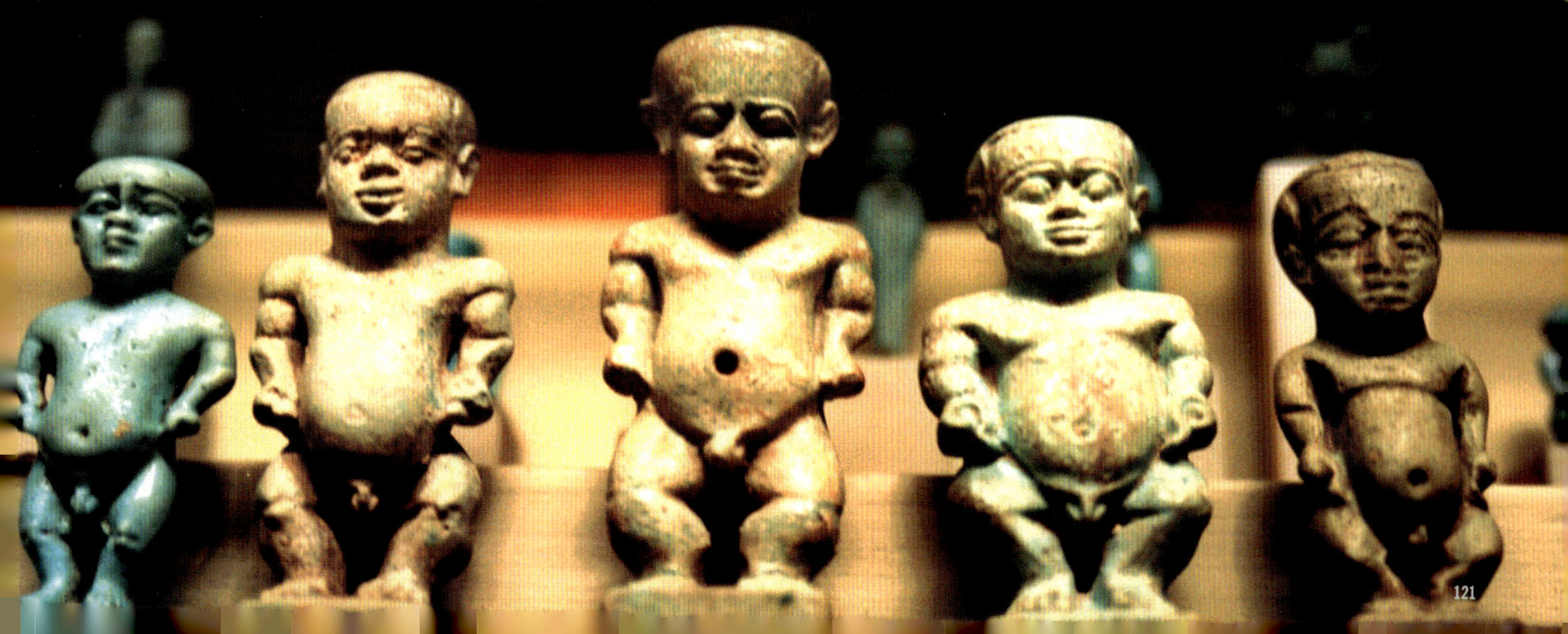

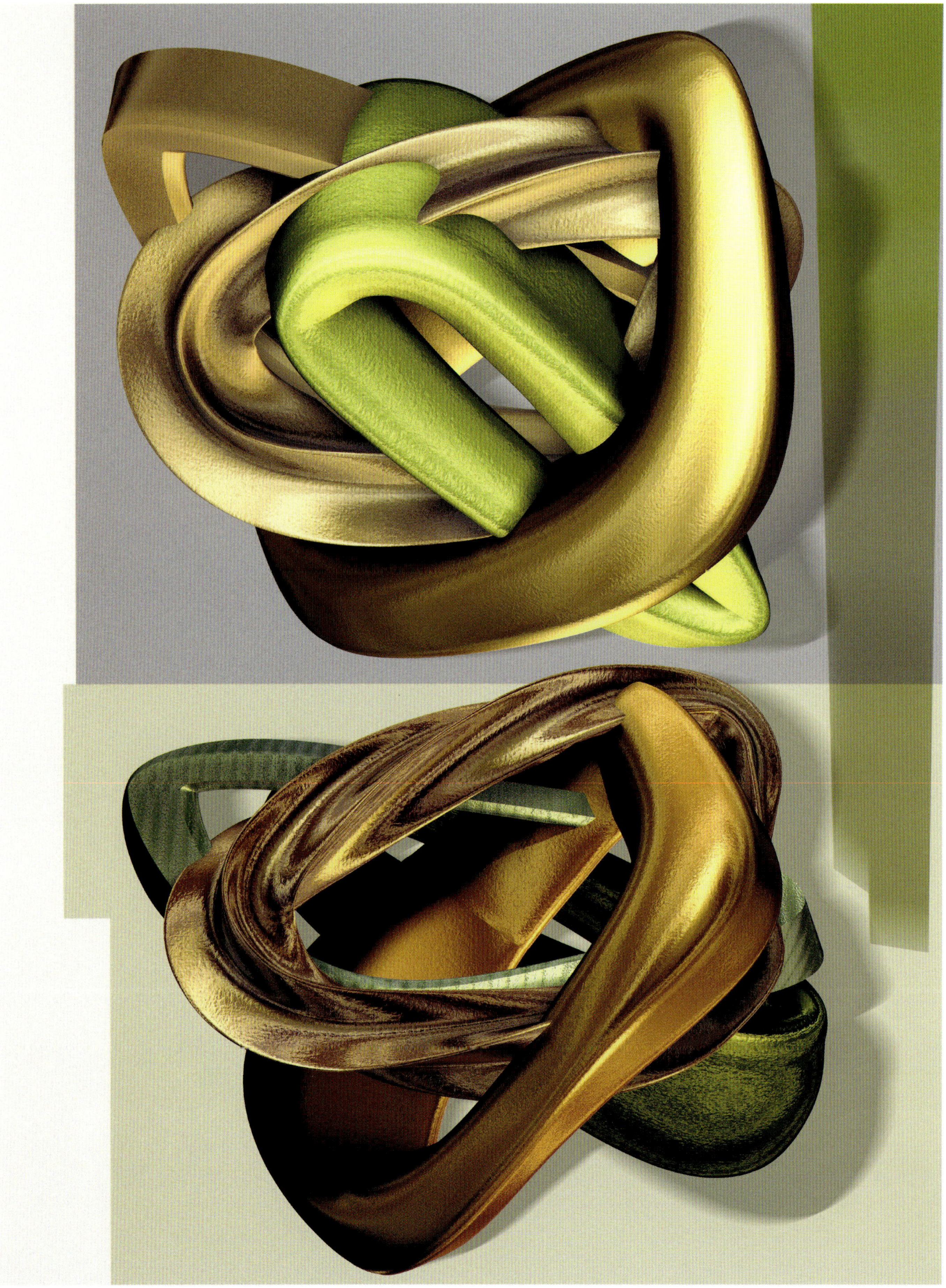

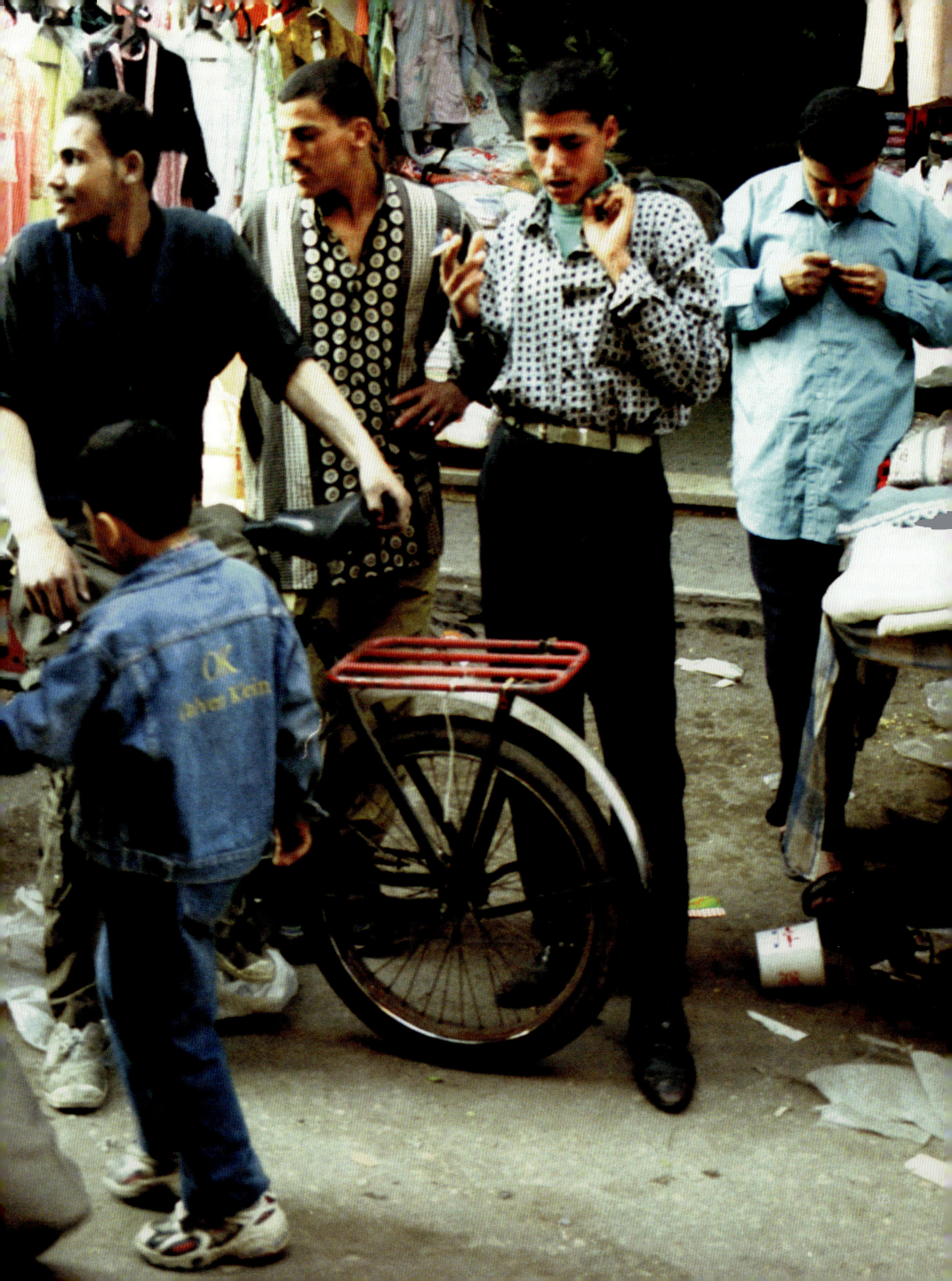
CK
Calvin Klein

CT Team: By extension, the locus of hyperbolic language distances, those with the greatest ranges of distribution globally, were teamed with density: the breadth, tightness and strength of the references within the same frame.

3.653º SSE: The immediacy of the range, received as a single unit of reference. Conceived as being the product of the immediacy of the bonds between the smaller elemental compositions within the popularized frame. Generating an image, whereby observation is the story and the feeling of history.

**He said that he had come up with a better idea than the "one graviton" notion. It goes roughly like this. Consider a superposition of 2 quantum states, and consider the difference of mass density functions in these two states. Calculate the gravitational self-energy of this difference (after taking absolute values, presumably—though he didn't say this); the reciprocal of this energy gives a time, and this time should represent the lifetime after which collapse should occur, leaving the system in one or the other state.**
**I should emphasize a number of qualifications he added. —J Baez**

Date Posted: 20:19:57 02/02/03 Sun
In reply to: Doppelgänger 's message, "Back again" on 19:48:19 02/01/03 Sat

I agree with you, David's point about immaterial is important. This is an issue of whether reductionism alone can explain phenomena. Is there something fundamentally different about a whole that is explainable purely in terms of the parts and the rules of interaction? Gell-Man thinks complexity is the obverse of reductionism. Perhaps
he's right. It's a subtle point, but I certainly agree David's point is very important.

I gotta take issue with this though, though not for the reasons you might expect.
...

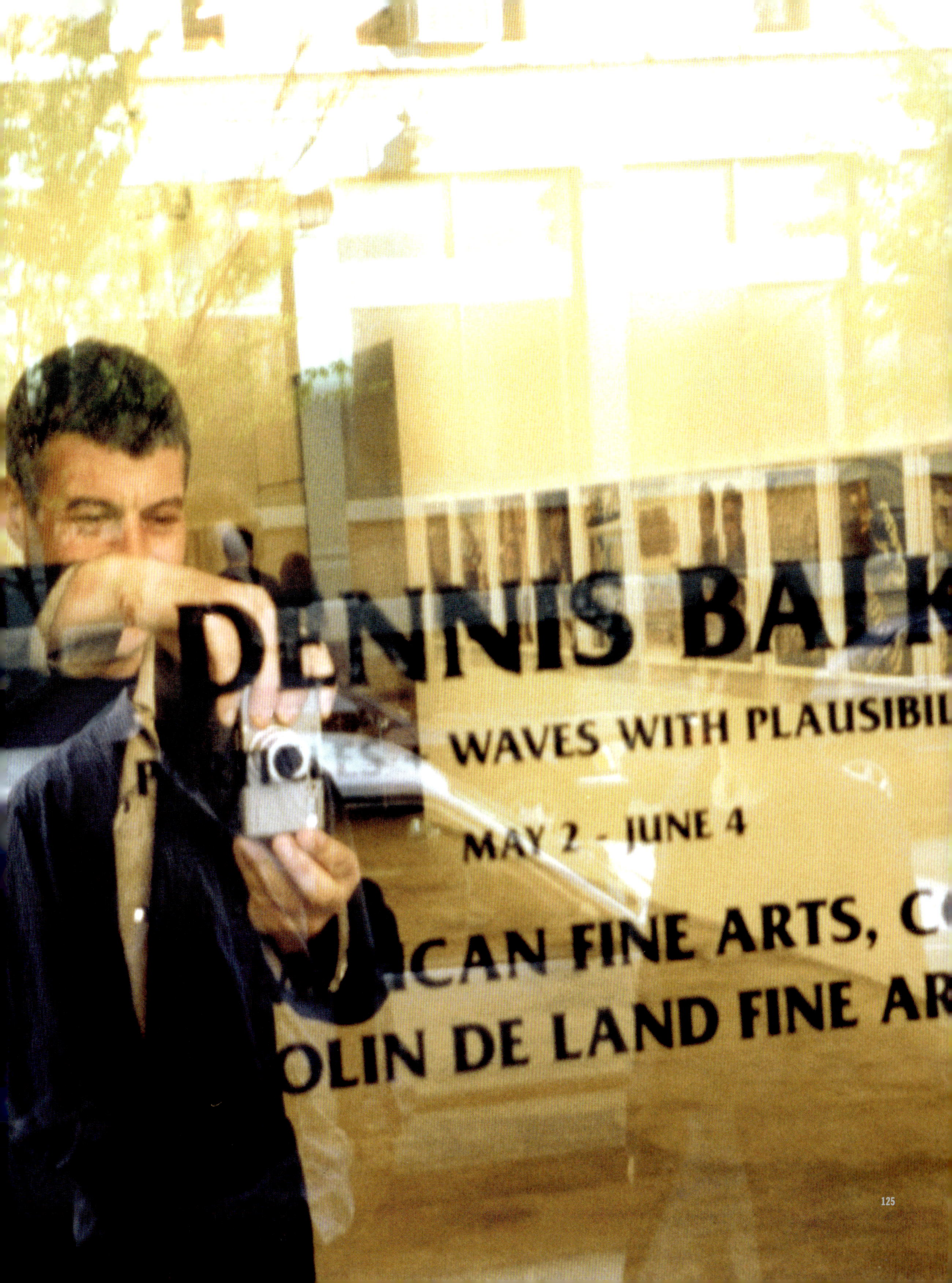
DENNIS BALK
WAVES WITH PLAUSIBIL
MAY 2 - JUNE 4
ICAN FINE ARTS, C
OLIN DE LAND FINE AR

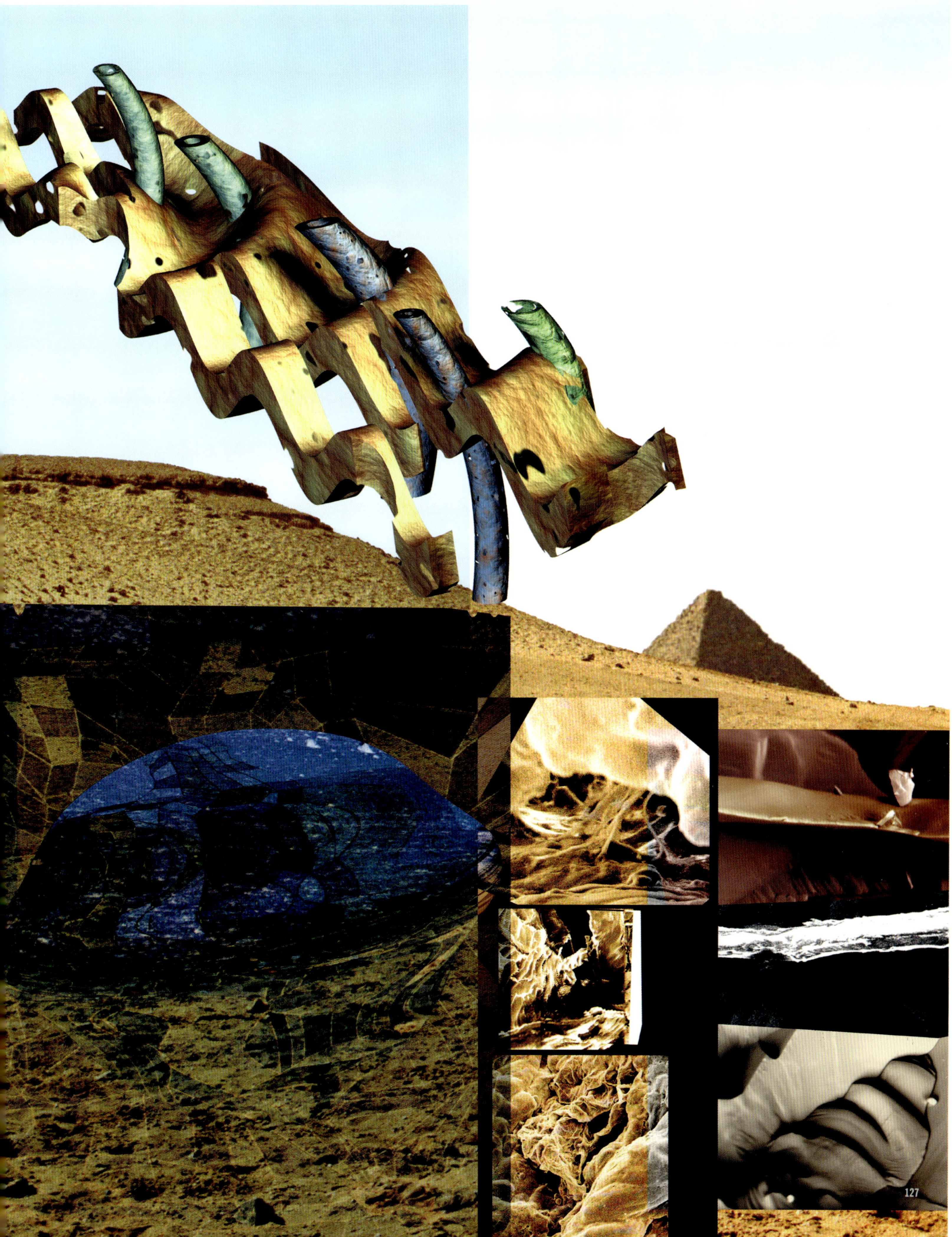

receptor group
instantaneous entropy of
post-Planck moment
debris shard composites

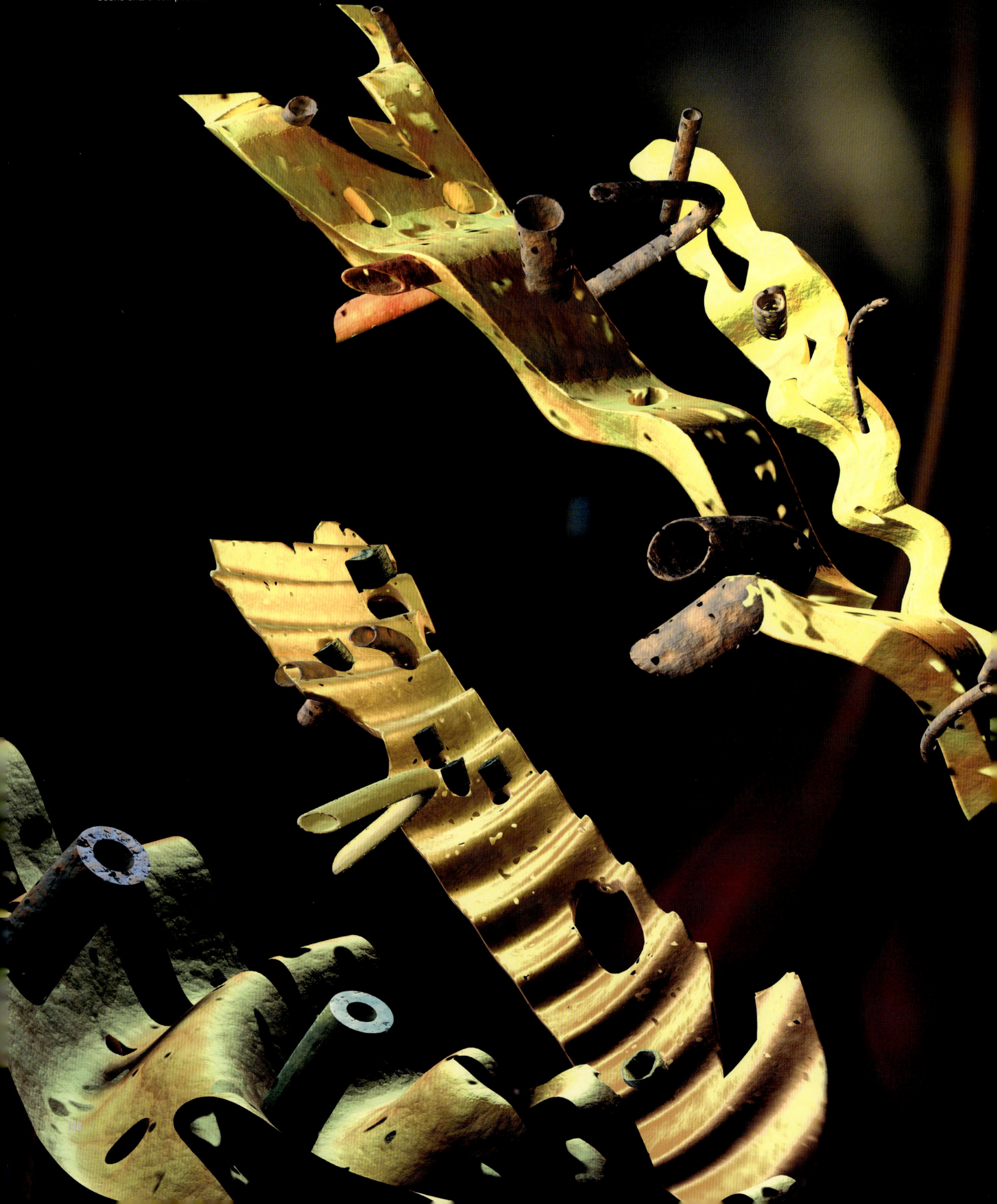

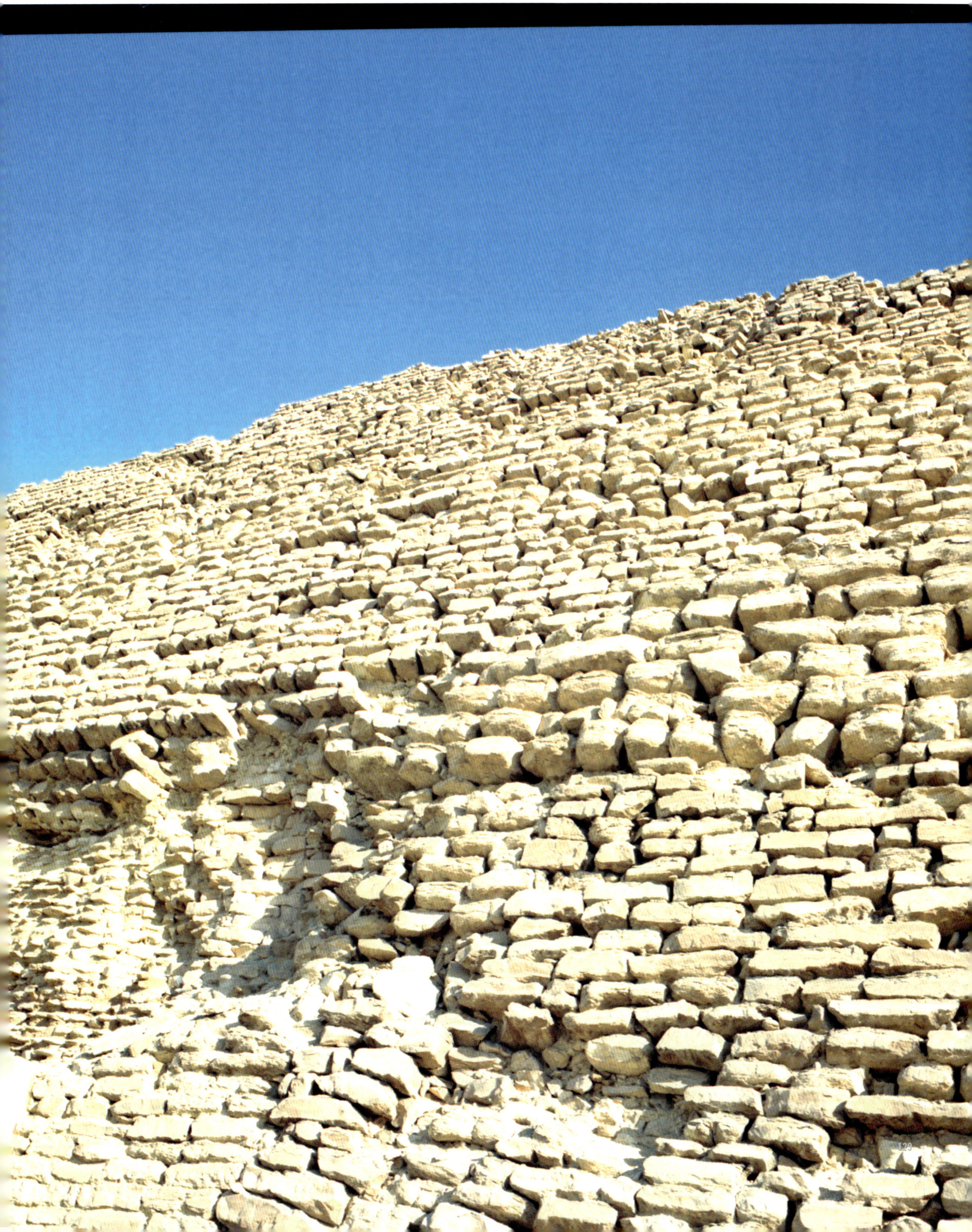

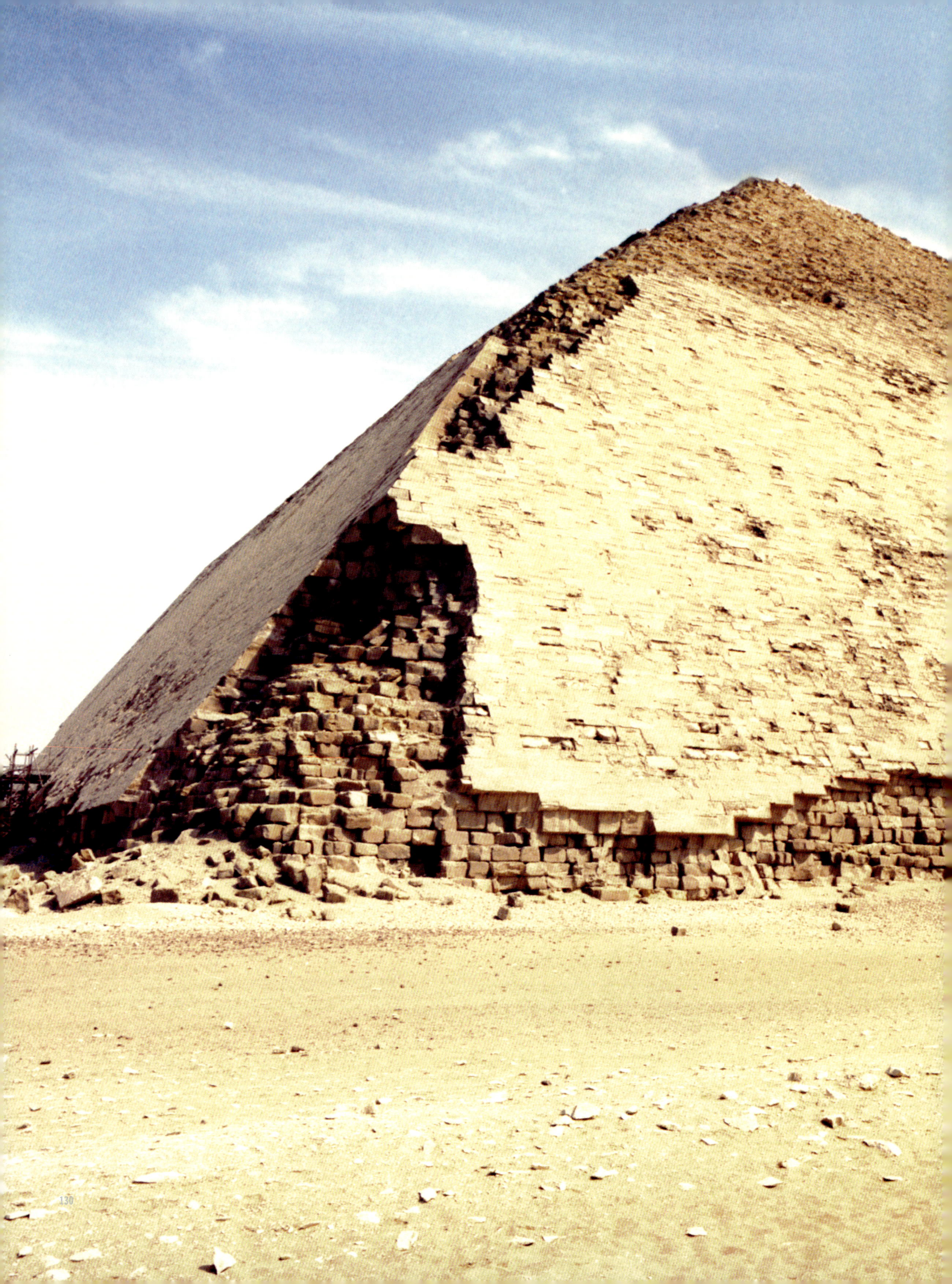

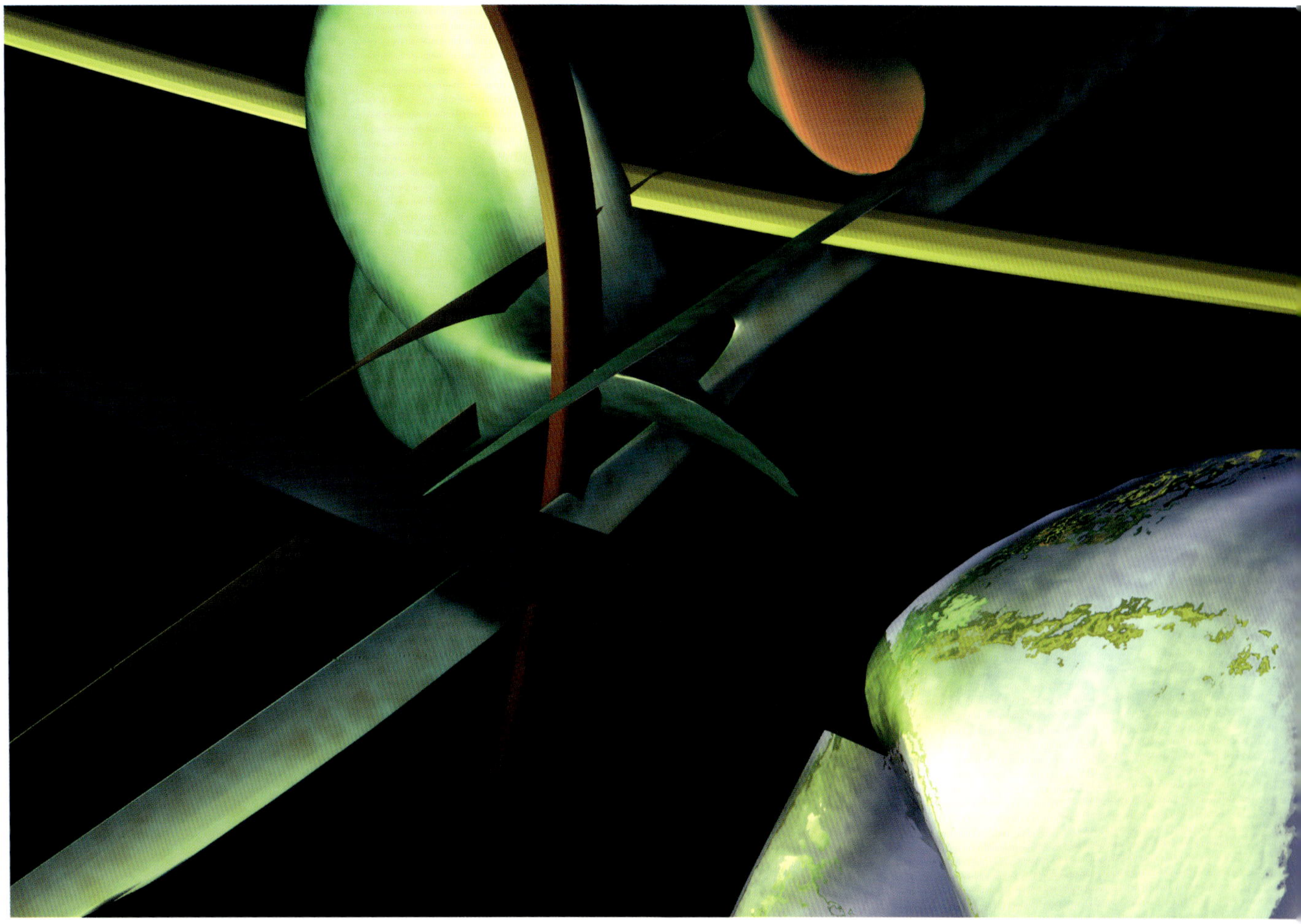

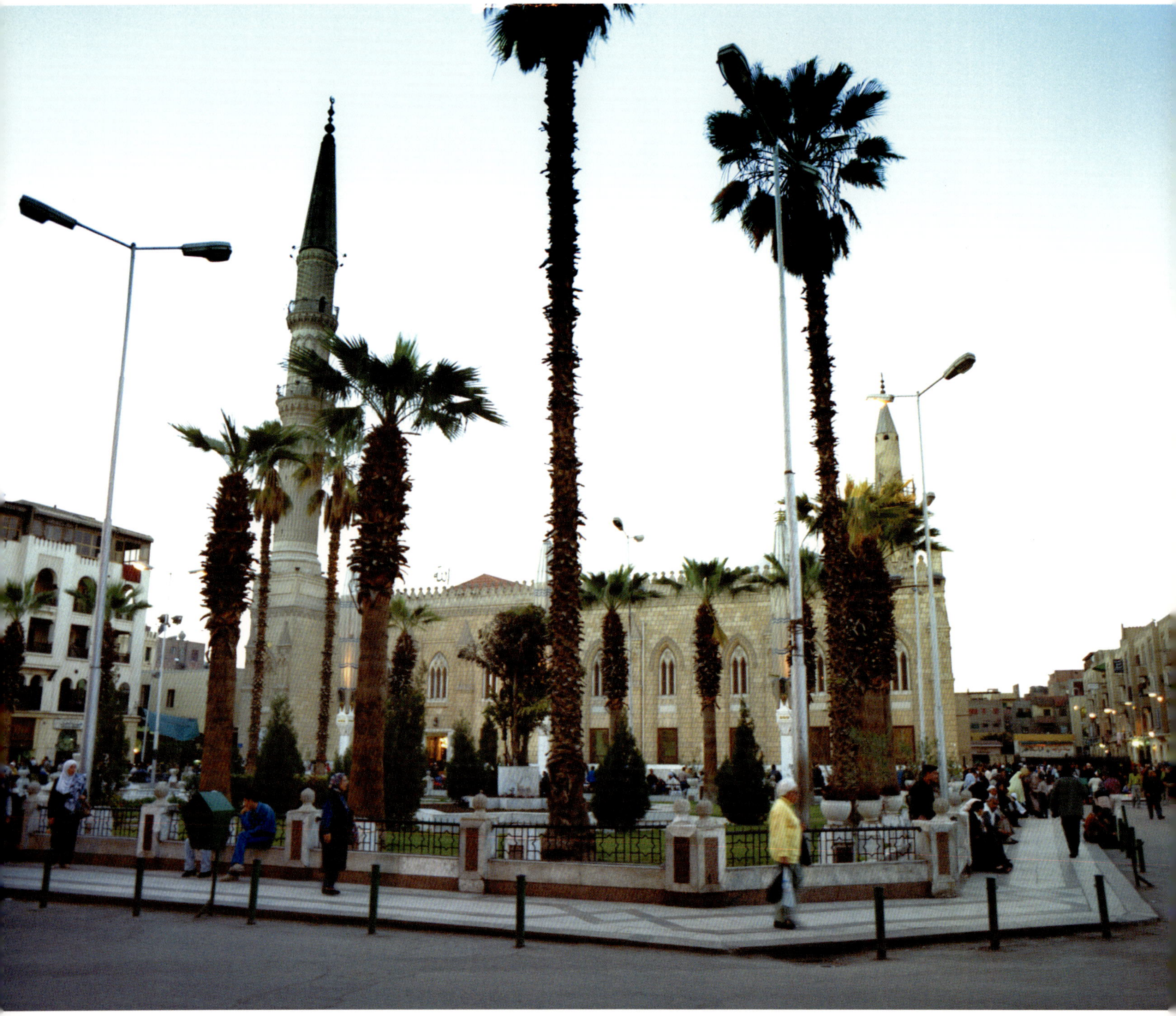

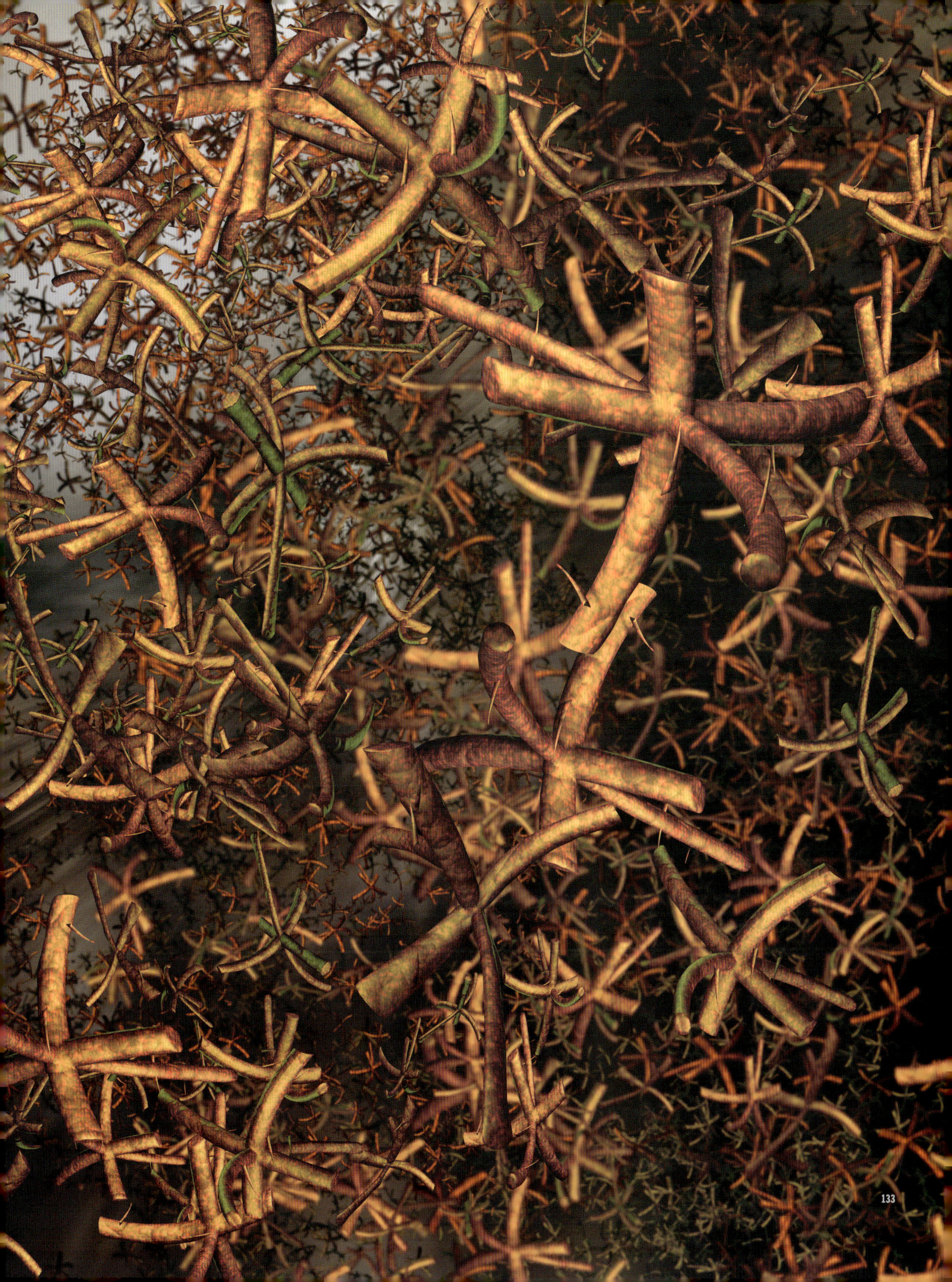

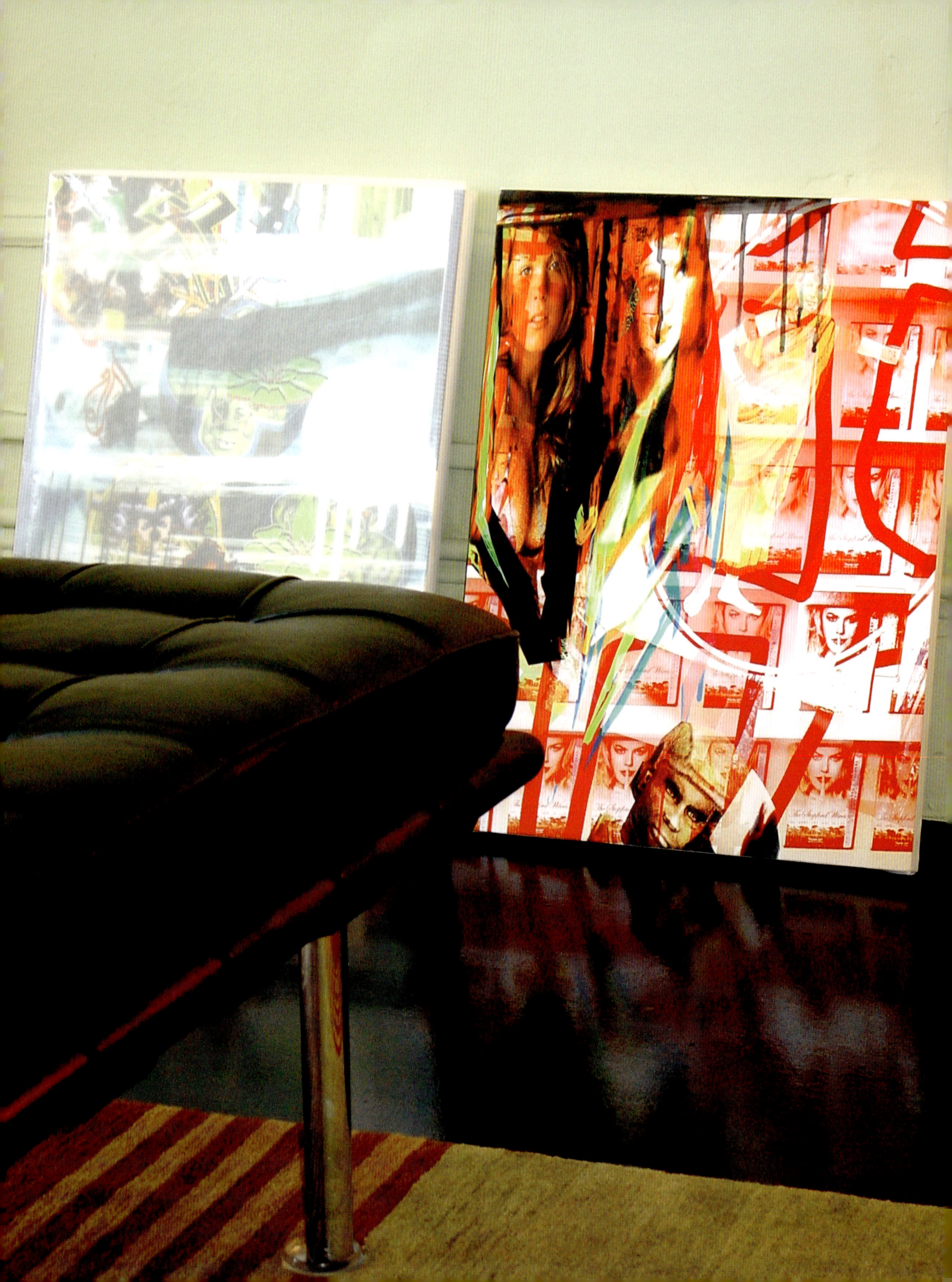

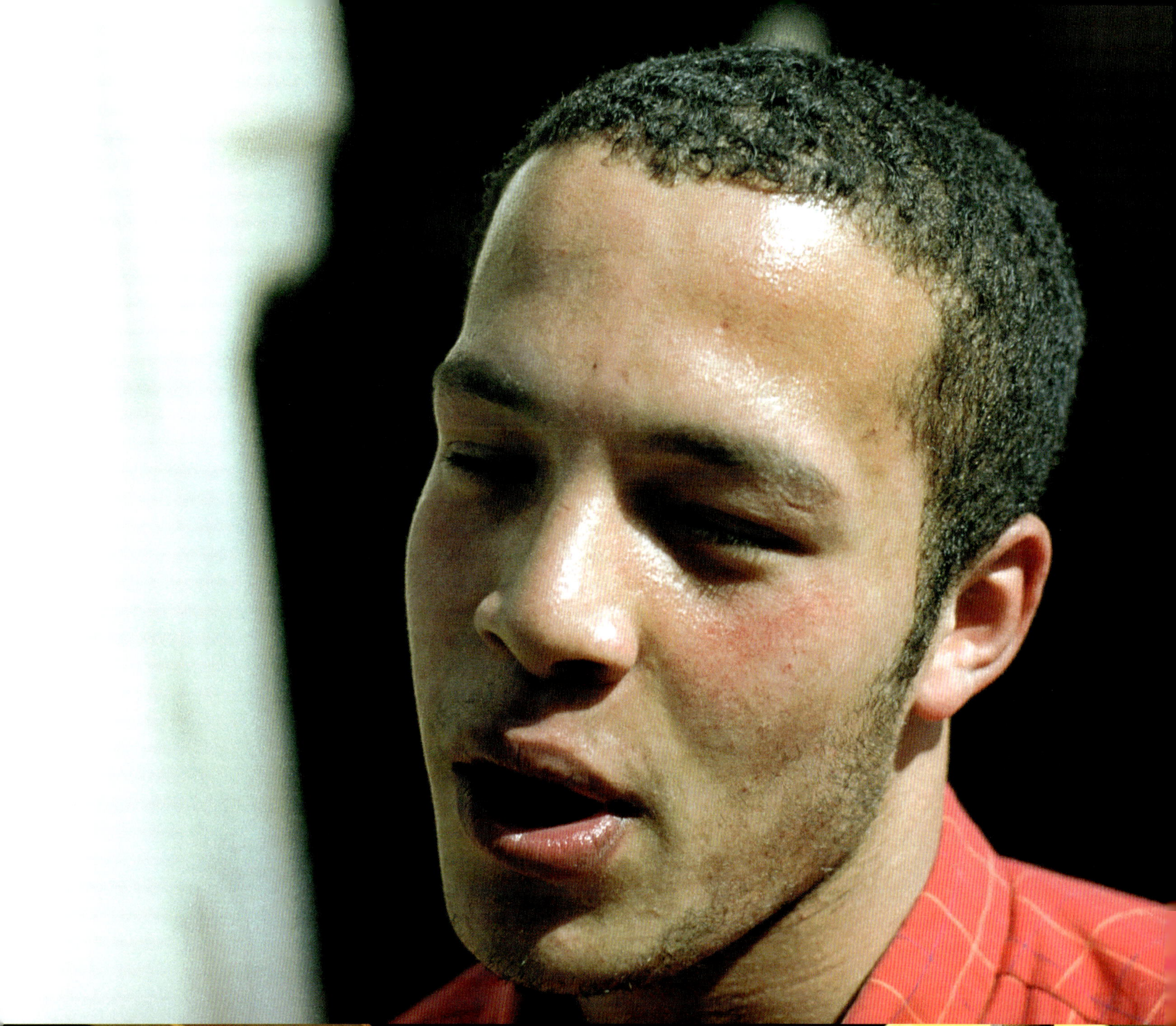
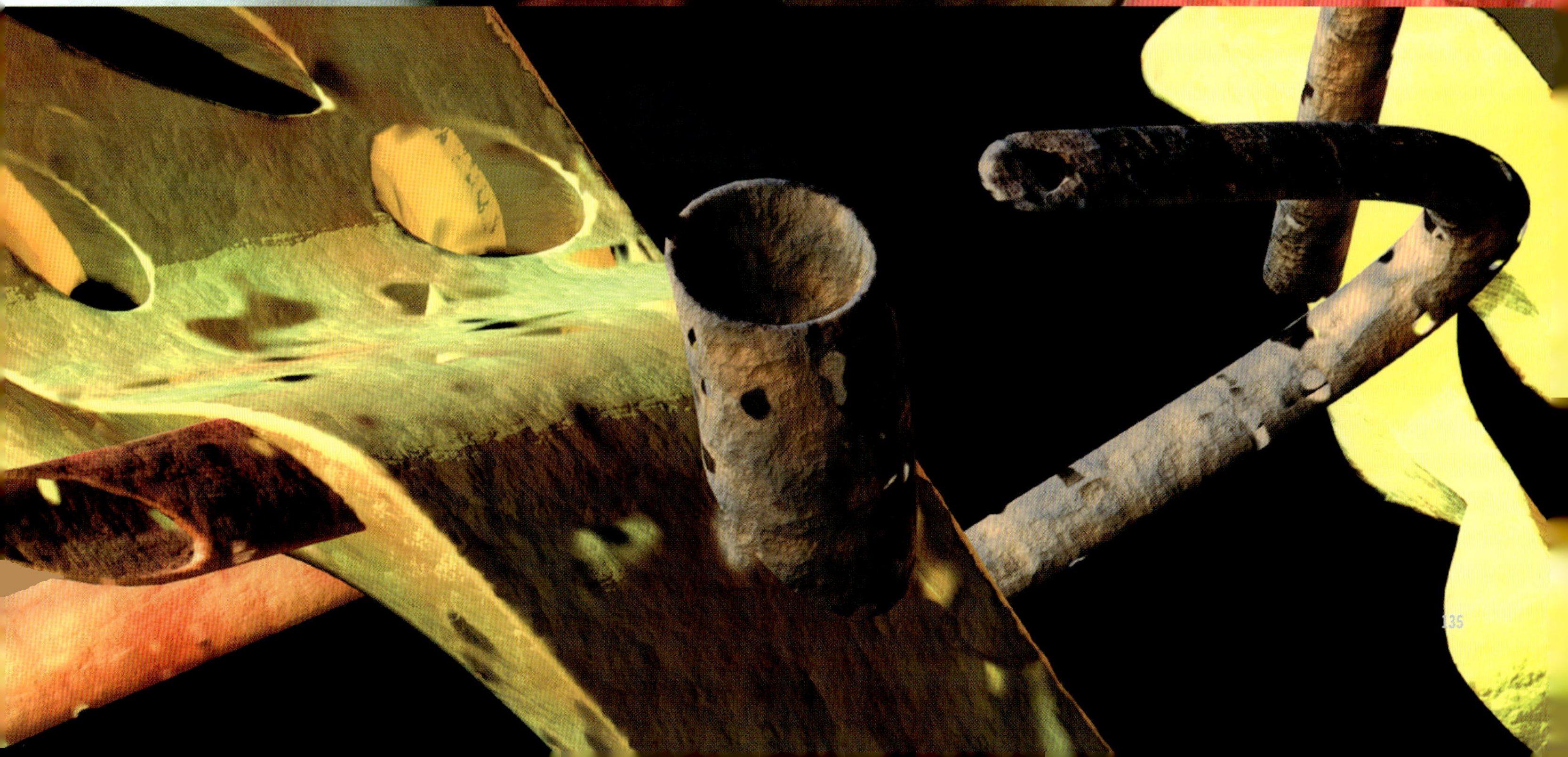

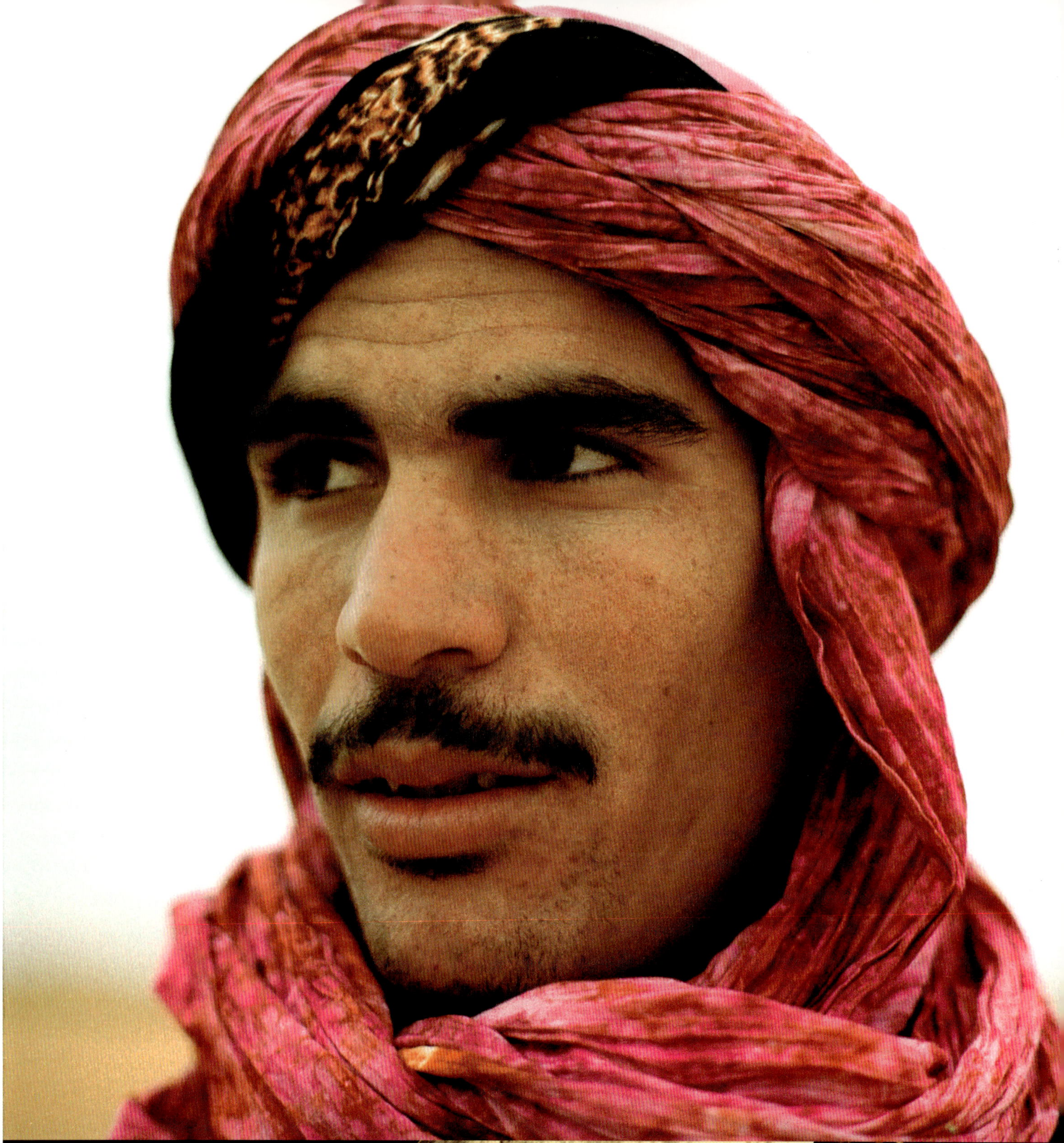

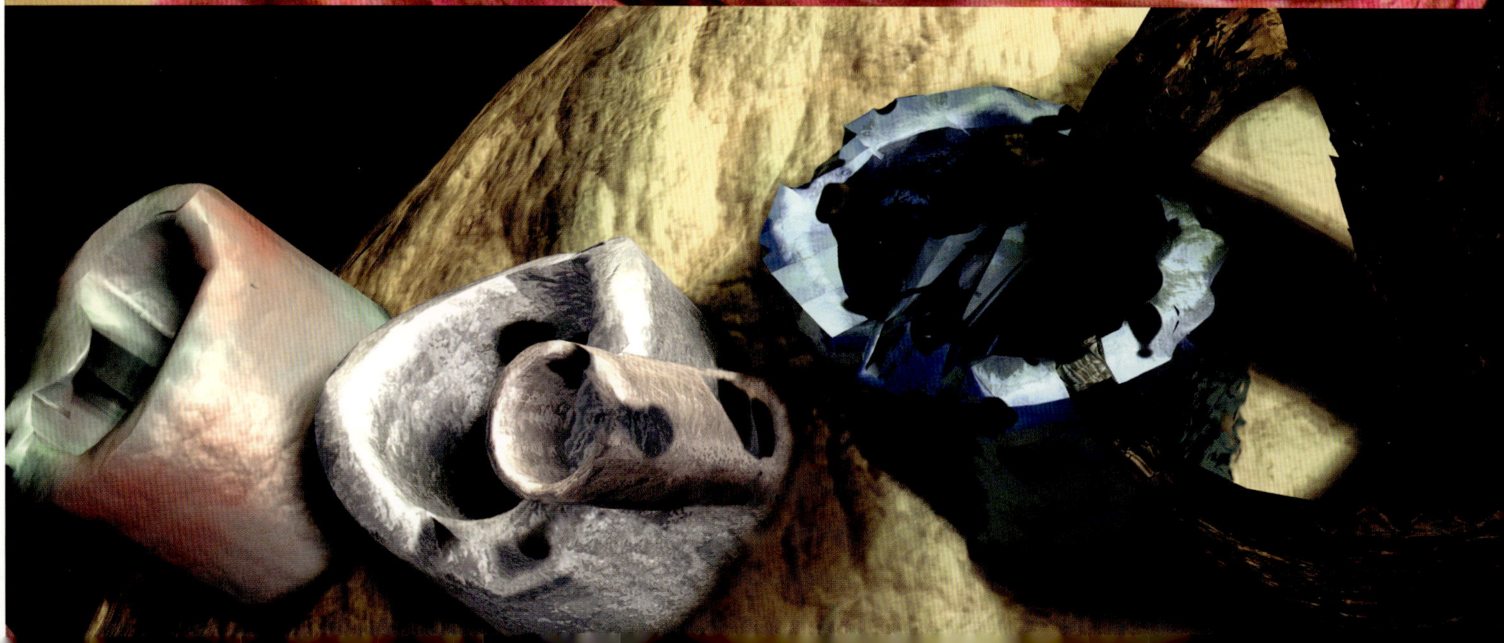

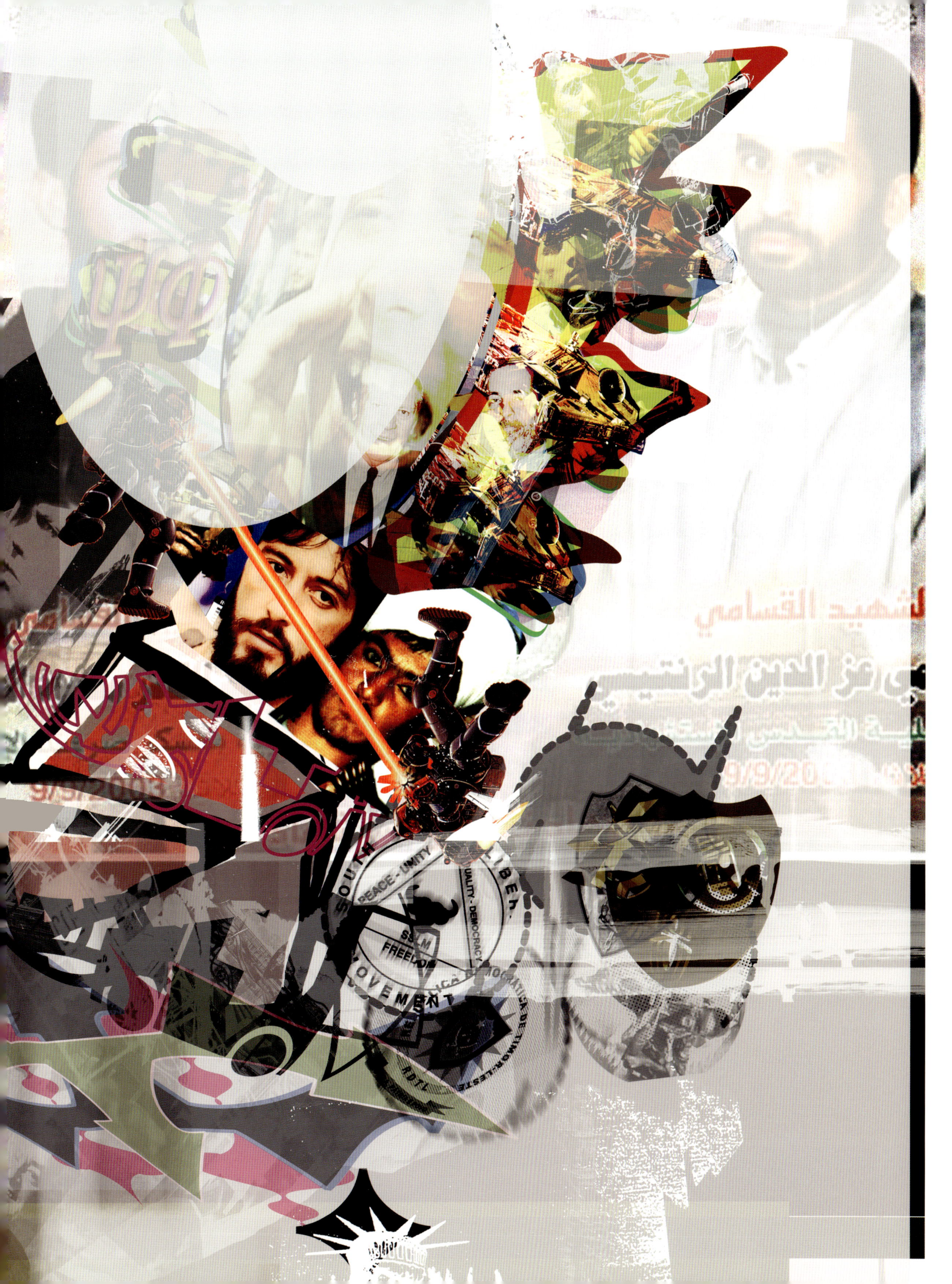
PEACE - UNITY
SSLM
9/9/2003

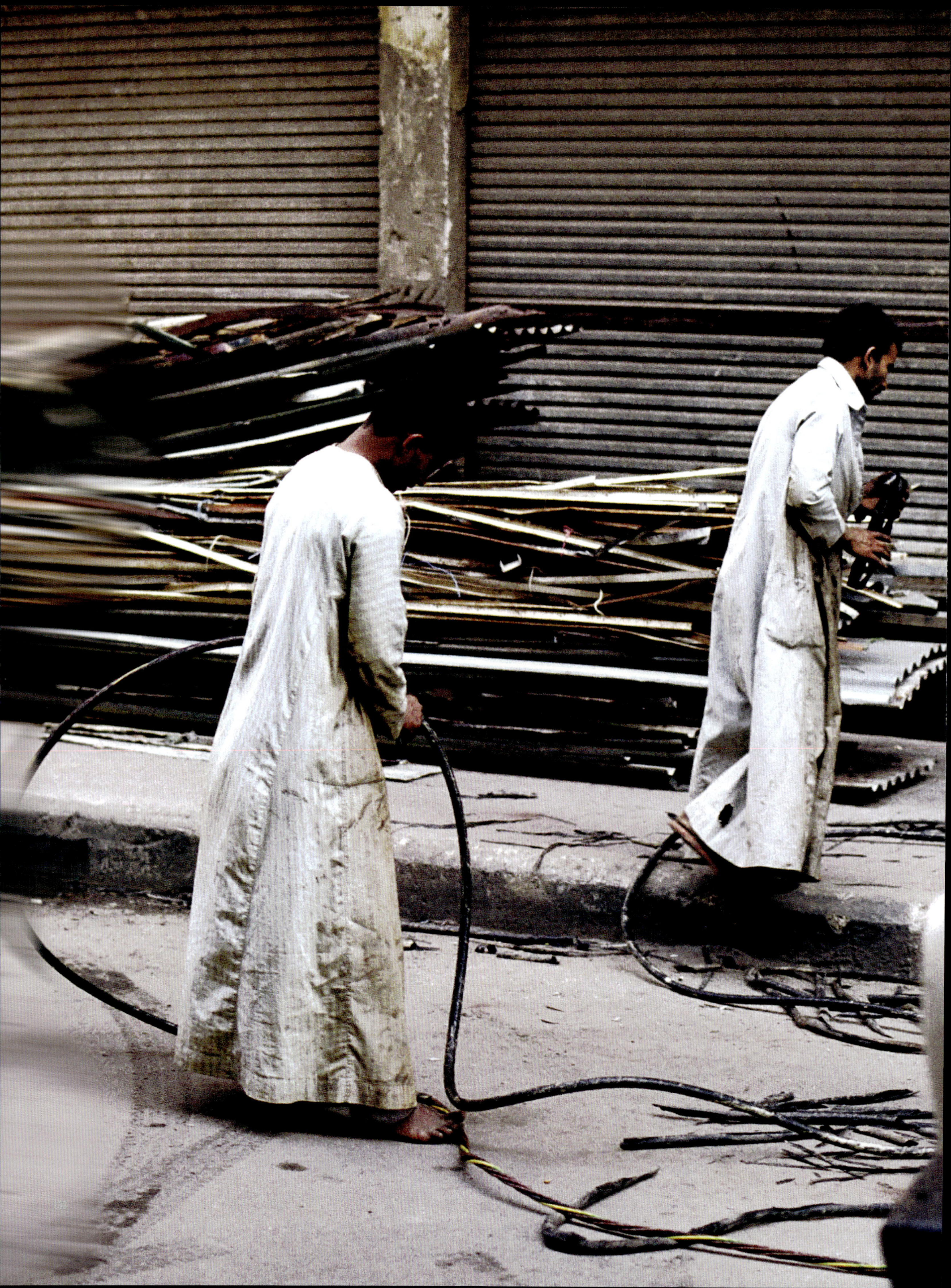

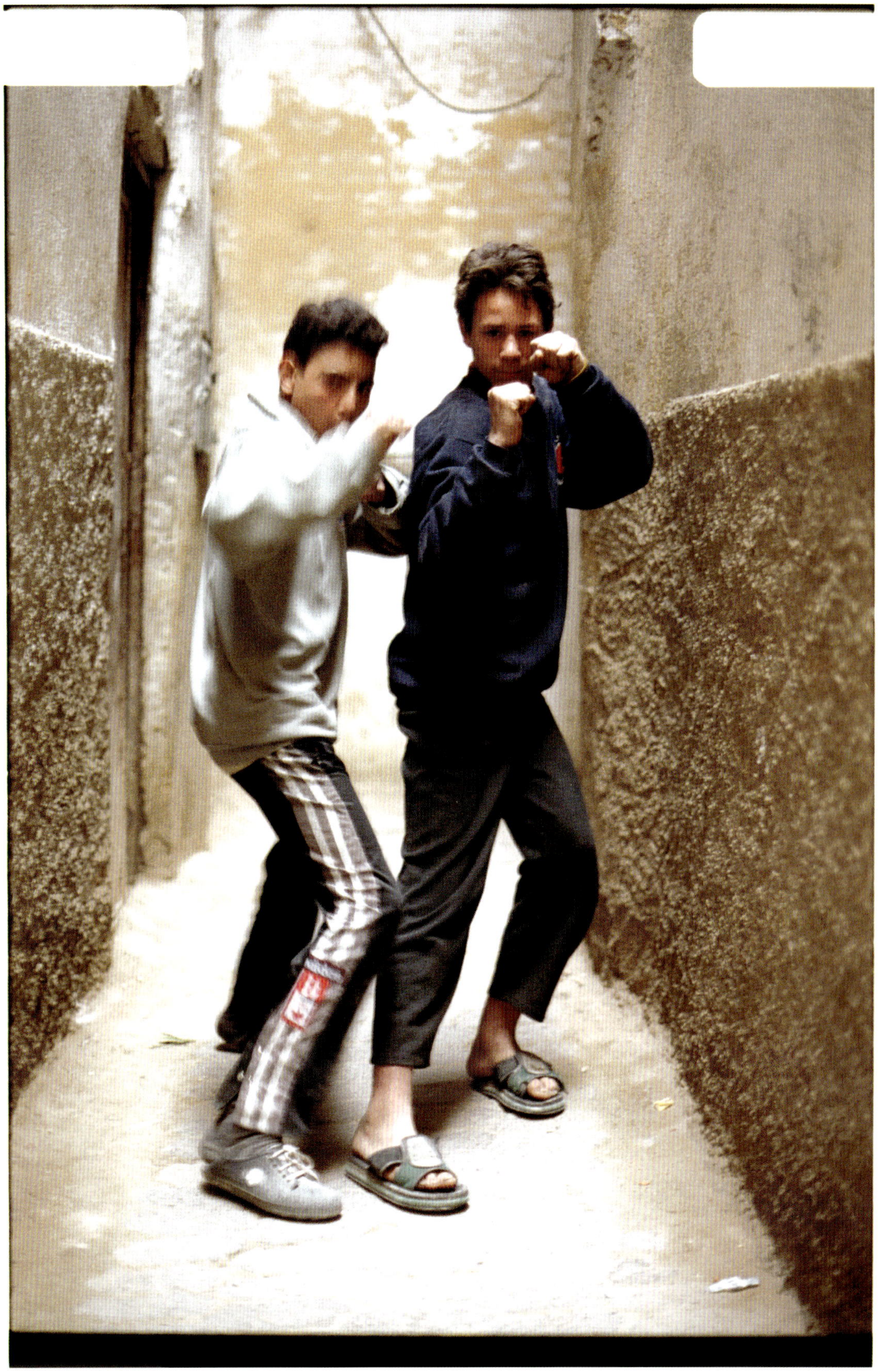

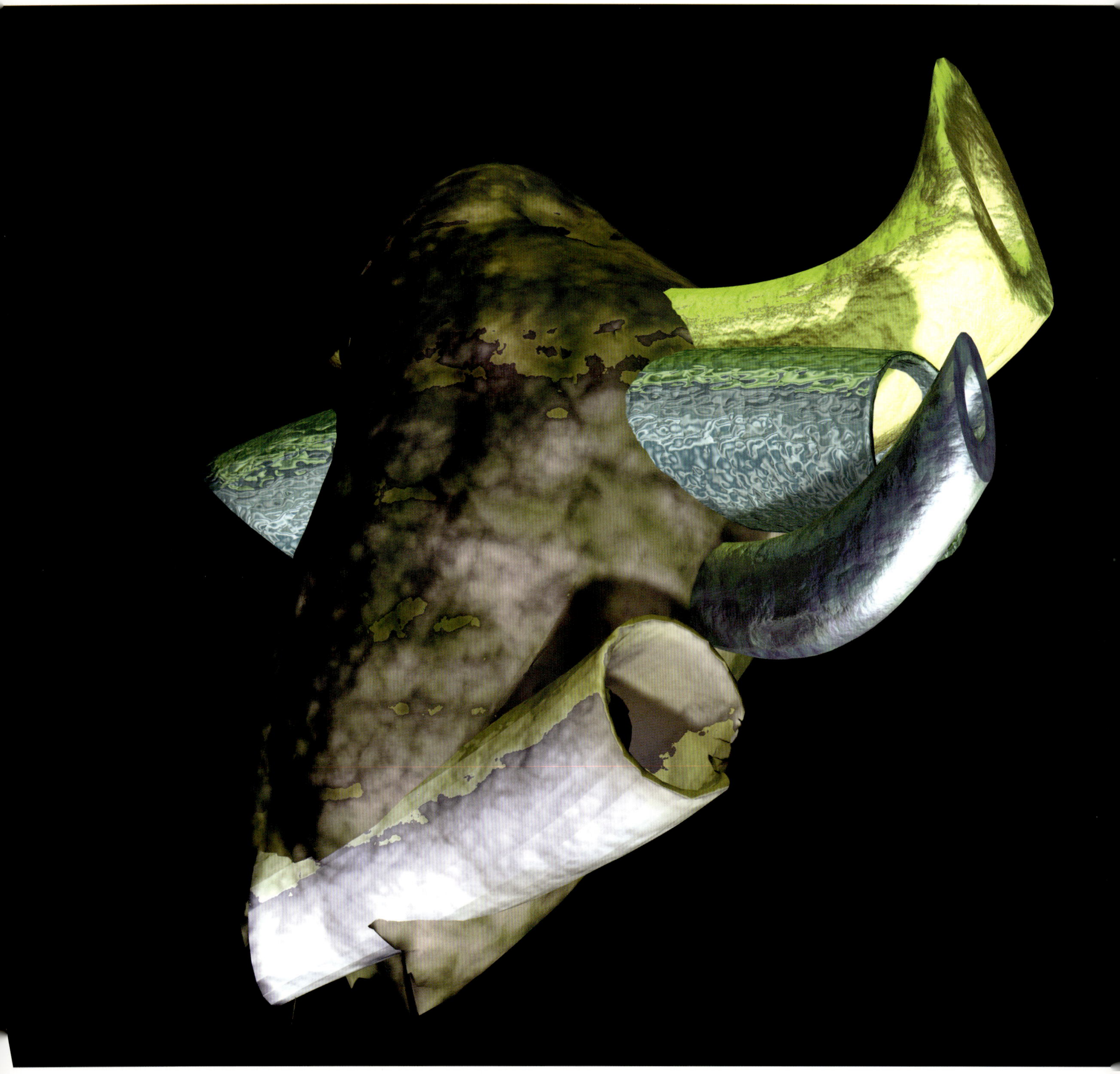

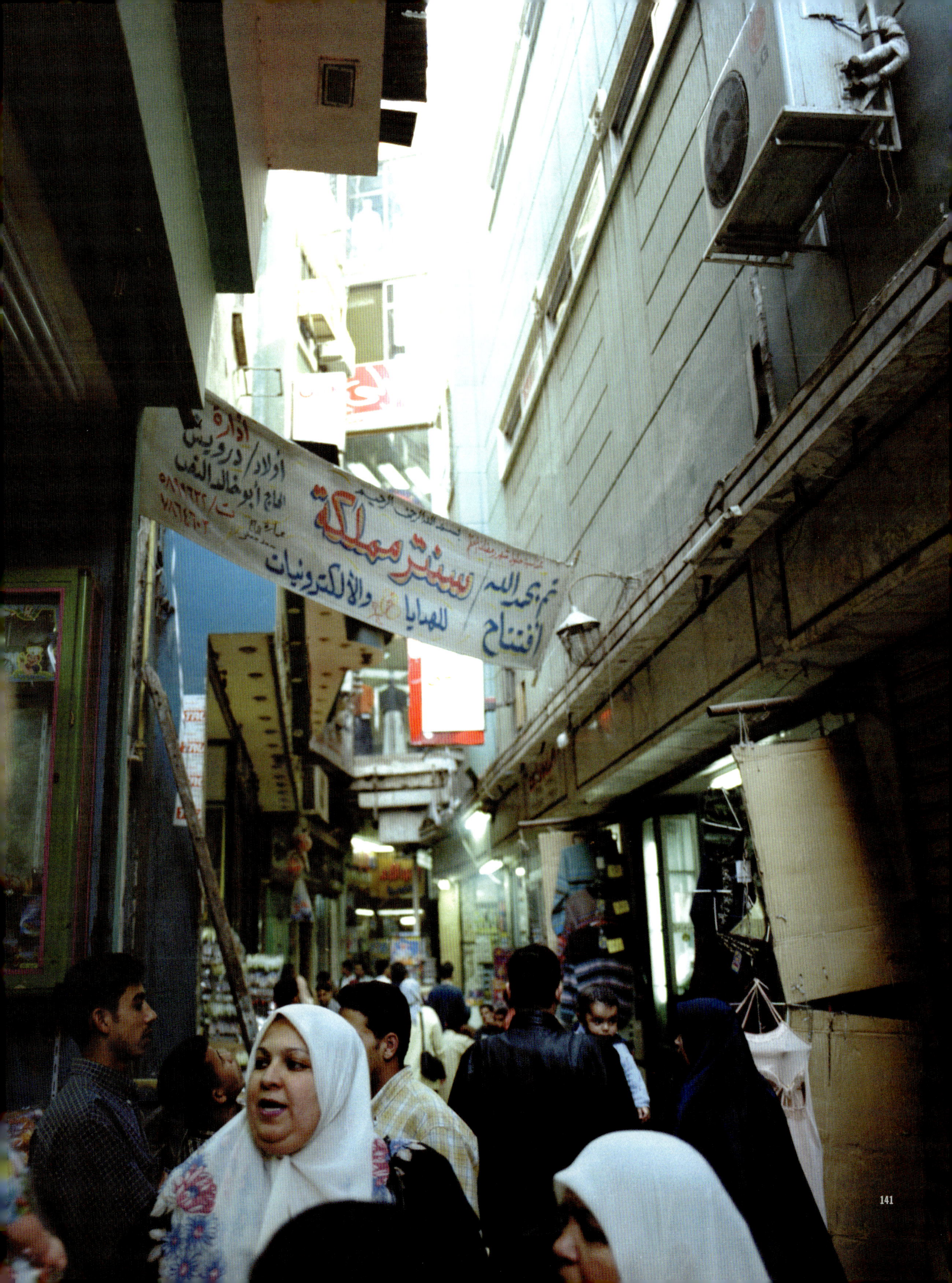
افتتاح
تم بحمد الله
سنتر مملكة
للهدايا والالكترونيات
ادارة
أولاد/ درويش

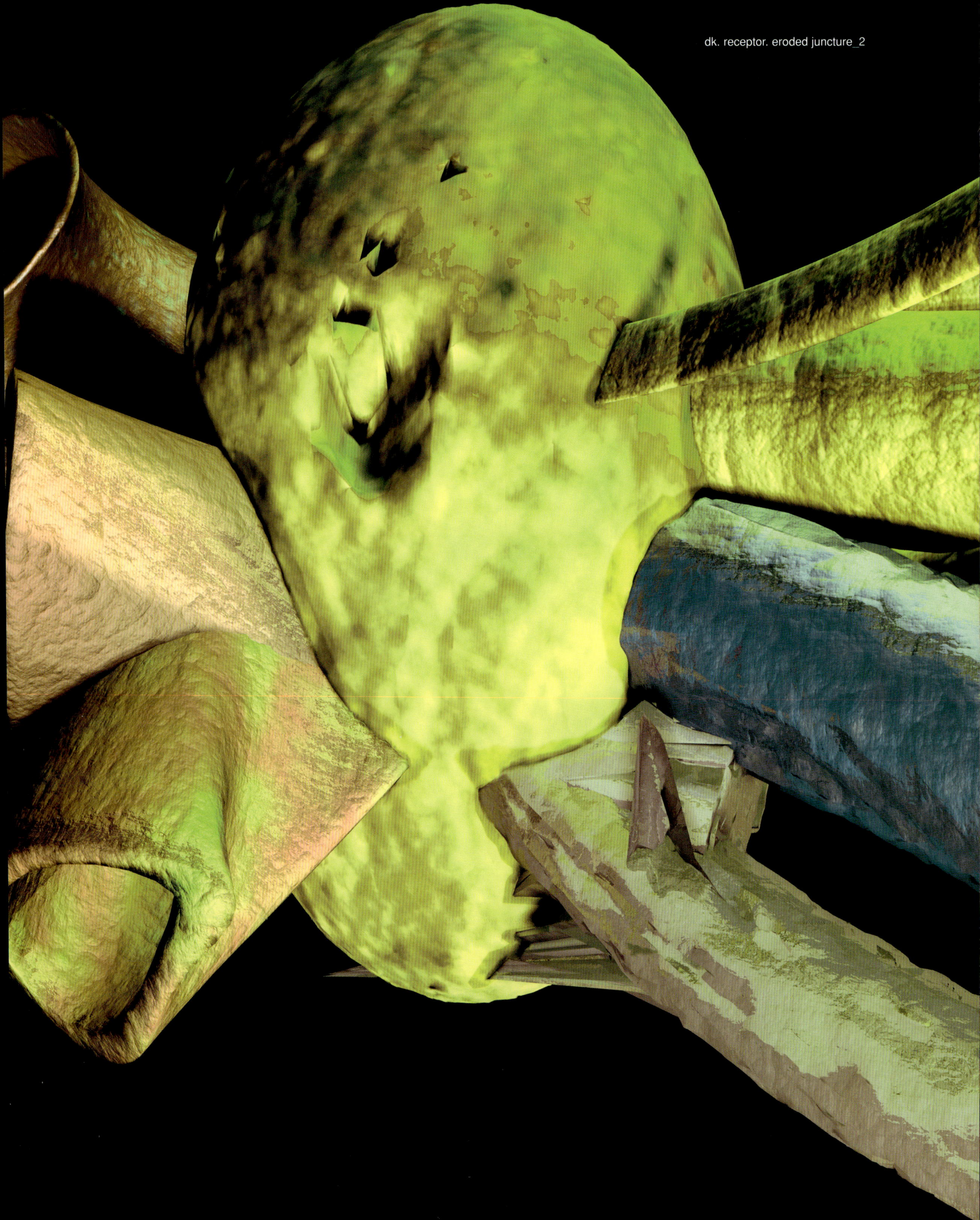

dk. receptor. eroded juncture_2

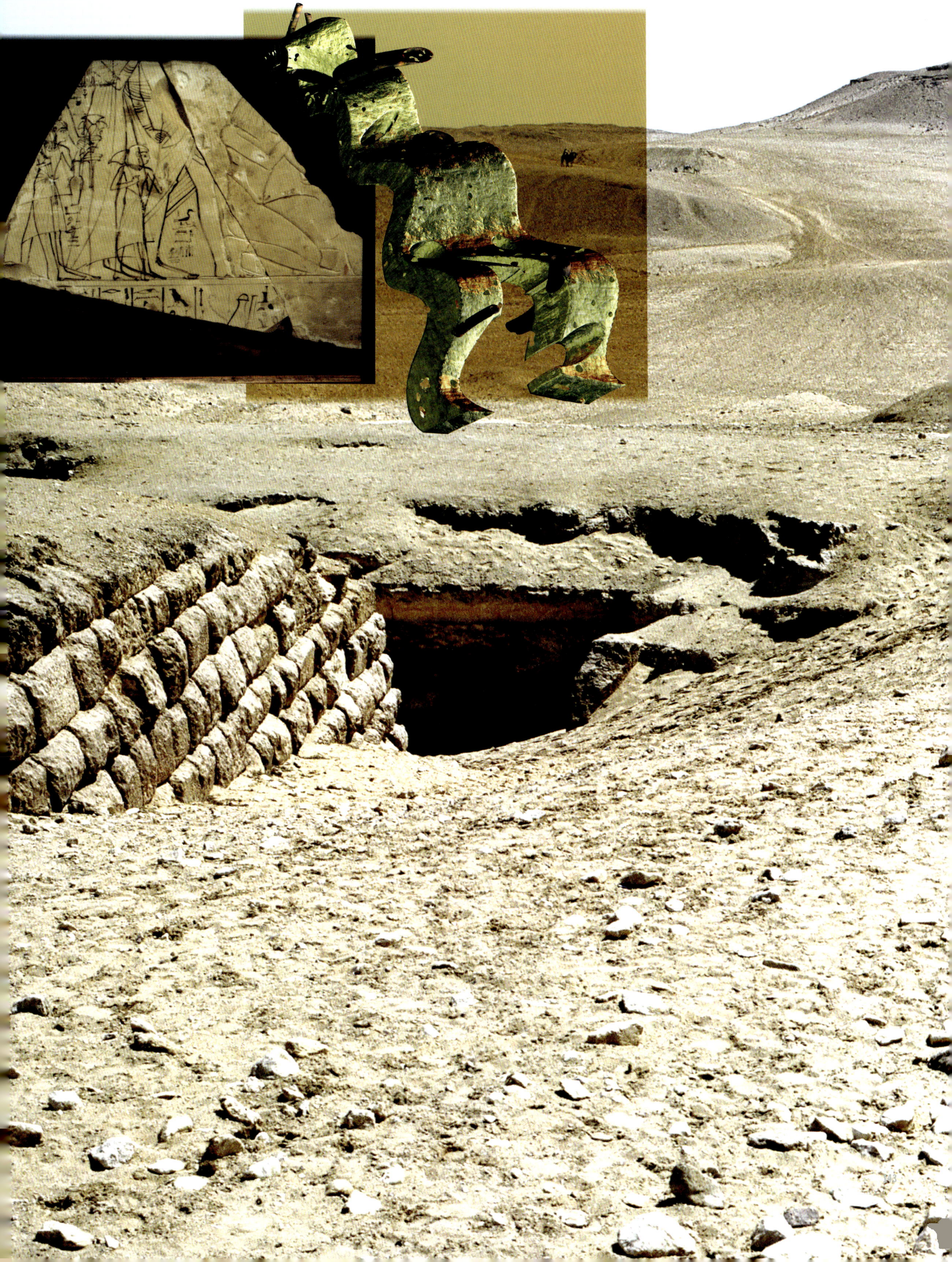

dk receptor /eroded junction 5_

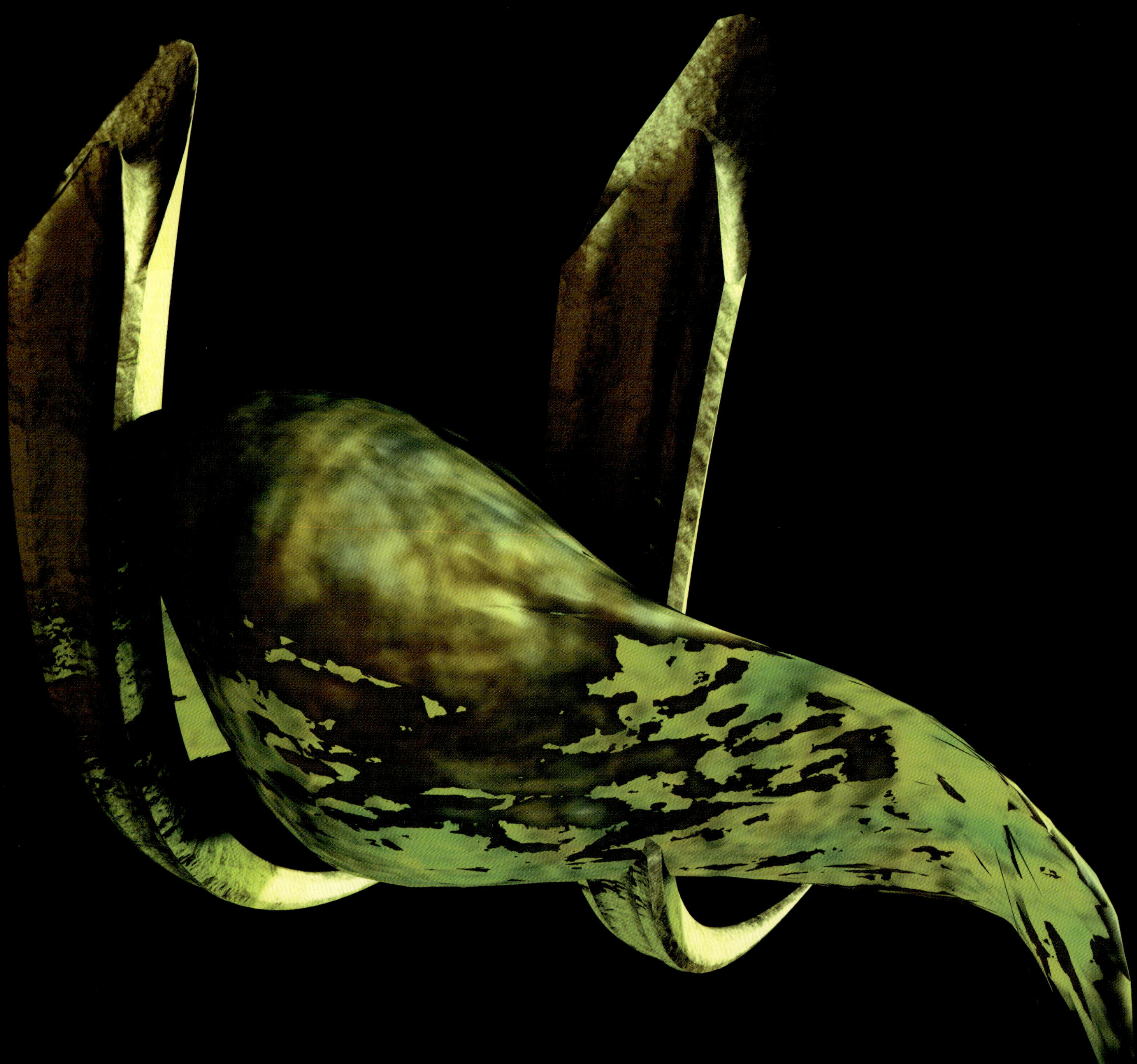

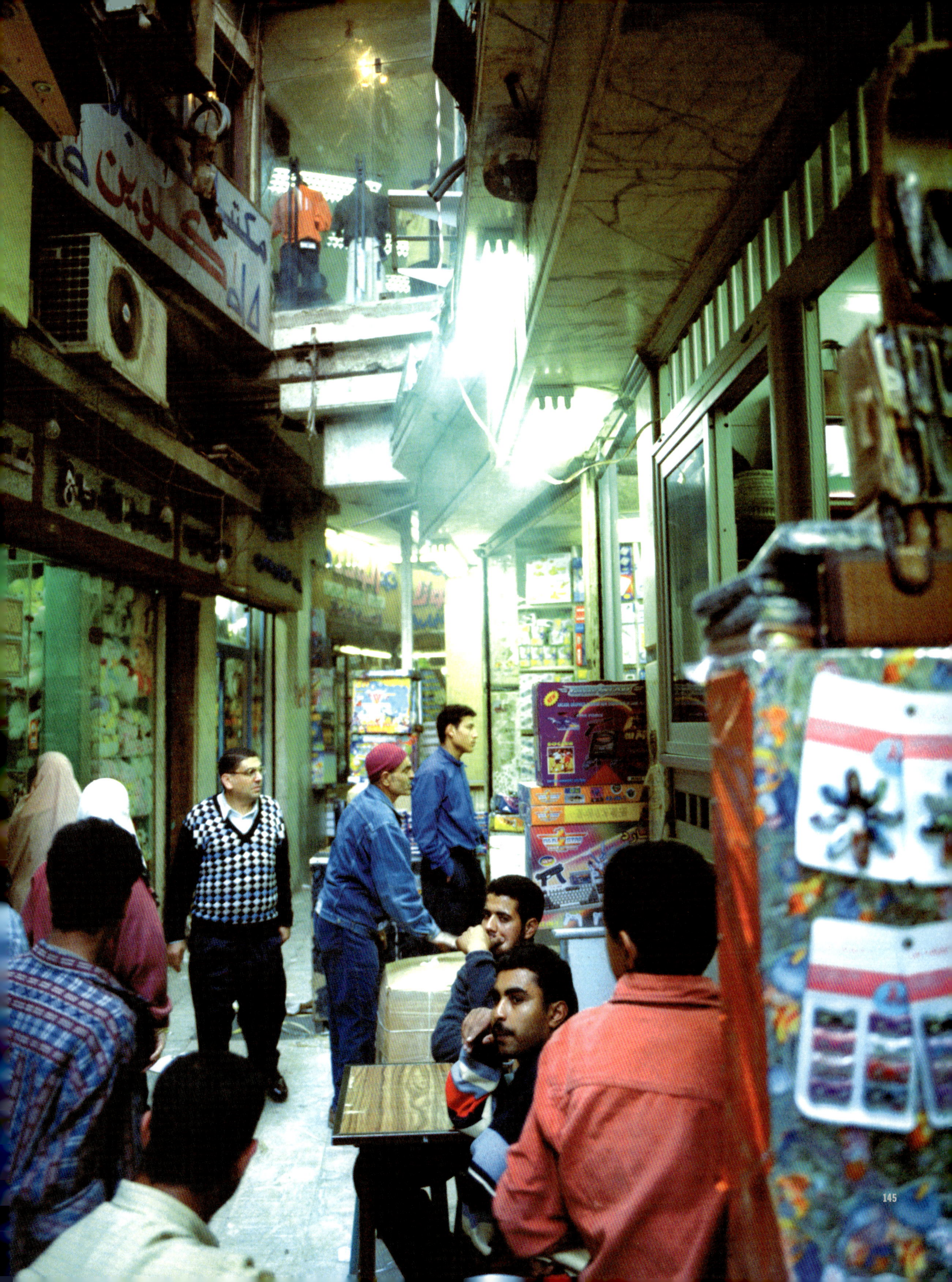

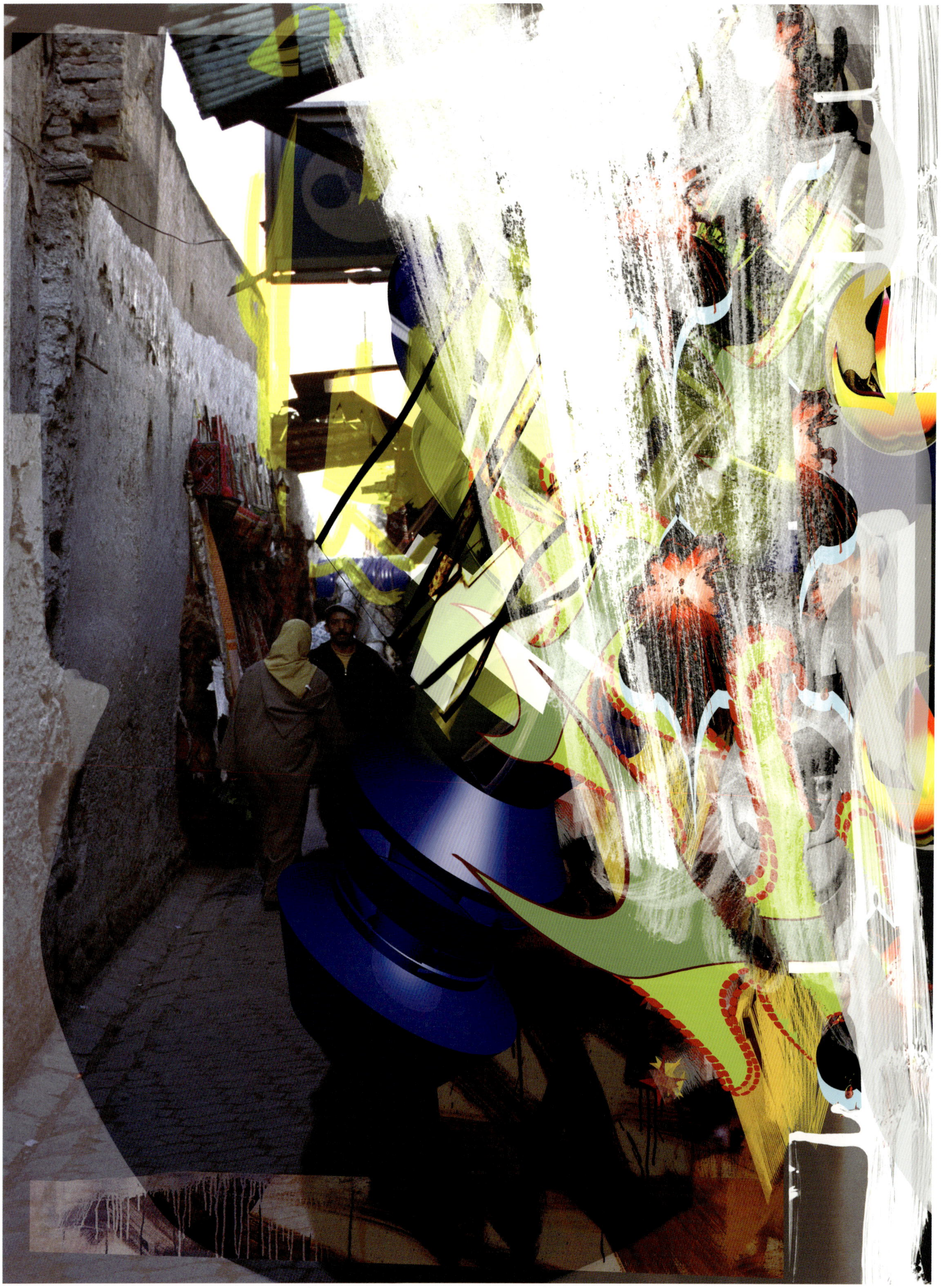

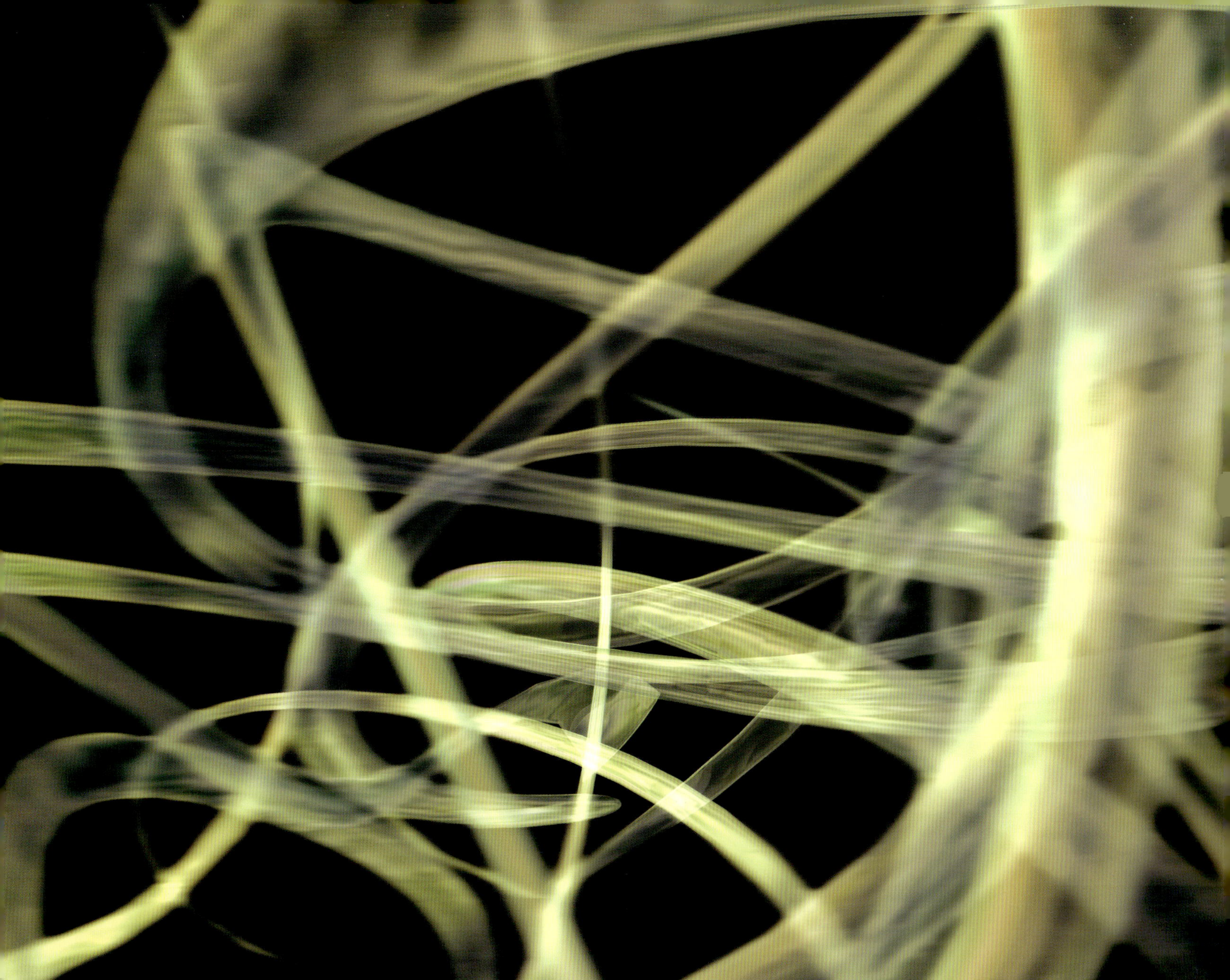

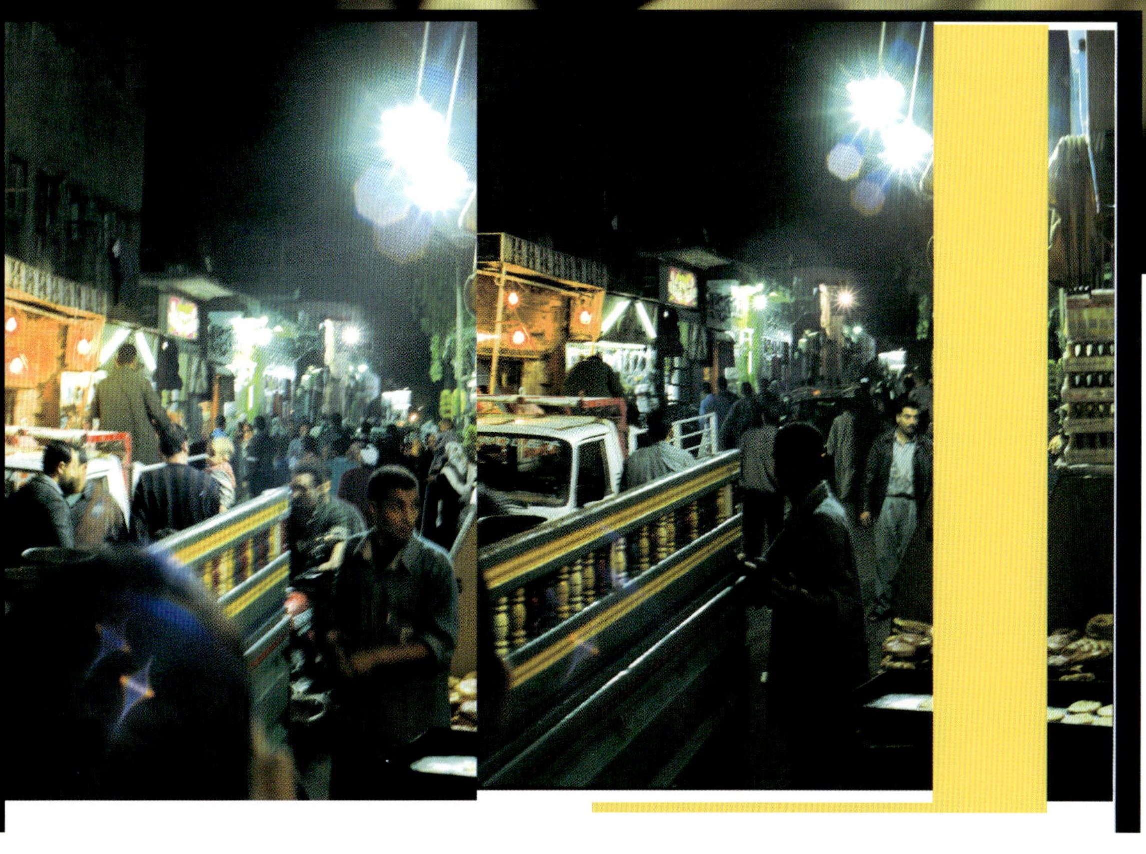

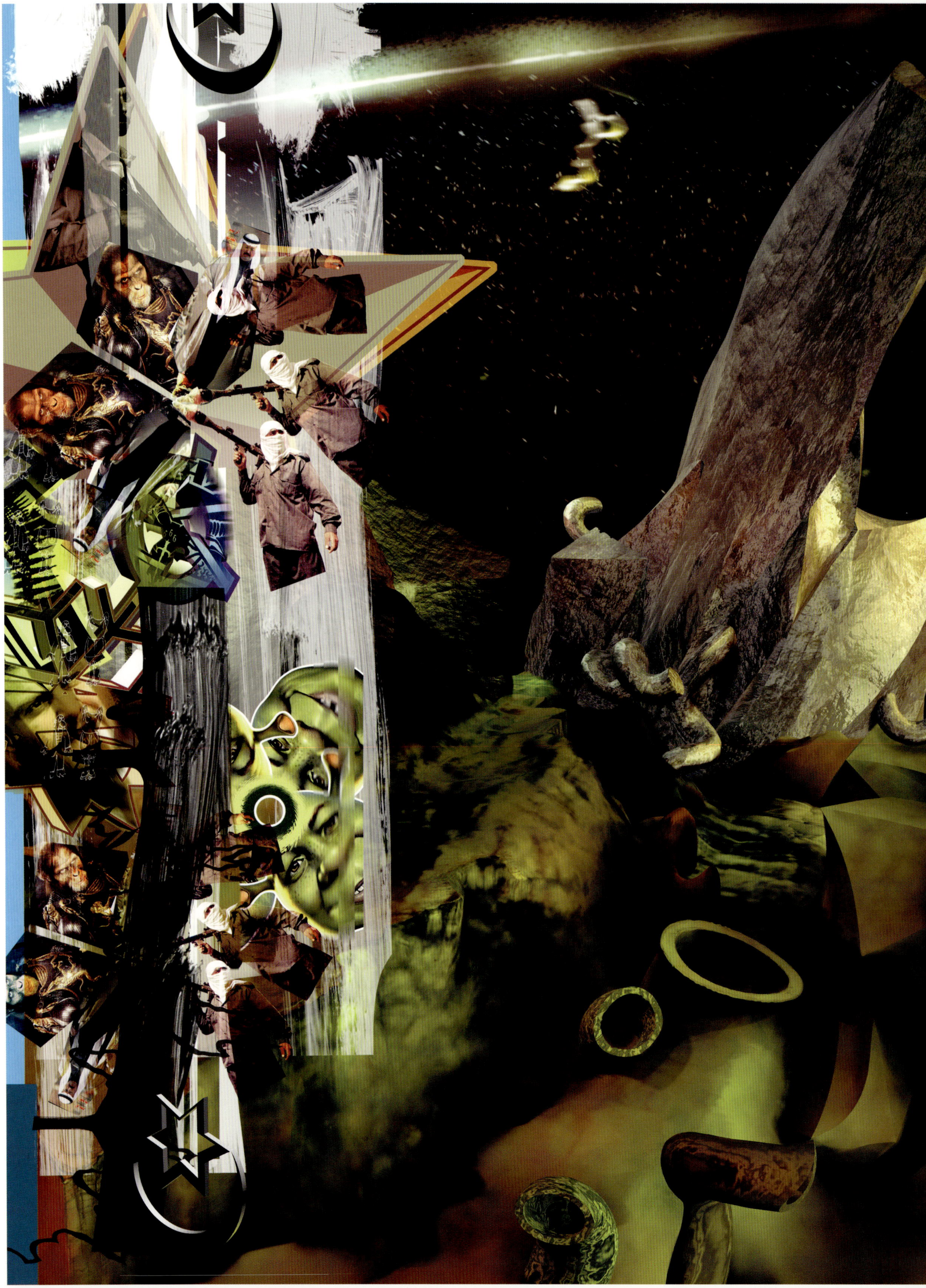

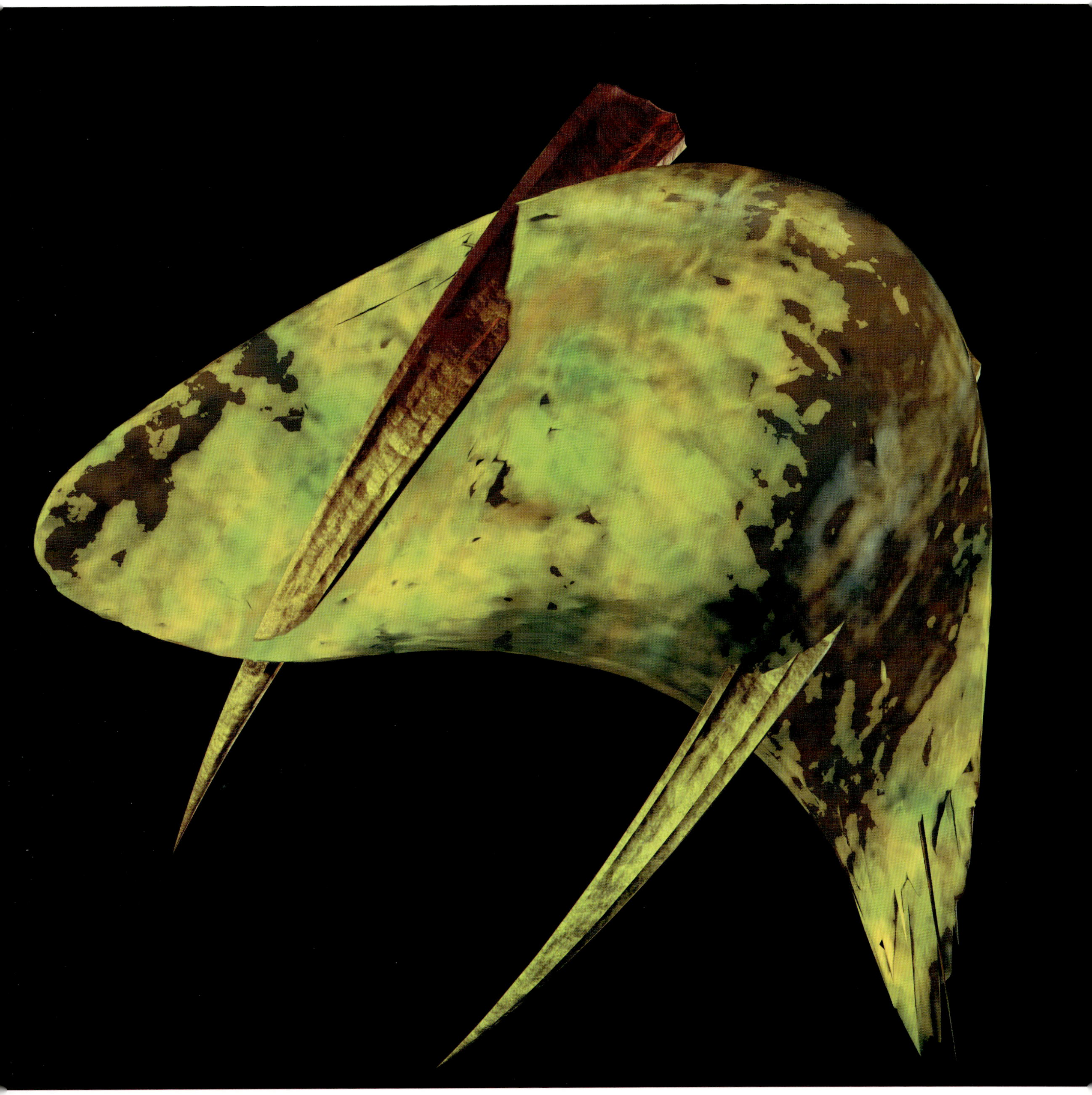

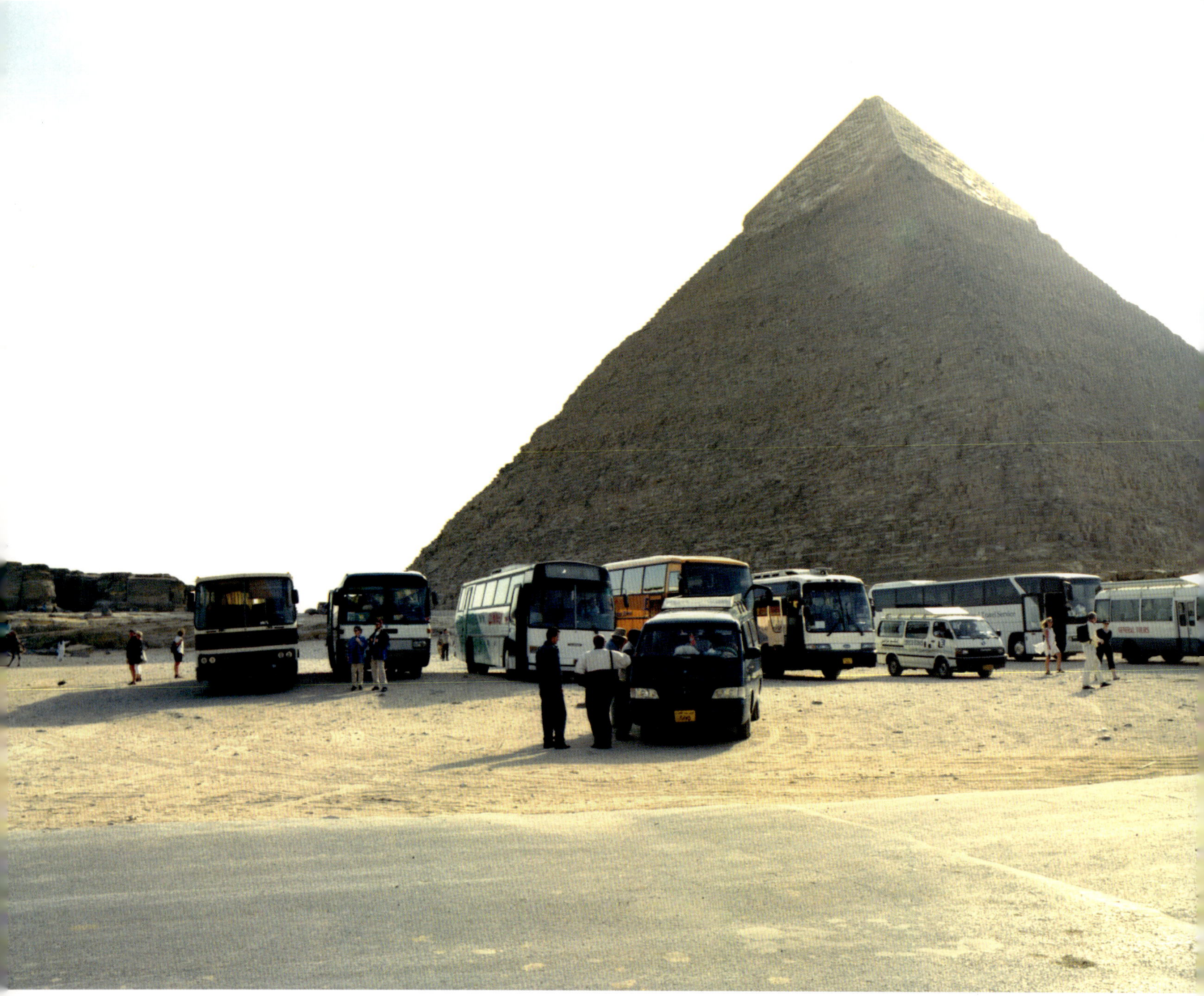

**MOST PLAUSIBLE FUTURIST MOVIES.** How likely is it that this set of circumstances might arise? Sometimes, for instance with meteor impacts, the rating is based on how plausibly the response to an unlikely event is depicted. Movies involving aliens are not penalized for the unknown but possibly very low probability of alien encounters. Movie Plausibility:

| Movie | Plausibility |
|---|---|
| Deep Impact | 9 |
| Rollerball (2002) | 8 |
| The Sum of All Fears | 8 |
| Close Encounters | 8 |
| Contact | 8 |
| Outbreak | 8 |
| Delicatessen | 8 |
| A Clockwork Orange | 8 |
| Fail-Safe | 8 |
| Gattaca | 7 |
| Jurassic Park | 7 |
| The Postman | 7 |
| Mad Max 2: The Road Warrior | 7 |
| The Boys from Brazil | 7 |
| Deep Blue Sea | 7 |
| The Island of Dr. Moreau | 7 |
| Seven Days in May | 7 |
| 2001 | 7 |
| 2010 | 7 |
| Escape from New York | 7 |
| Things to Come | 6 |
| Fahrenheit 451 | 6 |
| No Escape | 6 |
| On the Beach | 6 |
| Artificial Intelligence: A.I. | 6 |
| Mission to Mars | 6 |
| The Handmaid's Tale | 6 |
| Series 7 | 6 |
| Rollerball (1975) | 6 |
| 1984 | 6 |
| The Andromeda Strain | 6 |
| Starman | 6 |
| Solaris (1972) | 6 |
| Solaris (2002) | 6 |
| The Omega Man | 6 |
| Sleeper | 6 |
| The Abyss | 6 |
| Pitch Black | 6 |
| Red Planet | 5 |
| Vanilla Sky | 5 |
| Outland | 5 |
| Metropolis | 5 |
| THX 1138 | 5 |
| Z.P.G. | 5 |
| Ghost in the Shell | 5 |
| Back to the Future II | 5 |
| Colossus: The Forbidden Project | 5 |
| Soylent Green | 5 |
| Barb Wire | 5 |
| Stargate | 5 |
| Cocoon | 5 |
| Blade Runner | 4 |
| Westworld | 4 |
| Demolition Man | 4 |
| Bicentennial Man | 4 |
| Logan's Run | 4 |
| Predator | 4 |
| The Matrix | 4 |
| Johnny Mnemonic | 4 |
| Until the End of the World | 4 |
| Signs | 4 |
| Impostor | 4 |
| E.T. | 4 |
| The Stepford Wives | 4 |
| Silent Running | 4 |
| Escape from LA | 4 |
| Judge Dredd | 4 |
| Aliens | 4 |
| eXistenZ | 3 |
| Attack of the Clones | 3 |
| A Boy and His Dog | 3 |
| Strange Days | 3 |
| Species | 3 |
| Supernova | 3 |
| The Thing | 3 |
| Independence Day | 3 |
| The Thirteenth Floor | 3 |
| The Arrival | 3 |
| Armageddon | 3 |
| Fifth Element | 3 |
| Phantom Menace | 3 |
| RoboCop | 3 |
| Total Recall | 3 |
| The Time Machine (2002) | 3 |
| The Time Machine (1960) | 3 |
| Star Trek | 3 |
| Star Wars | 3 |
| Terminator | 3 |
| Virtuosity | 3 |
| Dune | 3 |
| The Sixth Day | 3 |
| Forbidden Planet | 3 |
| The Arrival | 3 |
| Dark City | 3 |
| Planet of the Apes (2001) | 3 |
| Zardoz | 3 |
| Planet of the Apes (1968) | 2 |
| Alphaville | 2 |
| Soldier | 2 |
| Starship Troopers | 2 |
| Twelve Monkeys | 2 |
| The Day the Earth Stood Still | 2 |
| Timecop | 2 |
| Millennium | 2 |
| Freejack | 2 |
| Battlefield Earth | 2 |
| Universal Soldier | 2 |
| Moontrap | 2 |
| Conquest of the Planet of the Apes | 2 |
| Moon 44 | 2 |
| Akira | 2 |
| Minority Report | 2 |
| The Quiet Earth | 2 |
| Sphere | 2 |
| The Lawnmower Man | 2 |
| X-Men | 1 |
| Waterworld | 1 |

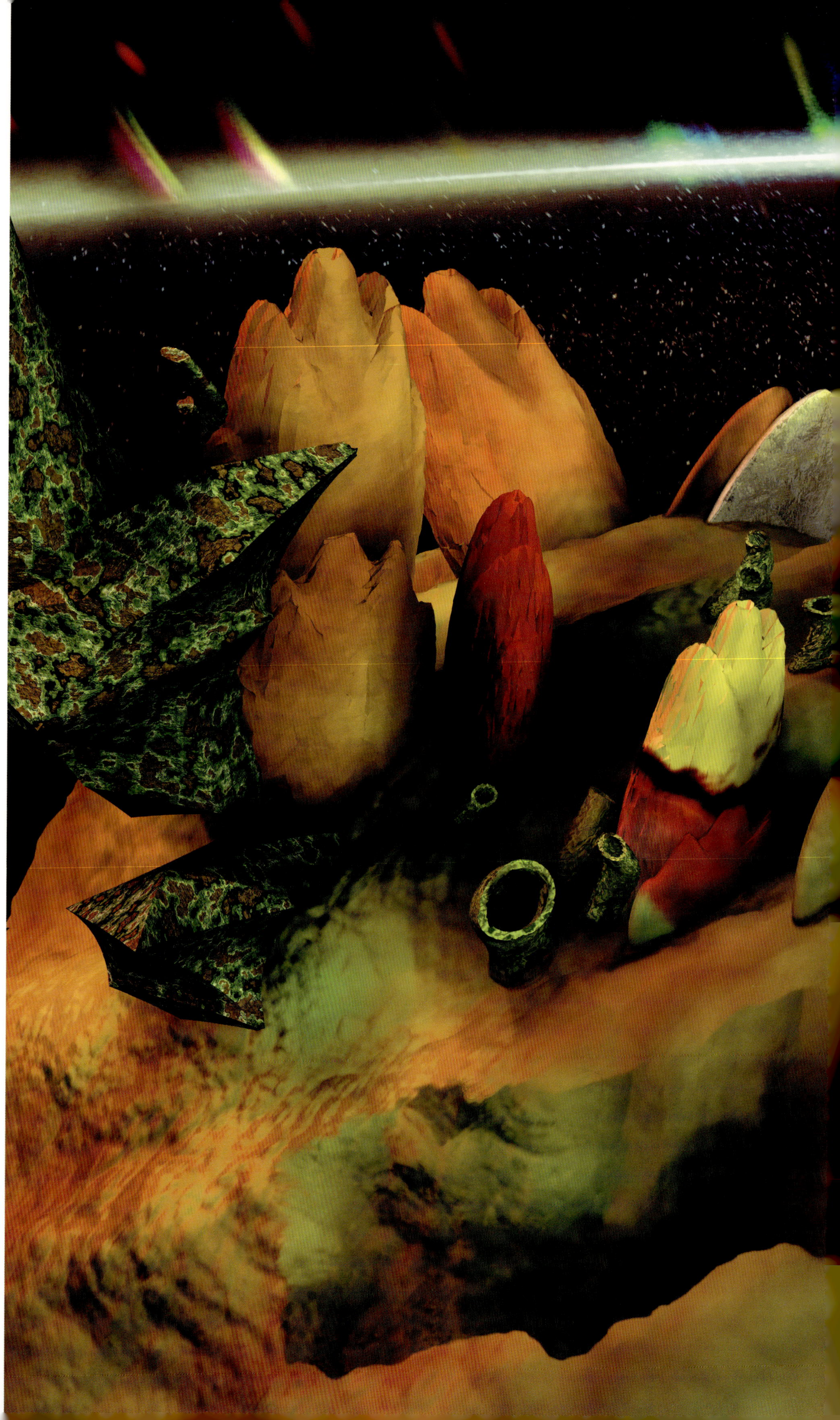

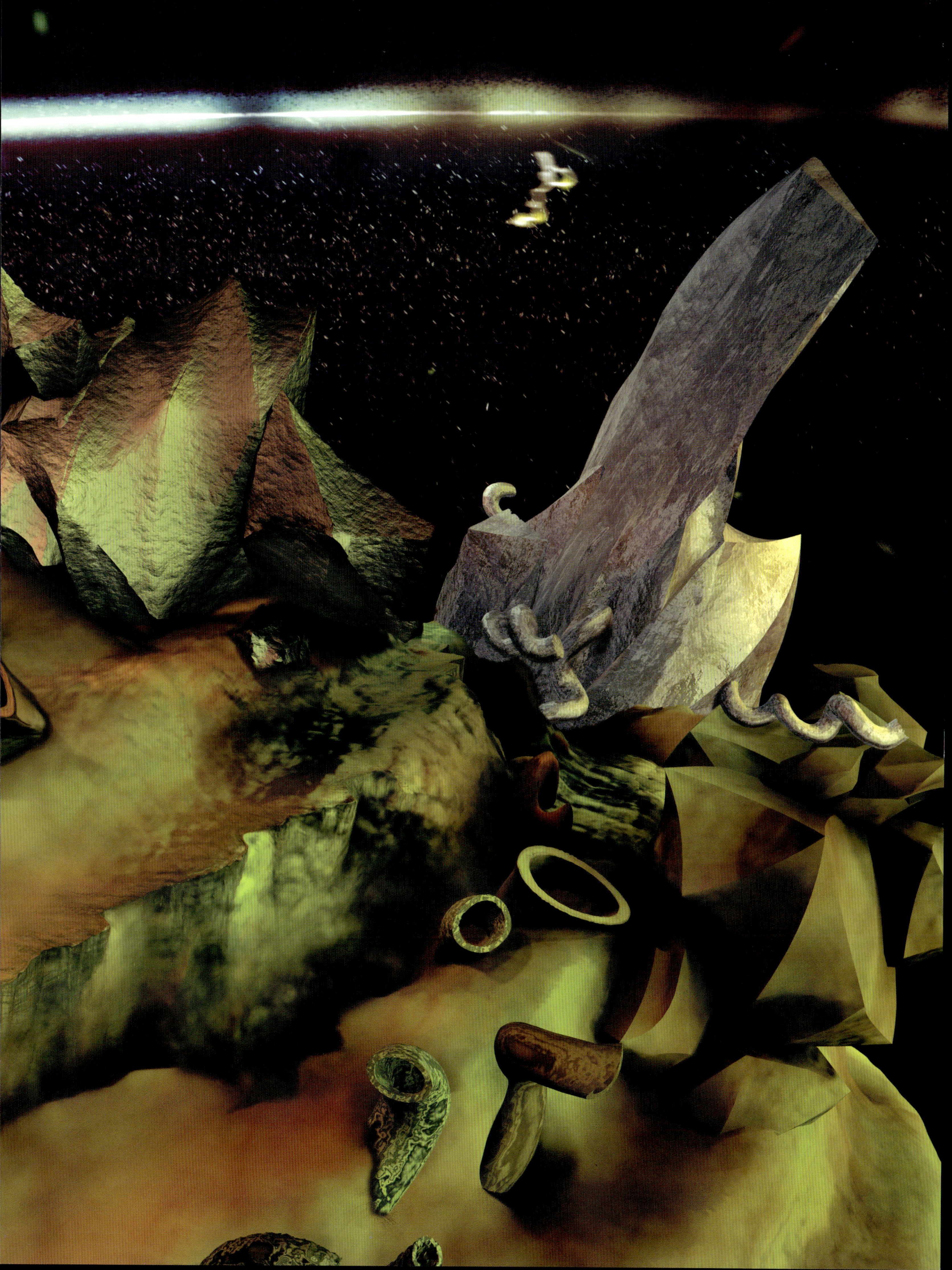

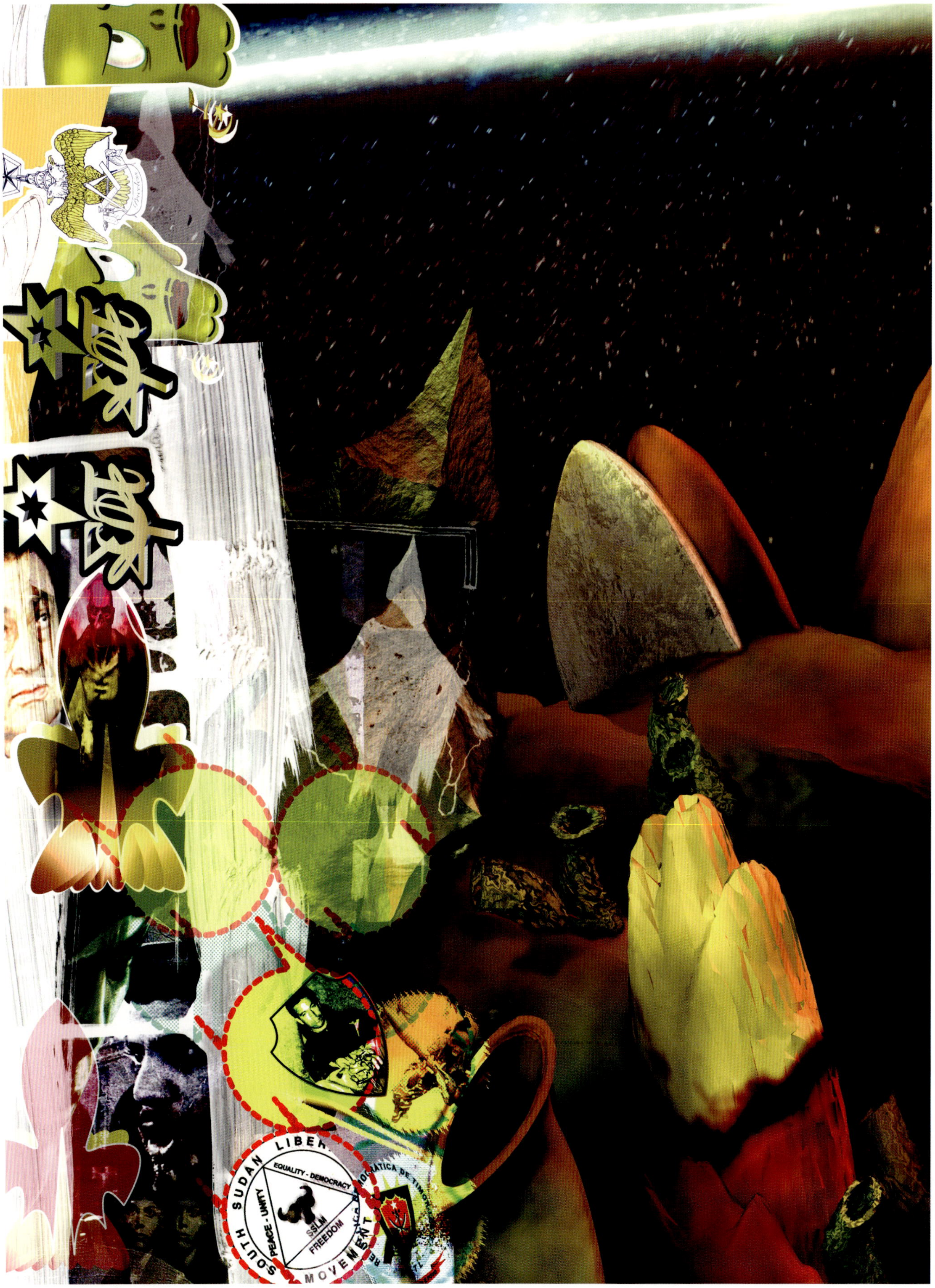
SOUTH SUDAN
EQUALITY - DEMOCRACY
PEACE - UNITY
SSLM
FREEDOM
MOVEMENT

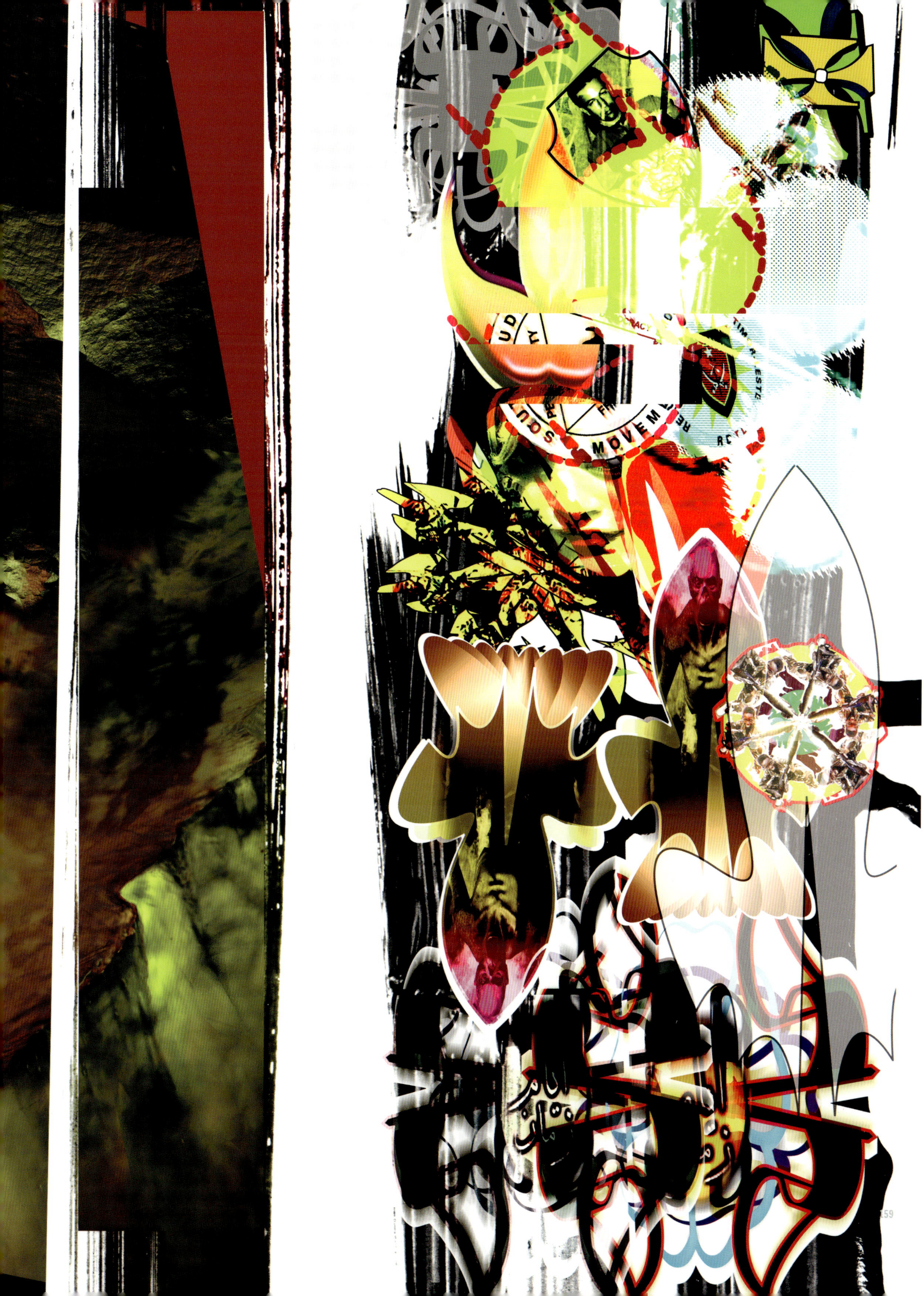

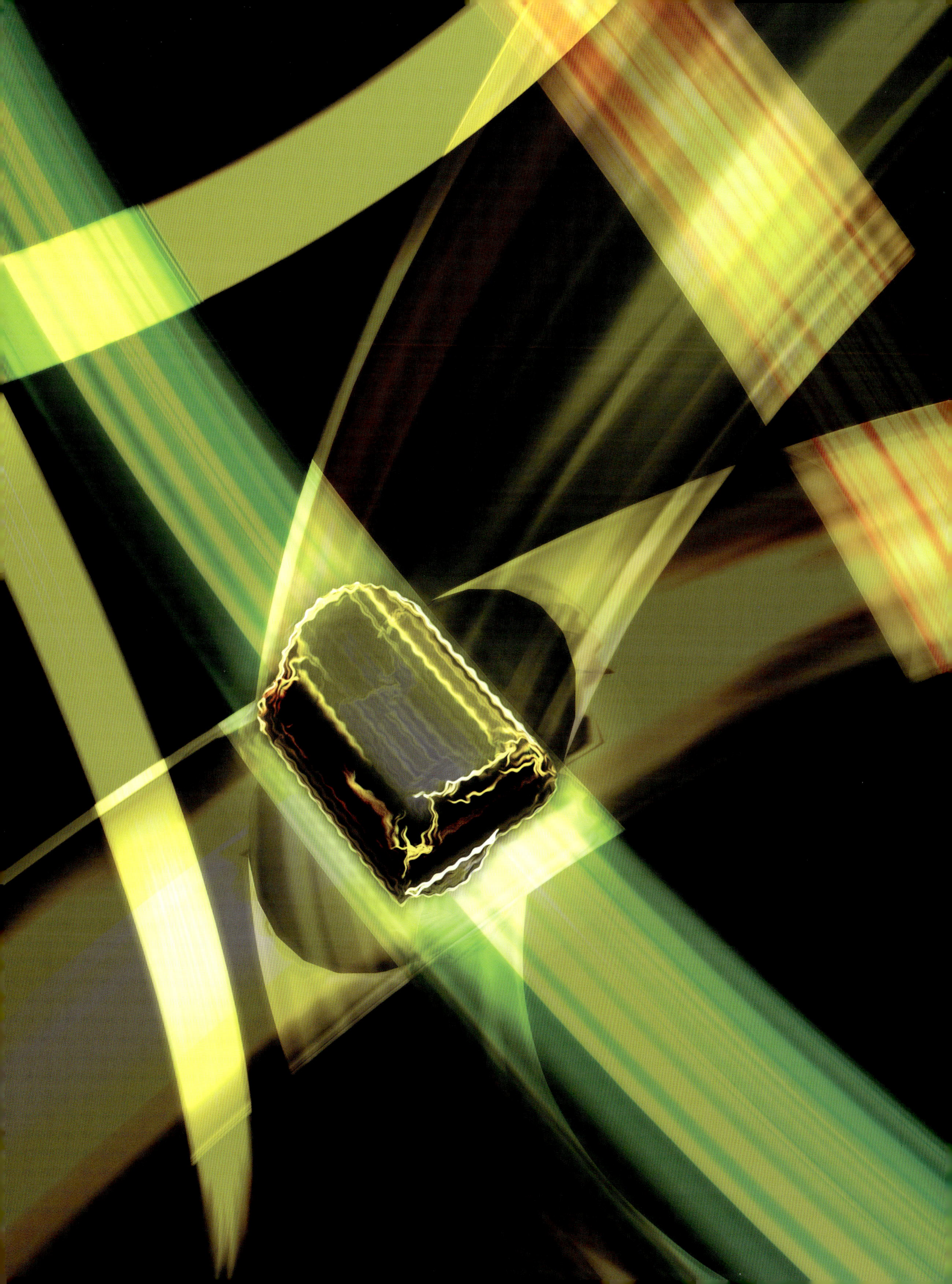

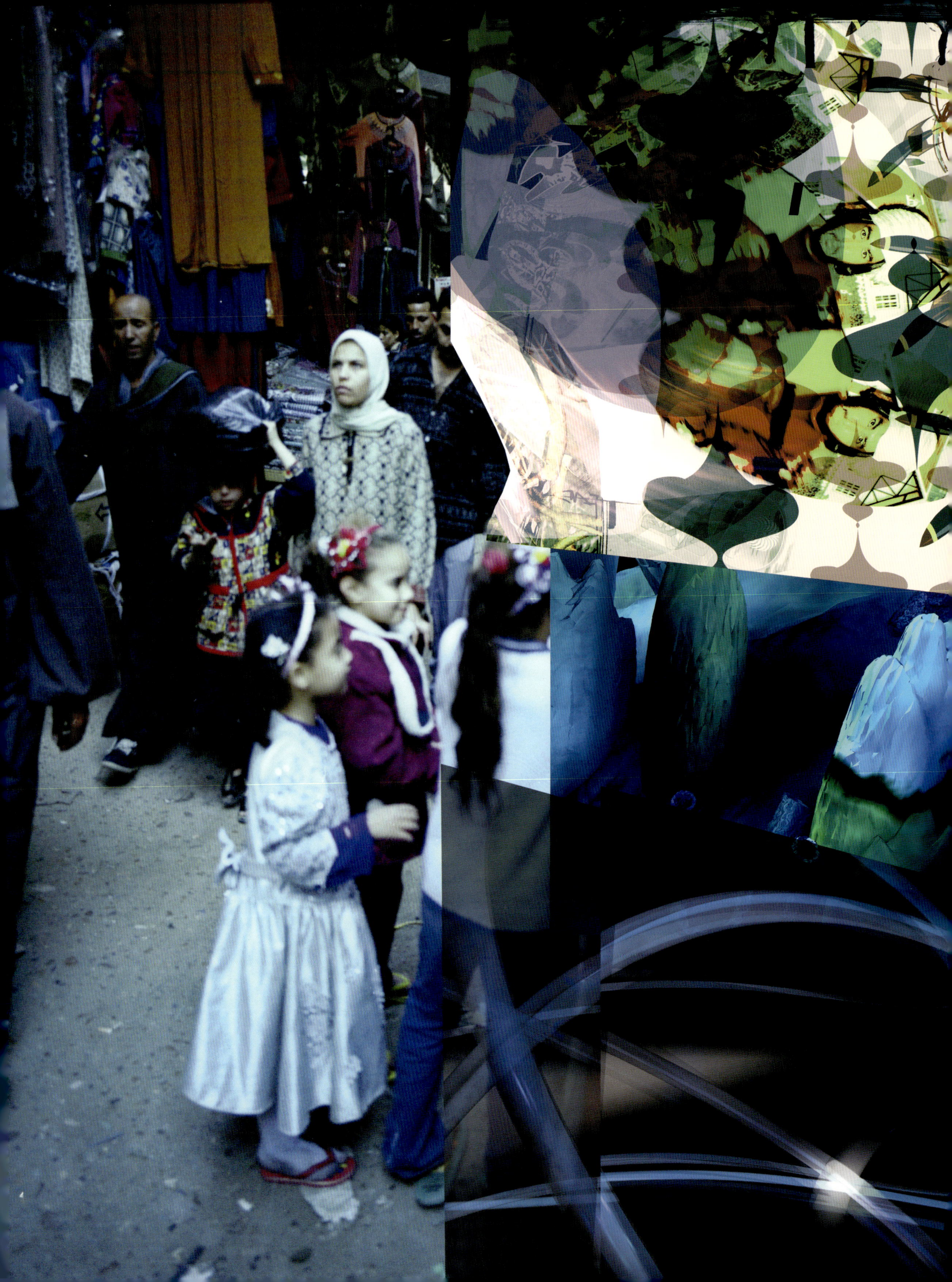

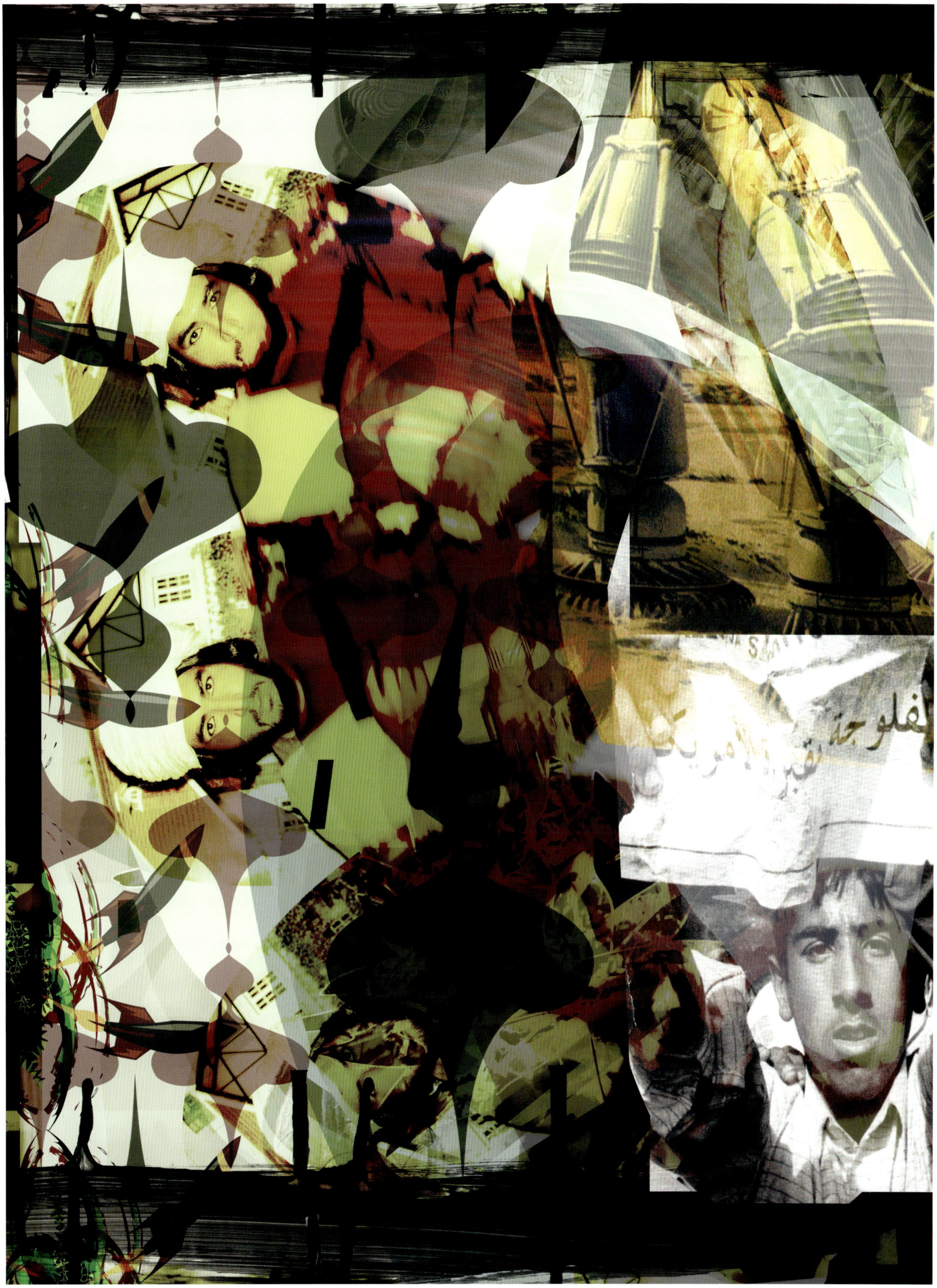

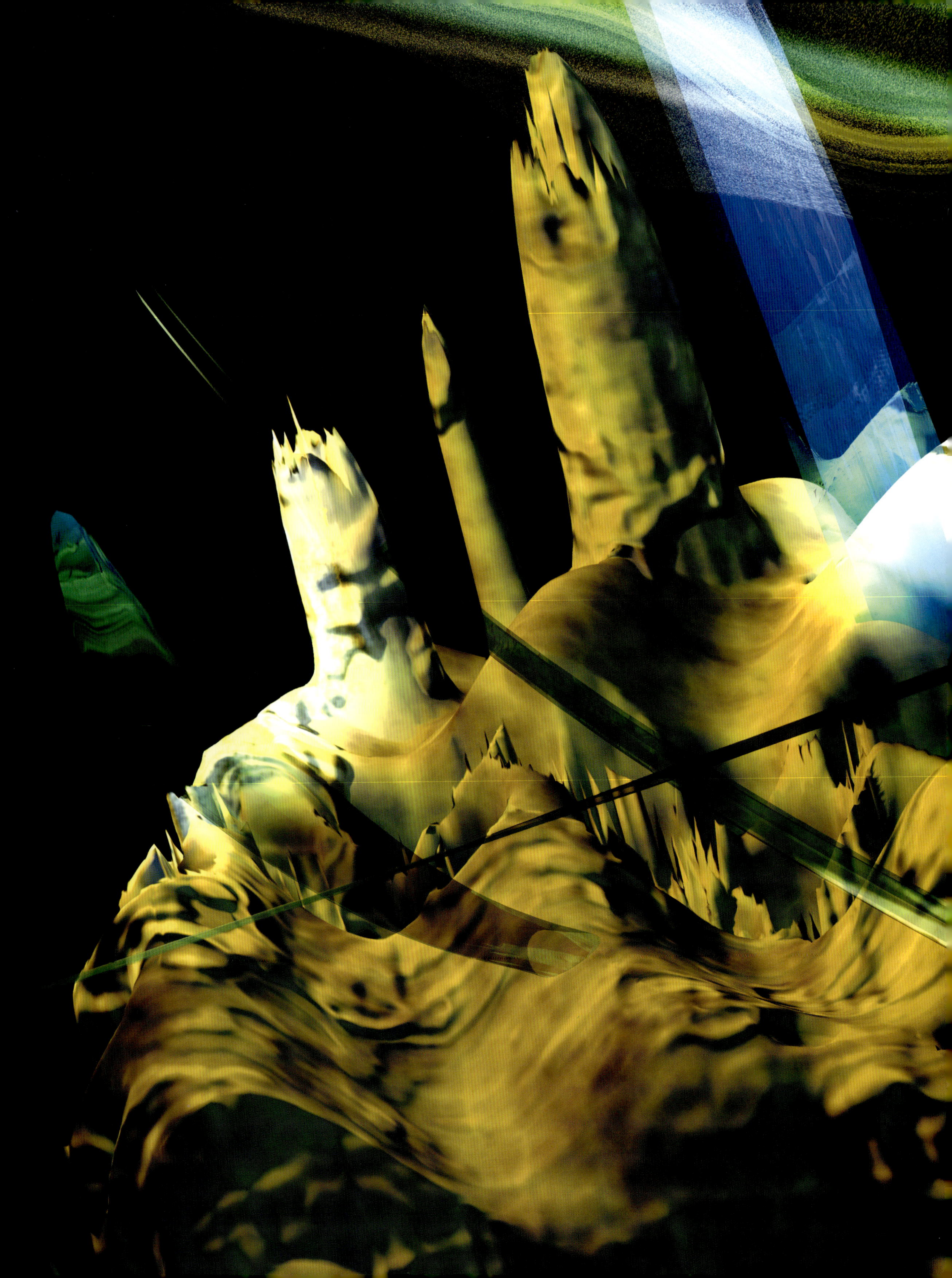

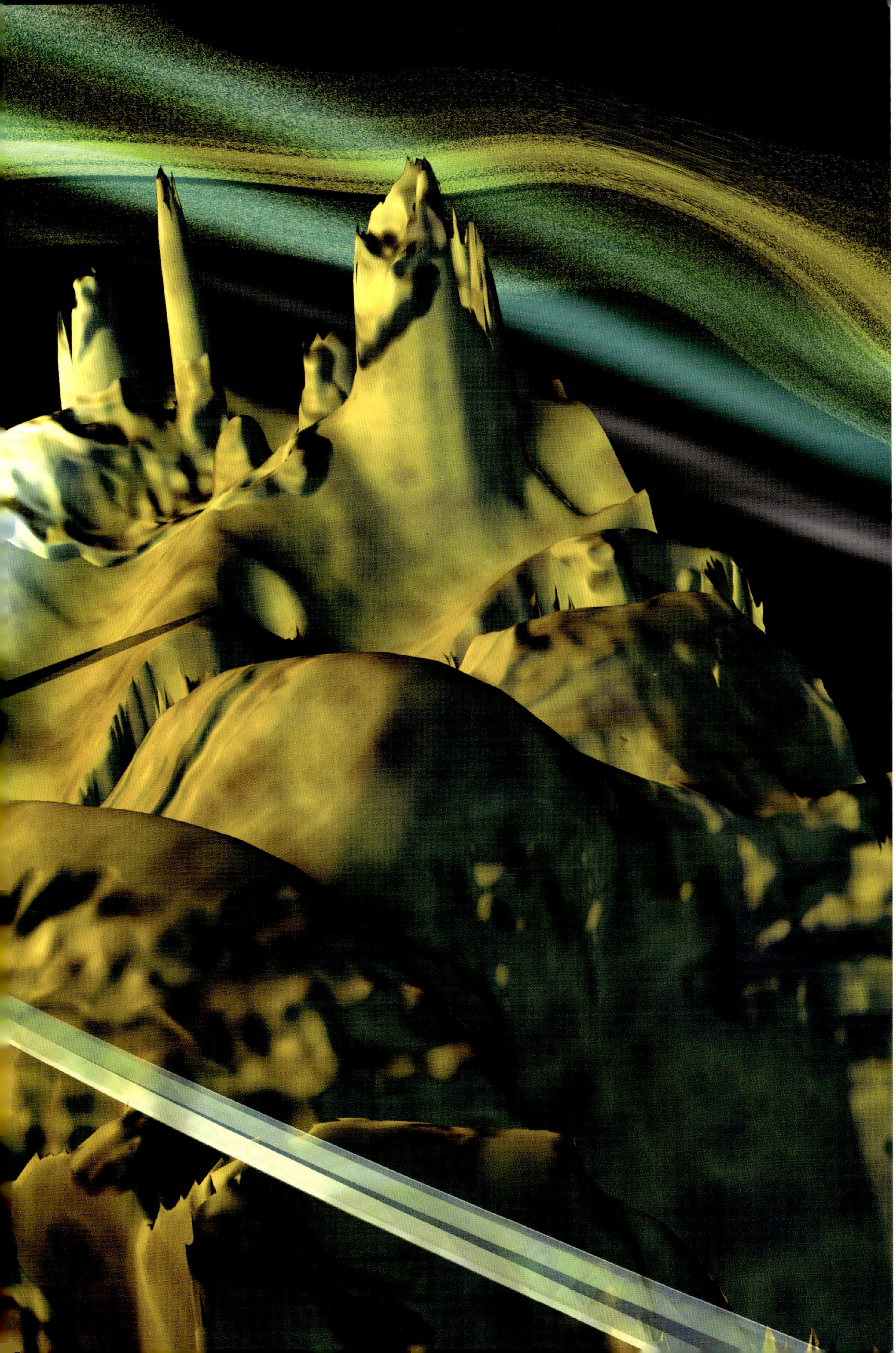

*Apache/1.3.33 Server at www.animal lawns.com Port 80*

LHC Progress
Dashboard

**AWAY Group: again, we followed the distant thunder Higgs field in. Outer quadrants did mirror, as you have said. Vectors showed no sign of irregularities—we are still moving toward 0°. Requesting procedure for entrance to the boson field. If such particle exists, the LHC will be able to make it detectable. over**

**PLEASE HAVE PICTURE BEFORE YOU SAY HELLO.**

PR: We're getting data sporadically. You guys awake?
On Monday 23 Ithaca confirmed the half-life source code.
McLeaod: tuned the entire physics-animation functions down to four lines of code. You guys need to fragment it and see what holds. We want six frames. No more blind alleys, we're on the last of it.

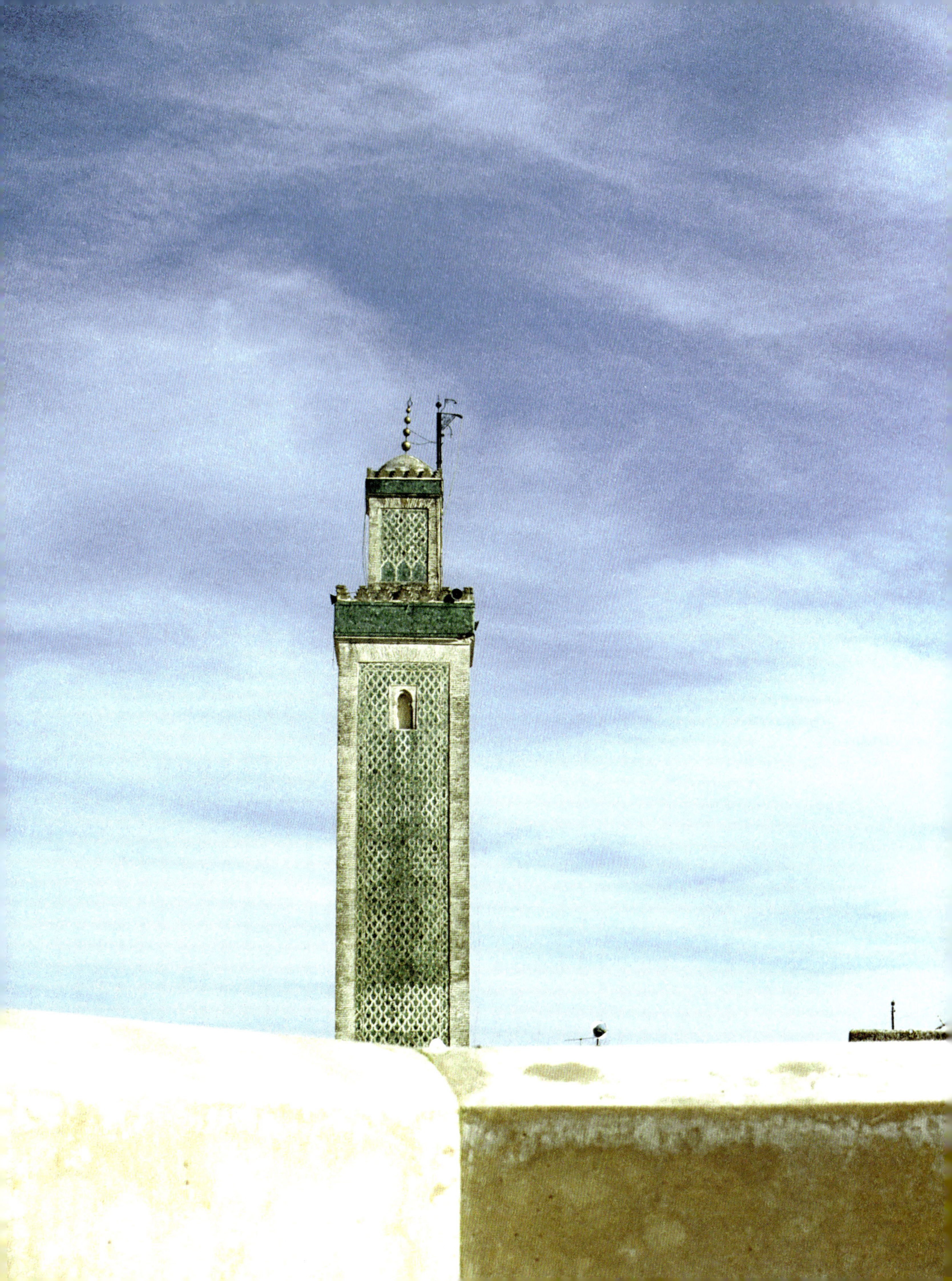

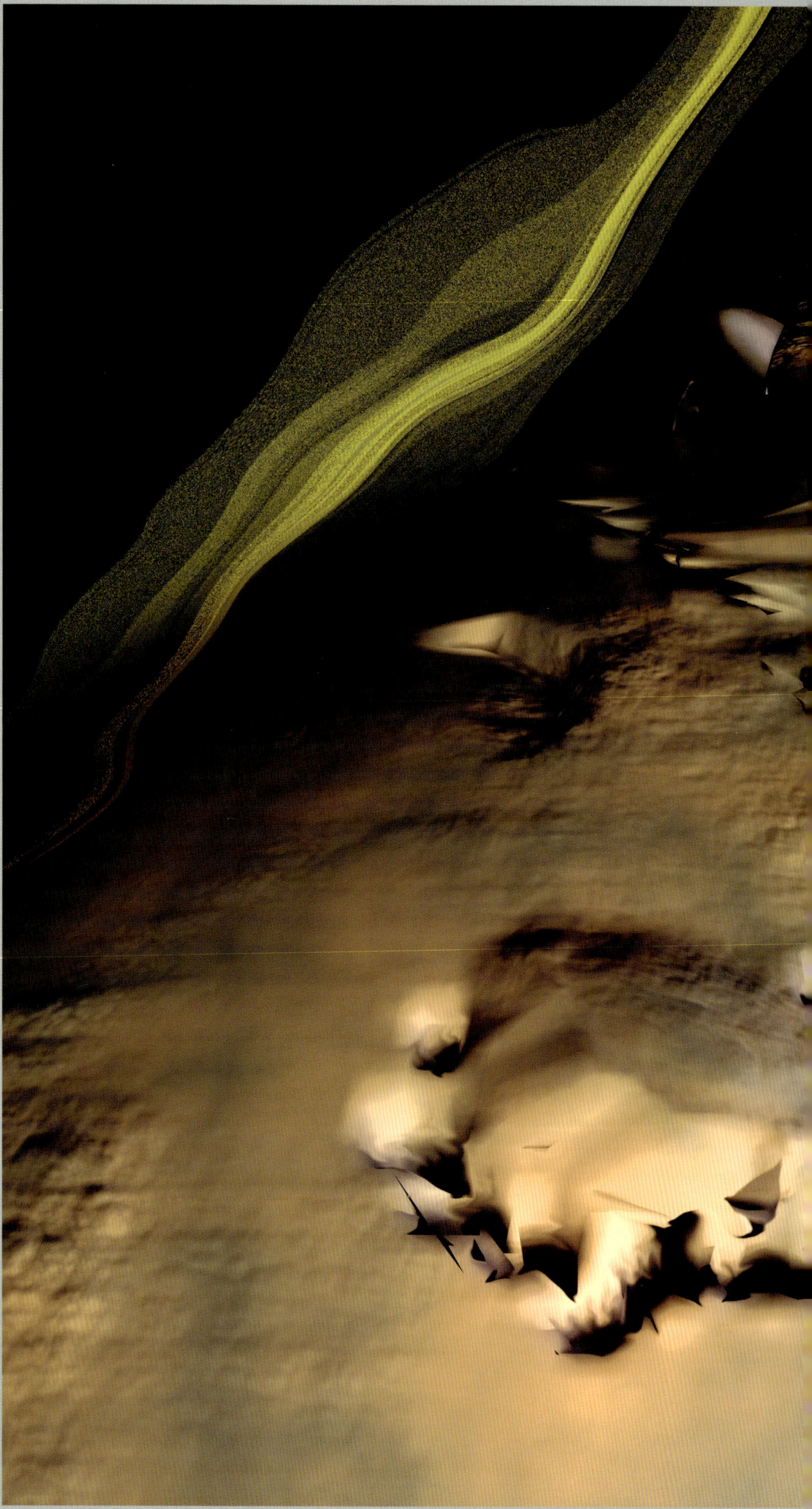

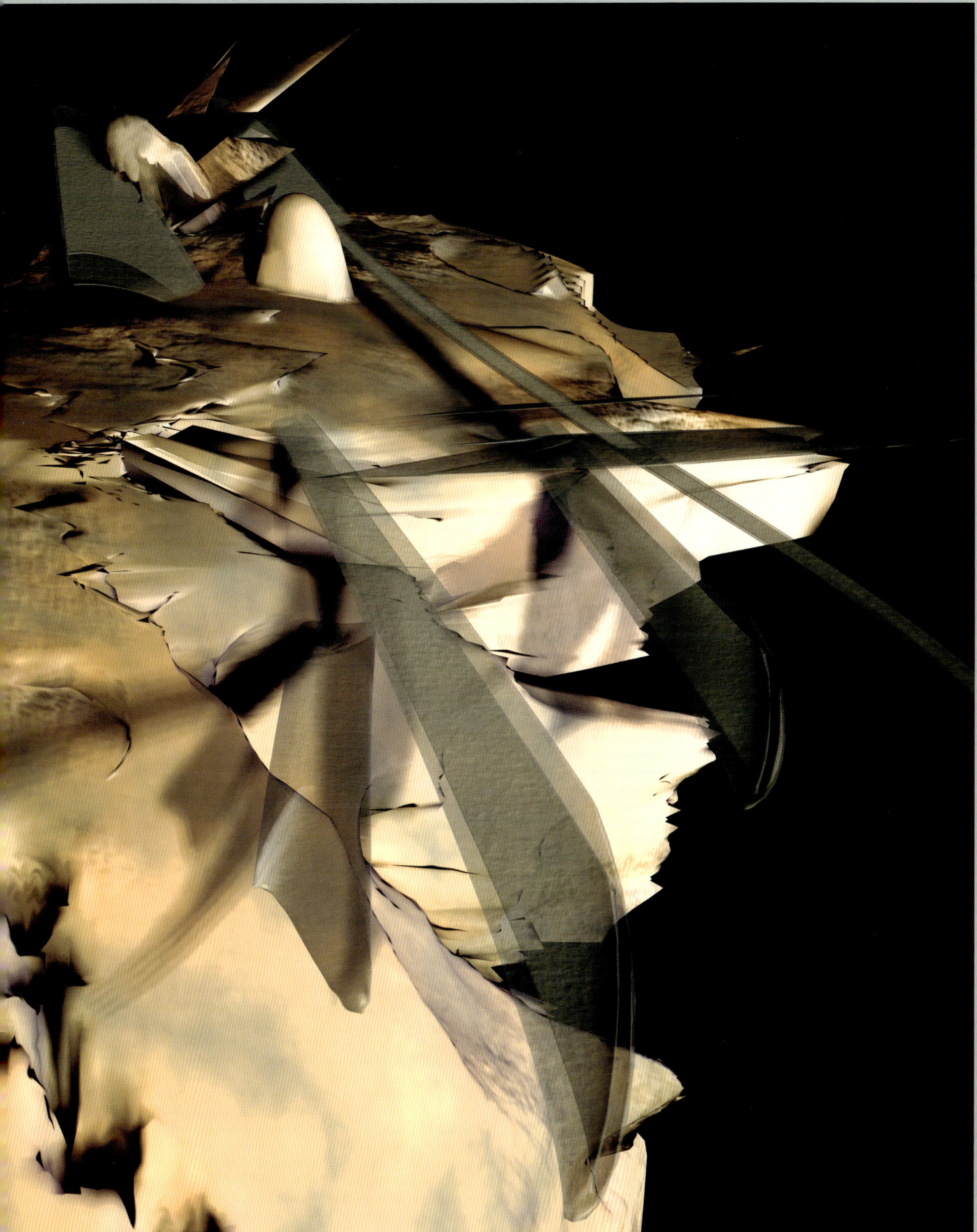

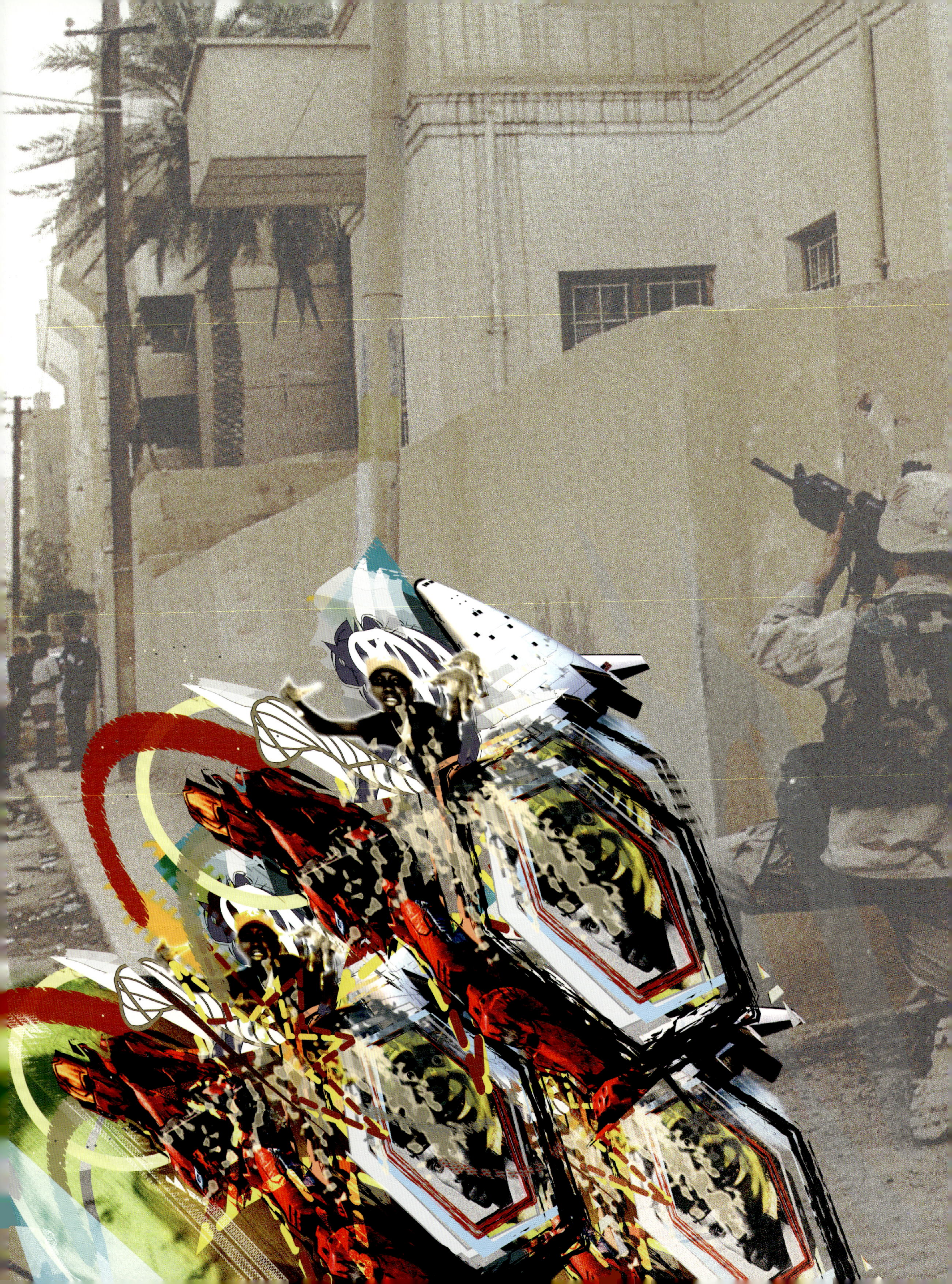

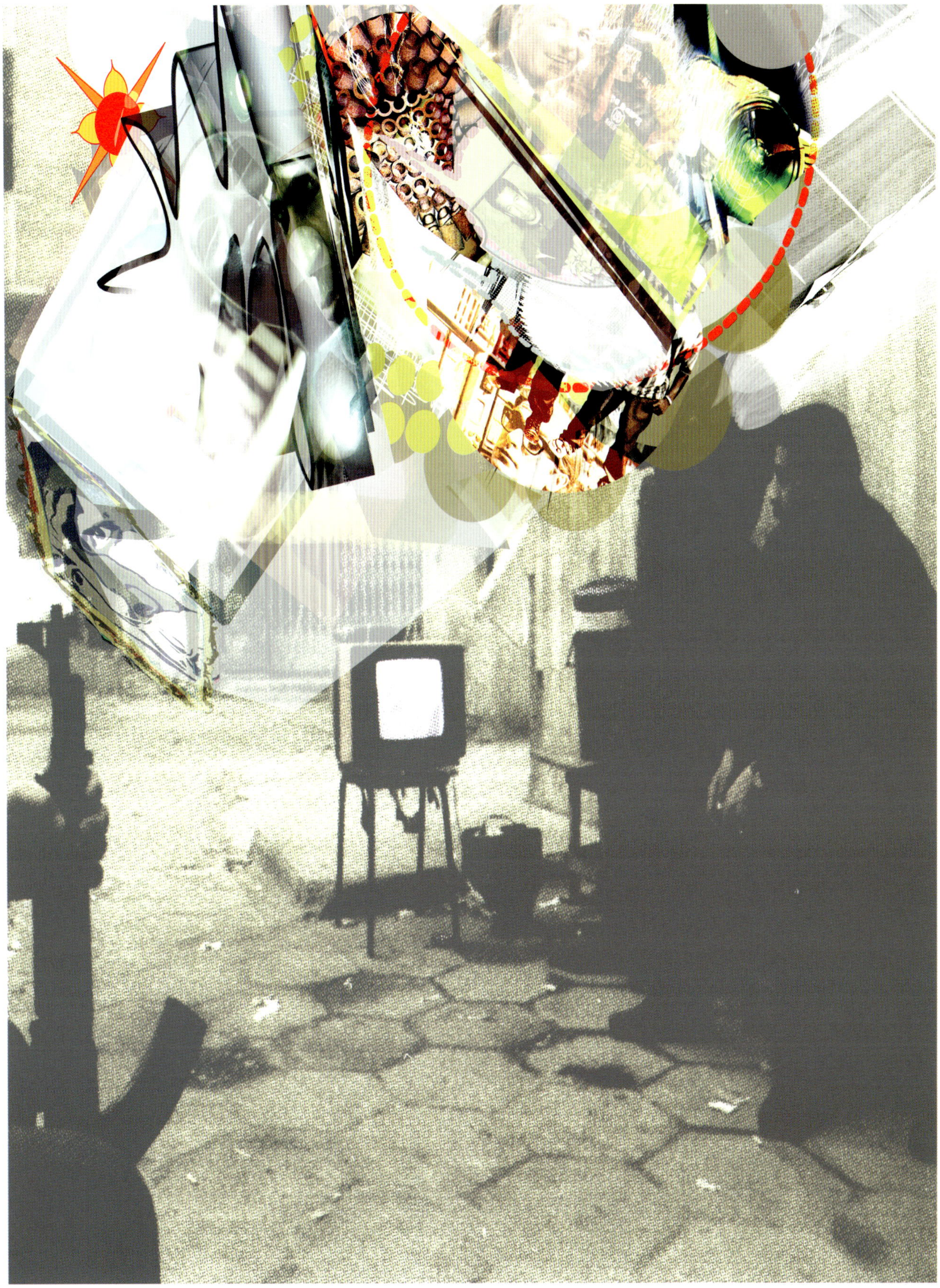

084_Gypsum 1300x, detail of gathering meme, and Giza sketch.

085_Untitled, Odd Spot. 18 x 24 inch digital canvas, 2004.

087_Meme Topology Unit Model. 2002. This is my master unit.

088_Meme Topology Unit Models/Gang in Theoretical Space #1. 20 x 30 inch ink jet print, 2003.

089_Necessary Relations #3. 20 x 30 inch ink jet print. This has the rough concept for an Islamic particle.

090_Meme Models Compared/Graffiti Revised. 20 x 30 inch ink jet print, 2003.

091_Untitled (Comic Explode). 18 x 24 inch digital canvas, 2004.

093_Merzouga Males. Morocco. These boys/men drove the Range Rovers and maintained the tent camp. It was the farthest out I had been in the western Sahara. I tried to convince them to let me go for a walk by myself during the day. My traveling partner Larry Barns freaked out, and said I would not make it back alive. In the end I didn't go.

094_Idle Background, Red/Green Version. 20 x 30 inch ink jet print, 2003.

095_Fez Kids. Morocco.

096_South to Djoser, with Meme Topology Unit Models. Also, Planck moment 5b (detail). Here I'm trying to model the concept of the instantaneousness of the Planck flash contributing to the ancient qualities of debris fill that occupy space at Planck scale (my concept).

097_Face and Time. Morocco.

098_Graffiti (detail).

099_Maturing Model Composite. 2001. This model has to do with fragmentation becoming articulate.

100_Egypt Stacked Koran. Cairo.

101_Untitled, Particle Sheet 6 (detail). 18 x 24 inch digital canvas, 2005.

103_Photo Magnetic Receiver (detail of array panel). Yerba Buena Center, San Francisco.

104_Swoon Spirit. 20 x 30 inch digital canvas, 2005.

105_Cairo Street.

106_Nut Dealers, Fez. 20 x 30 inch ink jet print, Morocco, 2003.

107_Nougat Dealers, Fez. 20 x 30 inch ink jet print, Morocco, 2003.

108-9_Cario Street.

110_Dark Receptor/Eroded Junction #3. This model incorporates the concept of a particle formed from the debris that collects on crossed remnant trajectories and sculpted by cosmic wind at Planck scale (my concept).

111_Horizons as Particle Surface, study/collage. Morocco.

112_Rocky Horizon #6, #7. Morocco.

113_Particle Landscape Horizon. Gouache and graphite on board, scanned for 20 x 30 digital print, 2003.

114_Rocky Horizon #5 with Ideographic Remnants. 20 x 30 inch ink jet print, 2004.

115_Ruins. Morocco, 2003.

116_Memphis cats. Egypt.

117_Rocky Horizon #4 With Ideographic Remnants (detail), with Particle Double Model #2.

118-9_Blind Alley. Morocco, 2004. Shard Field. 2005.

120_Untitled, Rocky Horizon. Morocco.

121_Dark Receptor/Eroded Junction within a Field, Figurines from the Egyptian Museum. Cairo.

122_Particle Double Model #3. 20 x 30 inch ink jet print, 2004.

123_Cairo Street.

125_American Fine Arts' front doors.

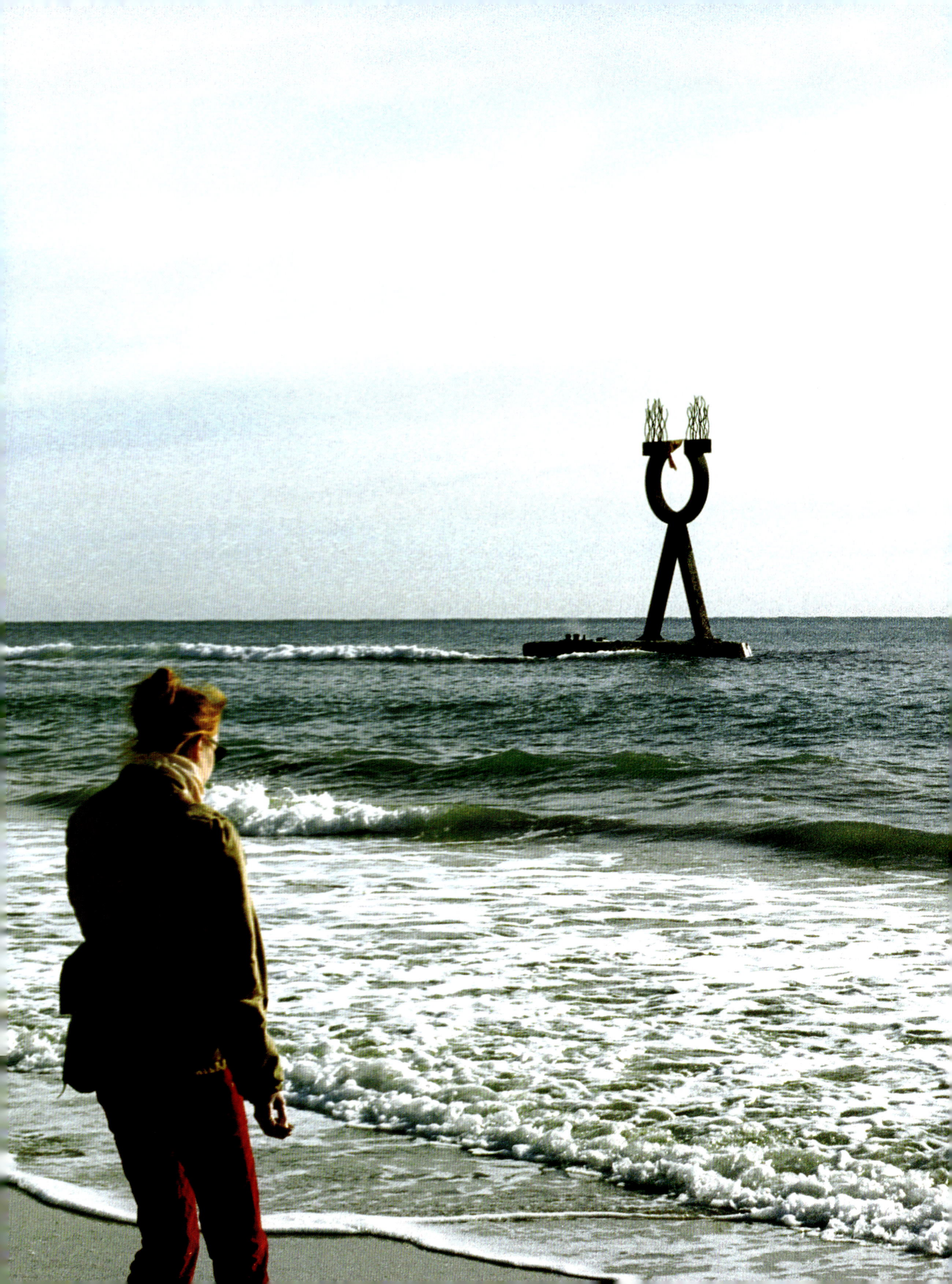

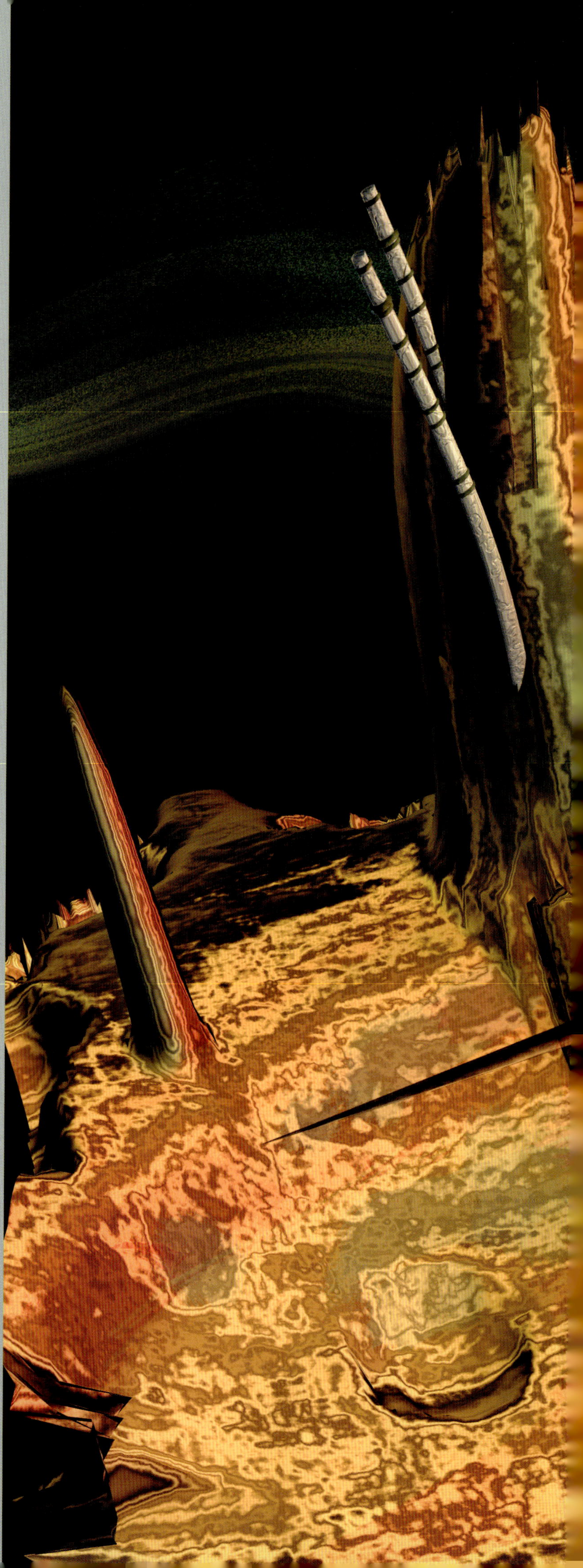

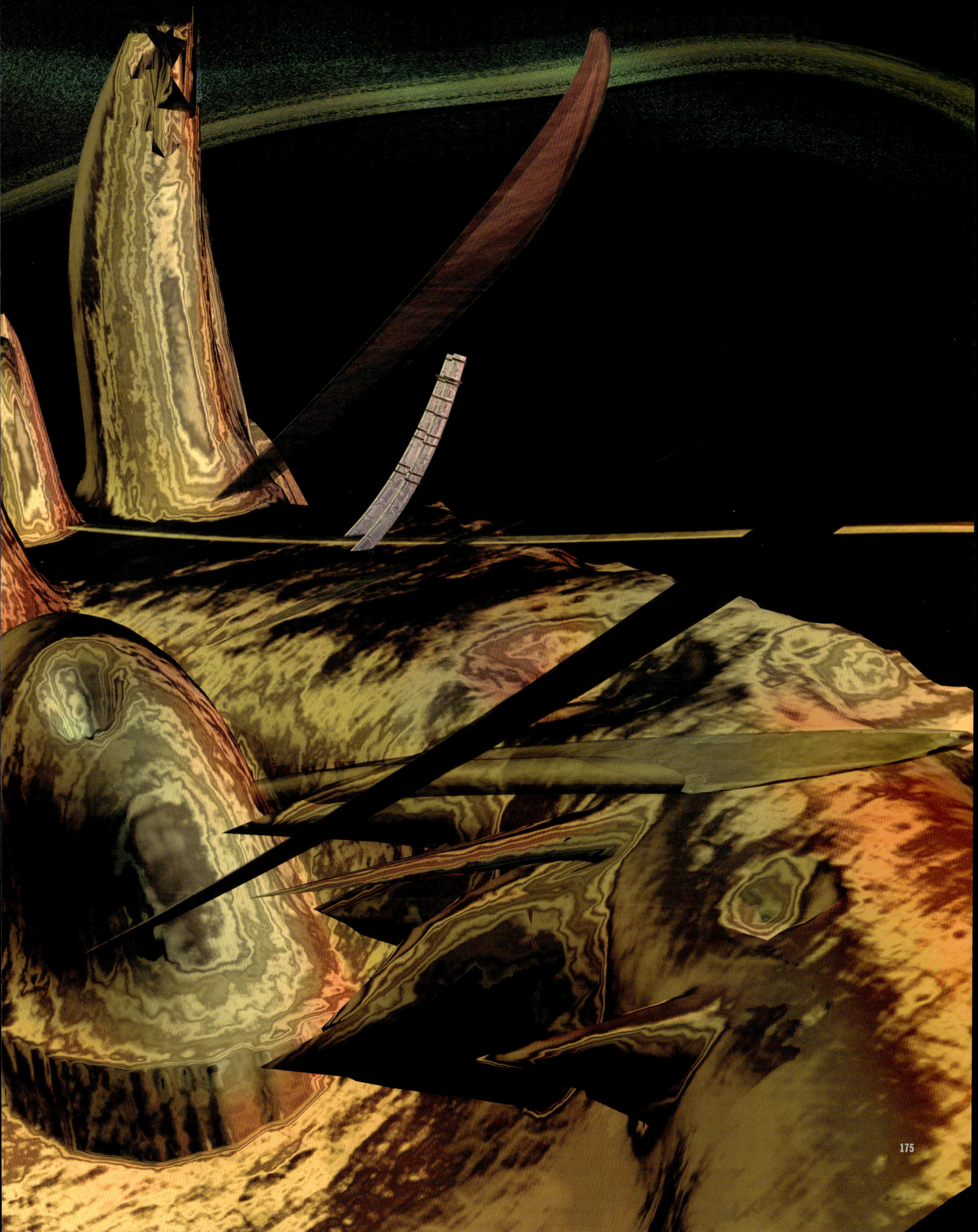

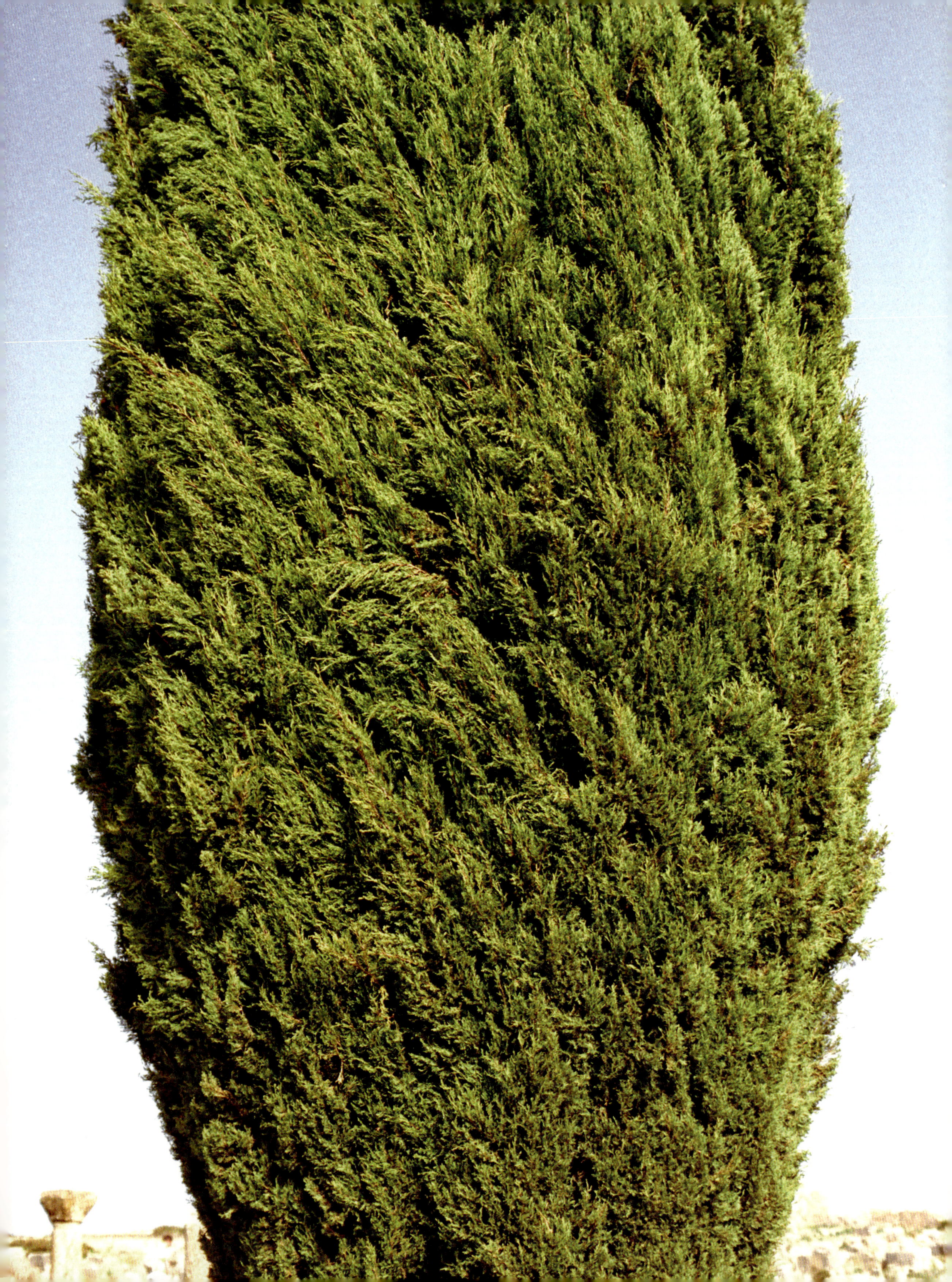

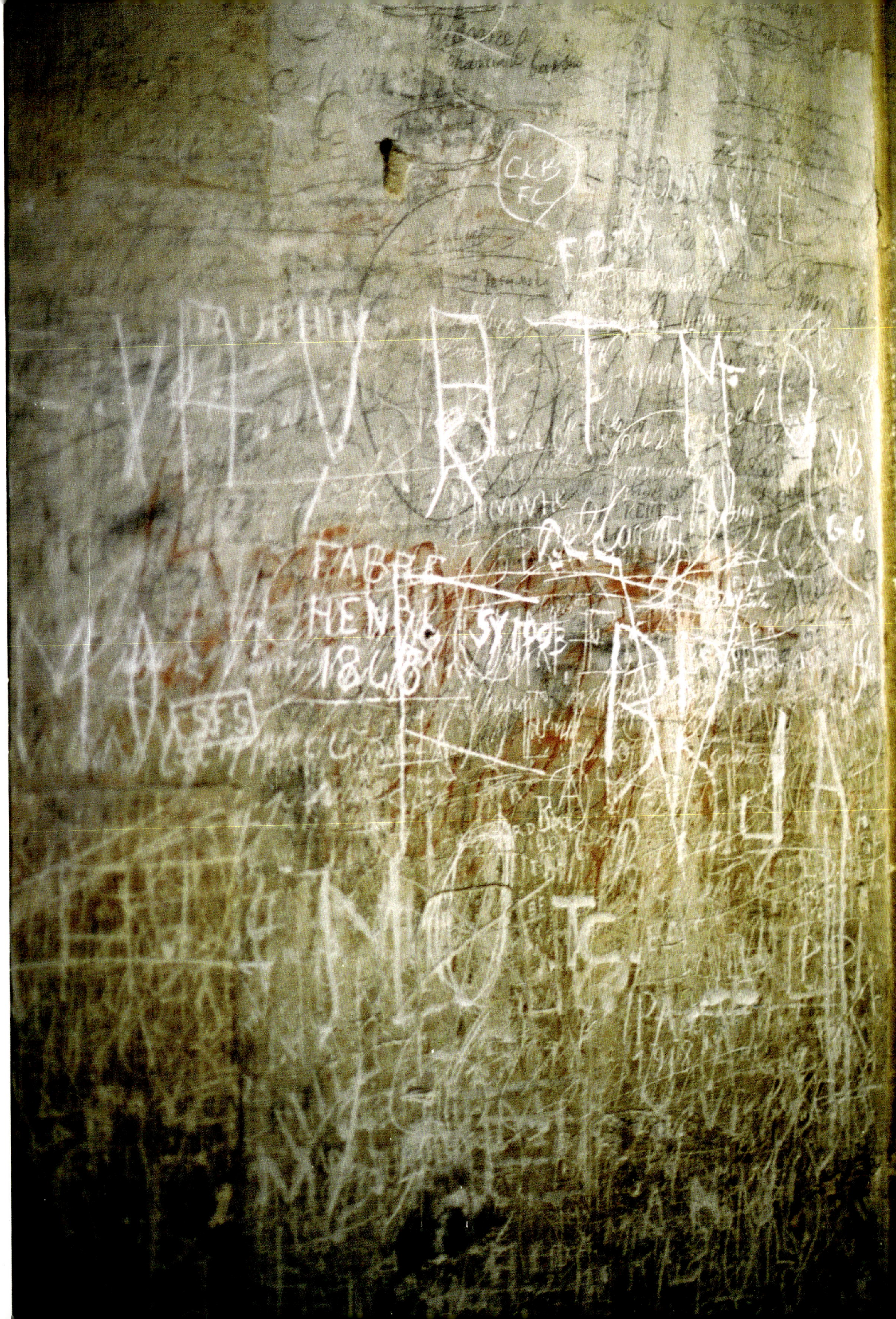

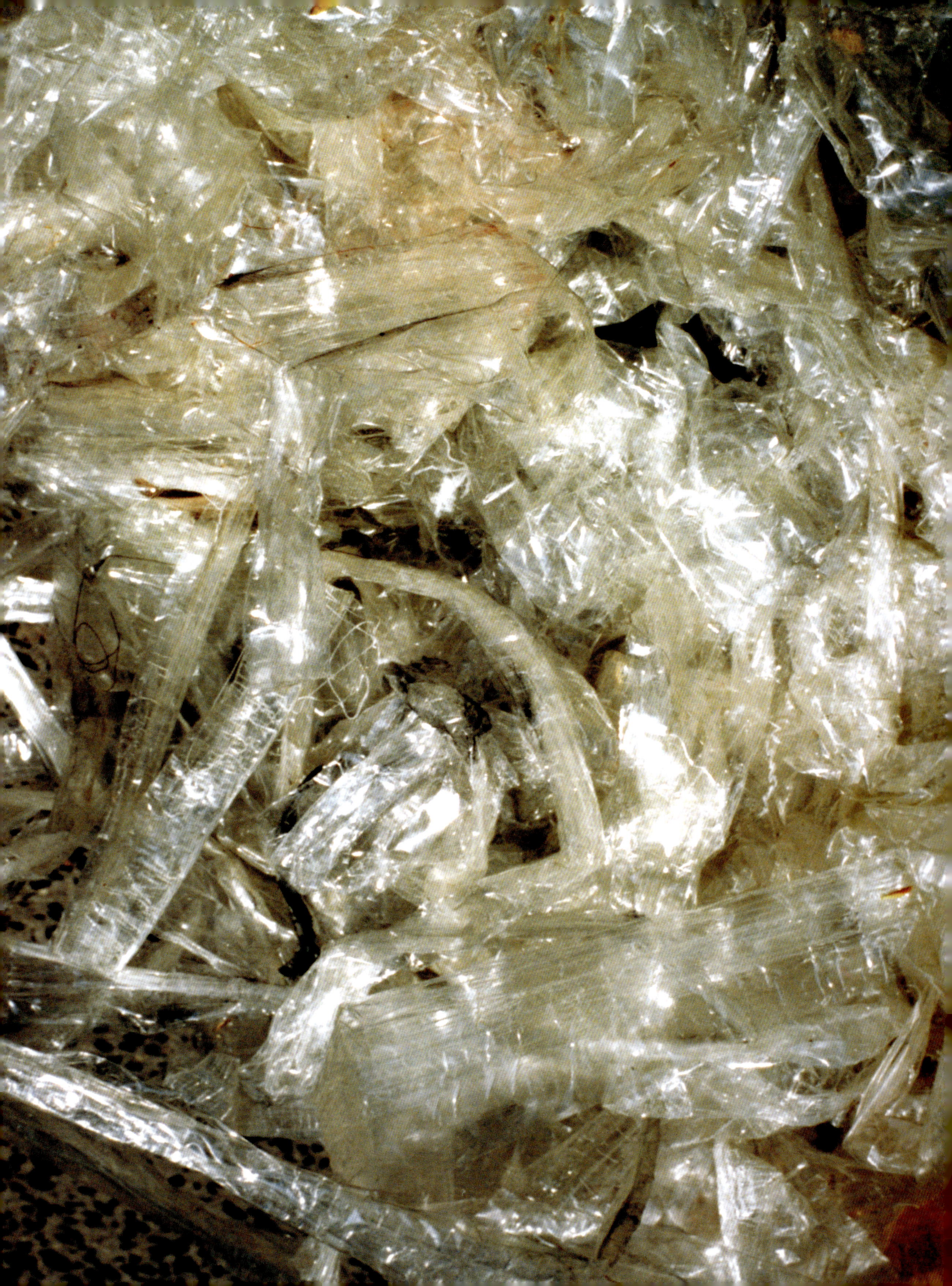

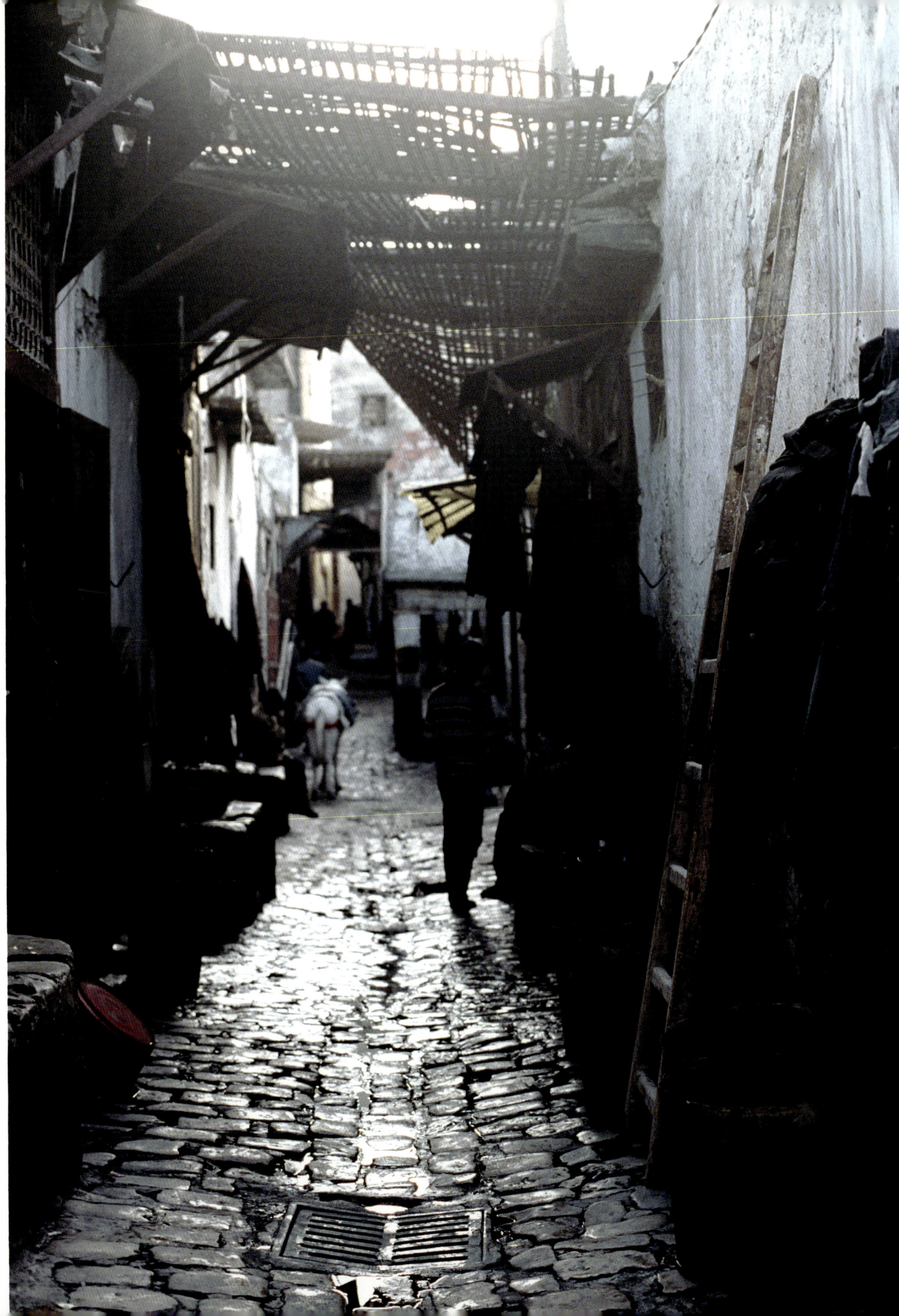

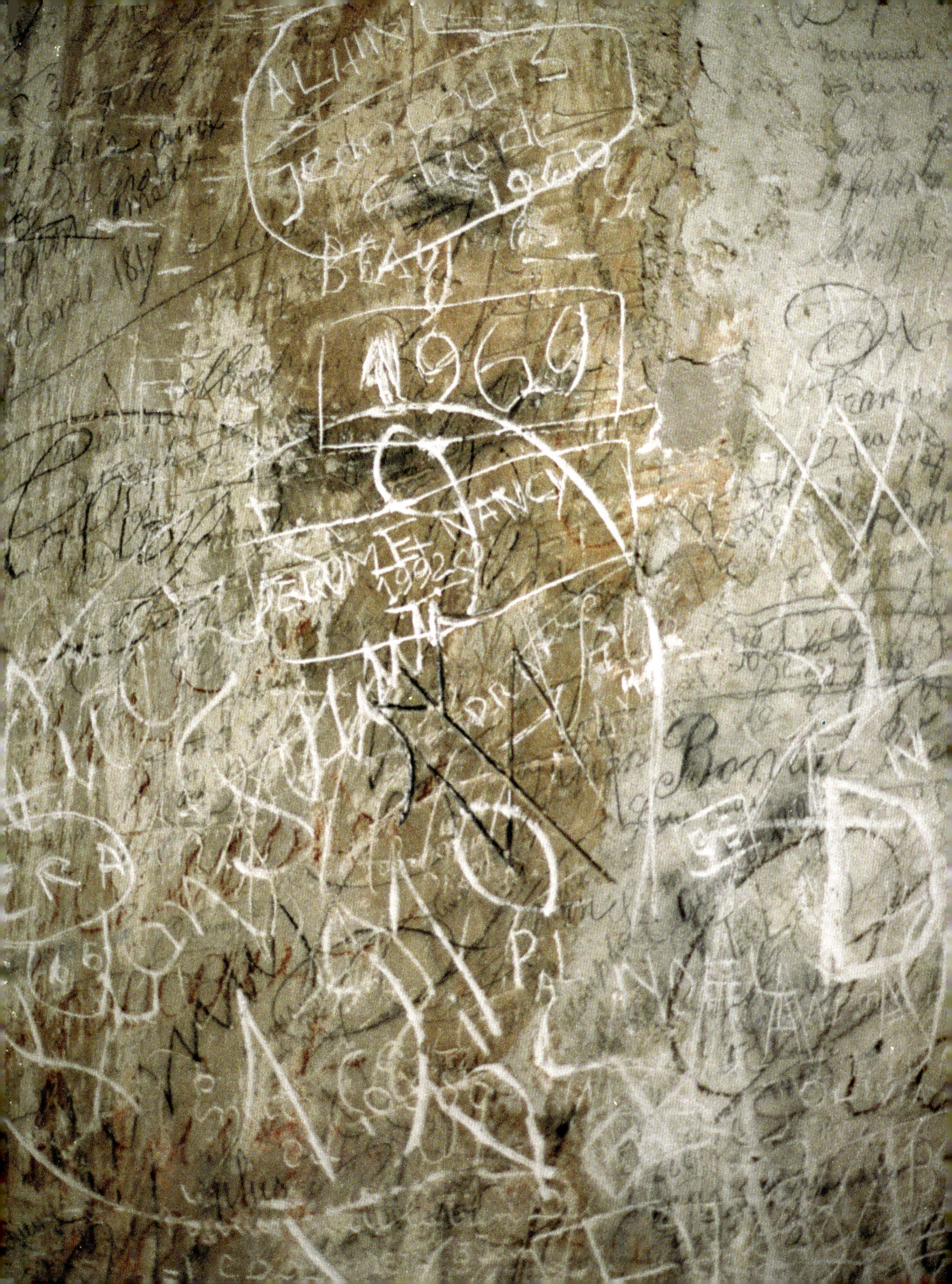

127_Giza looking SW, A Simple Unit, A Shard, and SEM Minerals and Muscle. 20 x 30 inch ink jet print, 2003.

128_Receptor Group (detail). Instantaneous entropy of post-Planck moment debris shard composites.

129_Egypt Lesser Pyramid, side wall. Giza.

130_Bent Pyramid. Egypt.

131_Juncture 4-stiff., dramatic version (detail). 2004.

132_Cairo Street.

133_Burnt Junctures Field. 2004.

134_Two Pictures at Residence. New York, 2005. Residence is a viewing space project produced by the MICA Foundation.

135_Receptor group #9 (detail), Fez Young Man. Morocco, 2003.

136_Merzouga man looks West with receptor group #3. 2003.

137_Untitled, Deniro-Pacino. 18 x 24 inch digital canvas, 2005.

138_Cairo wire strippers. 2001.

139_Fez young boys. 2002.

140_Receptor Master Model, a Juncture. 2004.

141_Cairo Street.

142_Dark Receptor/Eroded Juncture #2 (detail). 2004.

143_Egypt Giza Entrance with details. 20 x 30 inch ink jet print, 2003.

144_ Dark Receptor, Eroded Juncture #5. 2004.

145_Cairo Street.

146_Dark Alley, Fez, with Ideographic swipe. 2005.

147_String background #3 with time-lapse Cairo.

148_Untitled Particle with Ideographic Stroke. 18 x 24 inch digital canvas, 2004.

149_Cairo Street.

150_Cairo Street with Tea. 2003.

151_String Background #4m. 2005.

152_Minaret Night. Cairo.

153_Dark Receptor/Eroded Juncture #8. 2004. The Receptor series is an attempt to model essential collected mass.

154_Giza Tour Buses.

155_Sci-fi ranked by plausibility. Internet swipe.

156-7_Theoretical particle #6, book version. 2005. This addresses complexity at 0°.

158_Untitled Particle with Ideographic Stroke #2. 18 x 24 inch digital canvas, 2005.

159_Untitled Particle with Ideographic Stroke #7. 18 x 24 inch digital canvas, 2005.

161_Planck Juncture with Residual. 2004. This is from a group of string models showing the formation of an isolated unit of string.

162_Cairo Street with Ideographic and Particle Piece.

163_Untitled, layer piece. 18 x 24 inch digital canvas, 2005.

164-5_Planck Moment #5. 2005.

167_Fez Minarette. 2002.

168-9_Large Particle #2. 2005.

170_Untitled, corner sprite. 18 x 24 inch digital canvas, 2005.

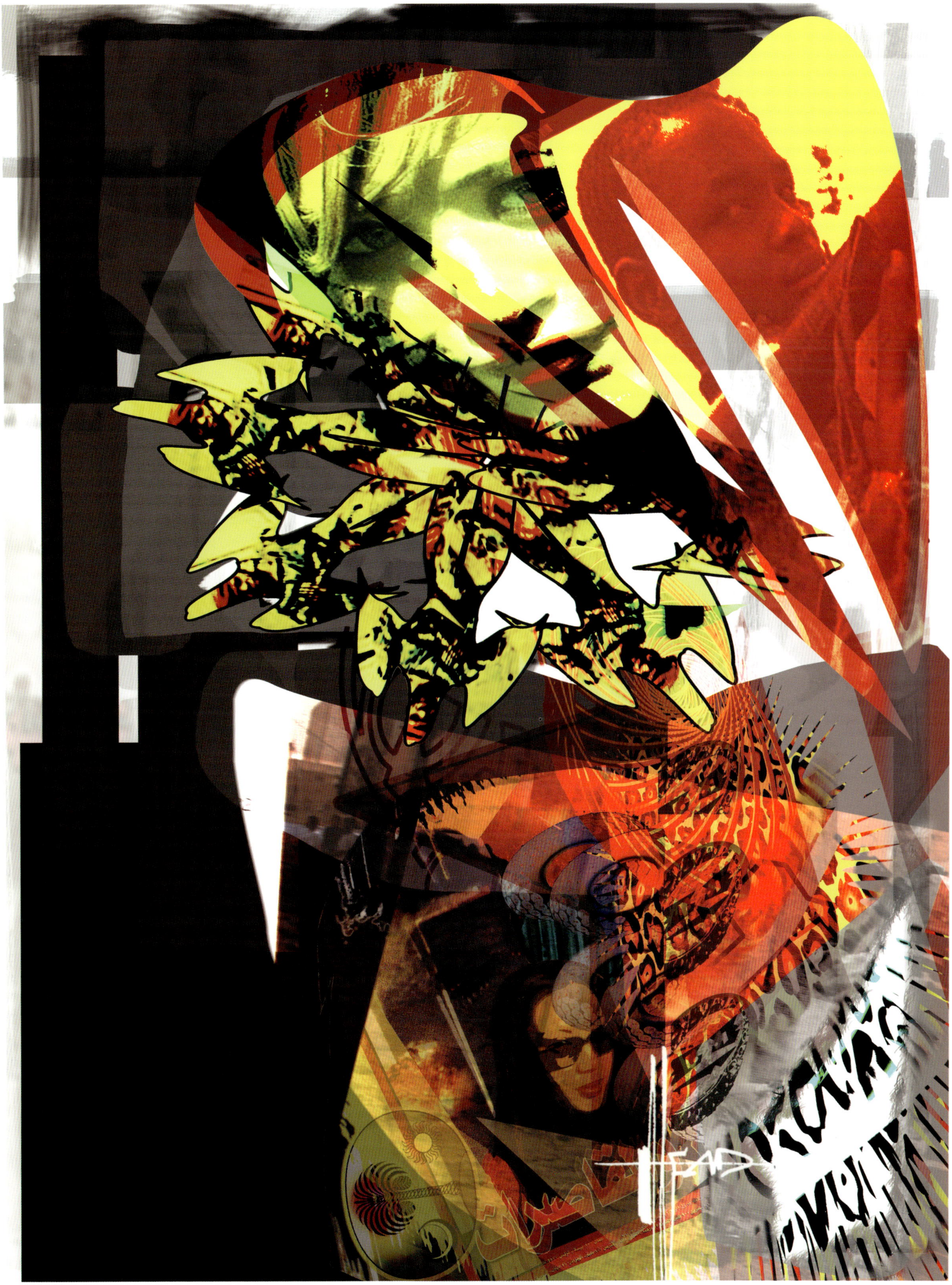

One evening in the spring of 2004, during the Sunni/Shiite insurgency that convulsed Iraq, an Arab satellite news network prepared to do a remote report from a hotel balcony overlooking central Baghdad. As is usual in such situations, the preparations themselves went out live over a satellite feed, observable to controllers back in the studio, and to anyone who happened to have a satellite dish tuning in. The crew busied themselves with checking light and audio levels, and since the actual reporter had yet to arrive on the scene, they did what TV technicians throughout the world normally do—they used a stand-in, in this case a fortysomething woman shrouded in a jetblack chador.

Just what her relationship was to the crew is unknown. She could have been a producer, perhaps, or the wife of cameraman, or simply a stranger, recruited on the spot. At first, the resulting footage is unremarkable, if not slightly comical, showing the woman as she stares blankly into the camera without uttering a sound, the deep frame of her garb tightly cropping her round middle-aged face so that floats like a small impassive moon against the glittering firmament of the Iraqi capital. To Western eyes unaccustomed to the juxtaposition of modern technology and medieval faith, her image probably looks surreal or at least exotic, a kind of 21st-century update of orientalism, with its romance of the Levant, and its condescending notions of Arabs as being at once noble and yet on some basic level, irredeemably savage.

Savagery, in fact, was in the offing that night, since unbeknownst to both the crew and the woman, their real subject was a carbomb about to detonate in the near distance. How the explosion unfolds in the clip is fascinating, and not just because as a species, we are easily mesmerized by carnage. The scene is an object lesson in physics, with the explosion appearing initially as a bright point of light flaring over the woman's shoulder, a singularity that suddenly inflates to halo her head for a nanosecond before filling the entire background. Since light travels faster than sound, all of this transpires before we—and of course the woman herself—hear the blast, or sense the effects of the aftershock. She suddenly bolts out of the frame and the camera, obviously now unattended, violently shakes for a few brief moments before returning to an eerie state of calm.

CNN picked up the footage, using it to lead in to its breaking reports on the bombing. It wouldn't be putting too fine of a point on the matter to say that for CNN execs, the clip was a godsend, easily worth a thousand times more in the currency of media coverage than if they had to rely on a laborious eyewitness description of the event, either on-camera or over the phone. (The last being the sort of thing typically accompanied by a static photo of the reporter and/or a map of the country he or she is reporting from.) The visual dynamism of the sequence, the serendipity of the camera being exactly in the right place at the right time, trumped both the content and context of the actual bombing, so that its record become an instant found object excluding any meaning beyond its own existence. CNN put it up on its website whereupon it was almost instantly linked to other sites like The Drudge Report, thus illustrating the physics of yet another universe—the Internet.

And yet almost as quickly as the effects of the bombing itself, the image seemed to dissipate—searchable, no doubt, within the infinite database of the World Wide Web, but otherwise withdrawn from public consciousness. Which is odd. One could argue that the sequence had the impact of similar records of death and destruction from the 20th century: newsreel of The Hindenburg explosion; Robert Capa's famous photograph capturing death of a soldier during the Spanish Civil War; the murder of Lee Harvey Oswald caught live on TV; the execution of a Vietcong suspect during the Tet offensive; the Challenger spaceship

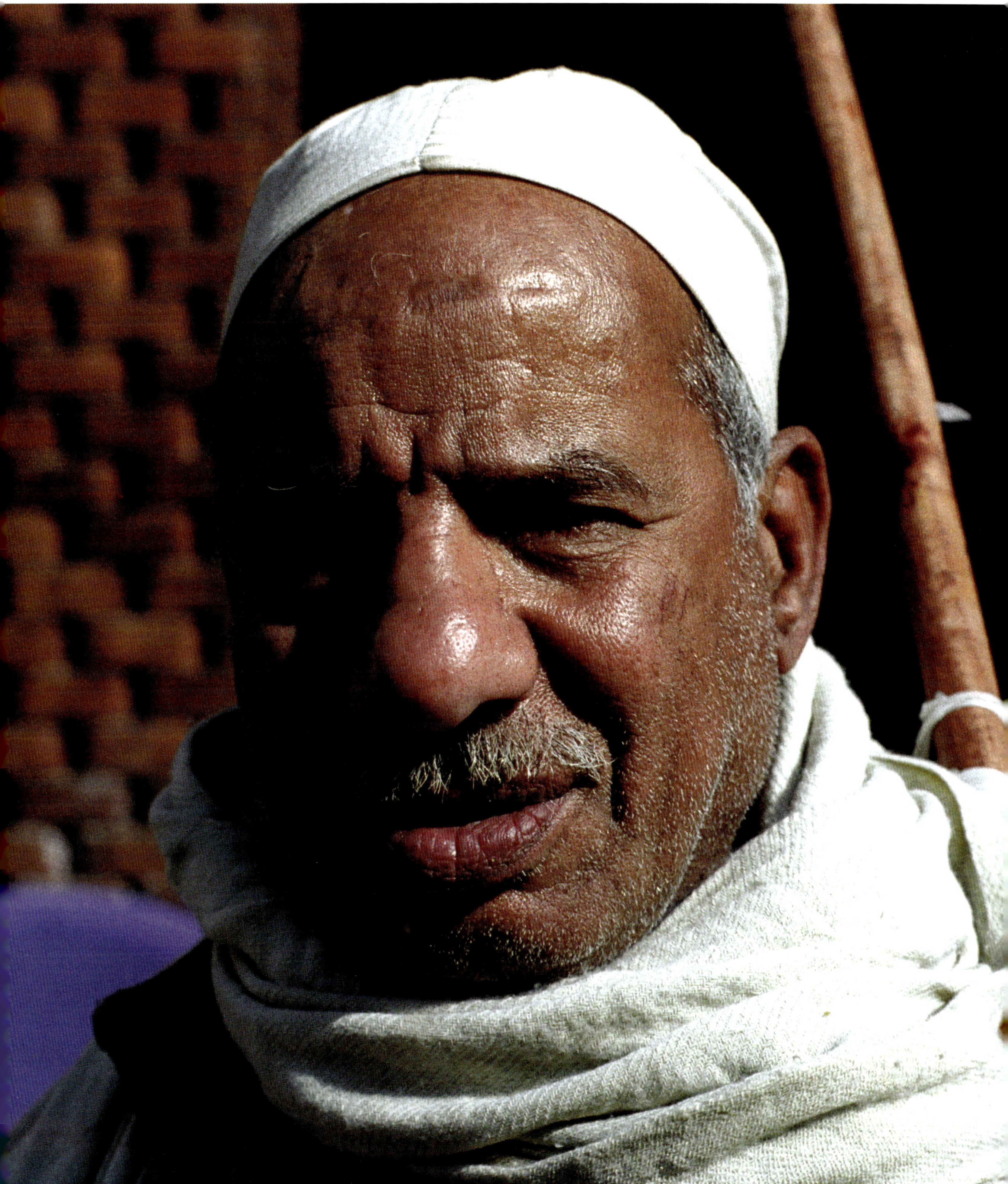

AMERICAN FINE ARTS, CO.
COLIN DE LAND FINE ART
22 WOOSTER STREET NYC 10013 (212)941-0401

TimeOut, January 28 - February 4, 1999

**American Fine Arts**
*22 Wooster St between Grand and Canal Sts (941-0401). Tue–Sat noon–6pm.* Dennis Balk, "Photo Magnetic Receiver." This could be dangerous gallery-going. Part of the show is a "body-accommodating device that magnetically aligns the viewer with a rare light phenomenon"—kind of like physically experiencing a James Turrell. The gallery warns that the electromagnetic current produced by the device could be harmful. Beam me up. Through Feb 13.

FAX: (212) 274-8706 

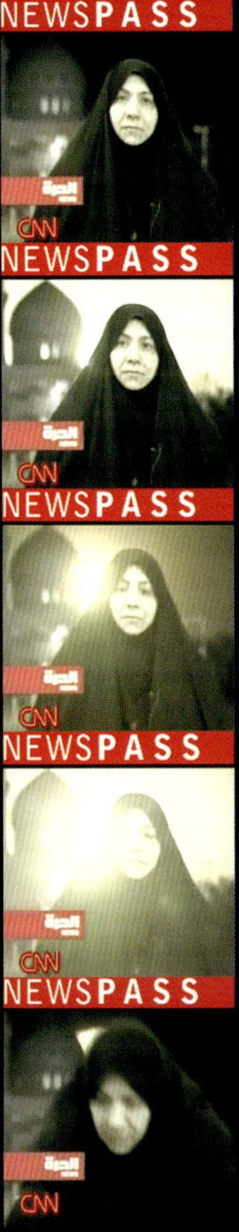


blowing up on lift-off, etc. One could further argue that the picture of the cowled woman with the light from behind is blessed with a certain art-historical resonance, suggesting as it does a Byzantine Madonna, replete with a gold-leafed penumbra. And yet the image seems to have had no cultural traction, in spite of being instantaneously accessible to potentially billions of pairs of eyes across several media platforms. One may reasonably wonder, Why? Is it that the image itself fails on some ineffable level of iconicity? Or is the reason the result of some form of prejudice—a pernicious bias that the suffering and vicissitudes of life in the Third World are somehow not worthy of First World consideration? Or is it perhaps due to some structural problem pertaining to the 24/7 nature of the contemporary media themselves—the fact that the linkage of cable TV and Internet creates a kind of global mobacracy of imagery so vast and unmanageable that no single representation can possibly break through and command our attention? Whatever the cause, this convergence of person and wartime outrage—this Madonna of instant history—did not succeed as meme, though that failure raises the interesting question of what, exactly, is a meme, and what is the meme's utility in explaining the operations of culture?

The common understanding of culture—that it is a uniquely human endeavor—is quite primitive, a view analogous, arguably, to the old cosmological notion that the earth was at the center of the universe. Indeed, as some naturalists suggest, culture, if it is defined as a set of learned, adaptive behaviors, may not be unique to humans at all. They point as an example to great apes like orangutans, where each generation of orang passes on certain forms of knowledge to the next—like a knack for probing for insects in the trunks of trees with the aid of a handy pointed stick. This is, at best, a Copernican level of understanding culture—but still an understanding which reveals, much as the Copernican revolution did, that the more humans know about their role in nature the less significant their role seems to be. Naturally, some people would refuse to abide by this paradox. The anthro-centric purists among us might argue, for instance, that the orang adaptation of insect-hunting with sticks is not acculturation at all, but simply instinctual. This argument, however, overlooks the problem of calling tool-use among great apes instinct, while accepting the adoption of stone hand-axes among our hominid ancestors as culture. Of course, the purest of the purists might then counter with the point that true culture began only with the arrival of Homo Sapiens some 100,000 years ago, since up until that point, technology had changed very little over the previous million or so years of hominid development. Contrast that, they would say, with the ensuing period, a relative blink in geological time in which humankind went from hunting and gathering to exploring the universe. By this reckoning, dynamism defines culture. But at what pace of change? And at what scale of time? And when does cultural dynamism cease to have any momentum, so that in effect cultural dynamism calcifies into cultural stasis?

These questions may be at the heart of our current conflict, a few brief seconds of which is illustrated by the image of the woman in Baghdad. Most commentators—and in the current all-pervasive mediascape they are legion—agree that the struggle is one between democracy and theology, materialism and spiritualism; between the values, in other words, of the Enlightenment and Modernism, and those of Islamic fundamentalism. But this clash of values, one might say, also signals something else again: a differential in cultural velocities, in the speeds with which cultural changes are effected and accepted. A free-market society based on hedonism, on the individual's pursuit of happiness or comfort or distraction, is all about the creative destruction of change as measured by how quickly ideas are circulated or marketed. A theocracy operates in the opposite fashion, propagating a set of cultural immutables as the only basis for organizing society. One might call this the difference between an internalized adherence to faith and an externalized pursuit of sensation. One might even call it the difference between moving and standing still. But the situation is more complicated than that because cultural transformations occur even at seemingly imperceptible rates of change.

Why? Think back for a moment to the notion that orangs have culture. If this idea is true—that is, if culture occurs in nature more frequently than we as humans believe—then culture is reducible to the following simple formula: Someone does something, and somebody else copies it. Studies in the wild show that orangs can imitate each other; less clear is whether they can improve upon one another's behaviors. On the other hand, we do know that ideas are not just copied in human culture, they are improved upon. Even if this means, in effect, that they are just being copied imperfectly. It is true that the frequency with which changes occur is an important determinant of cultural development, and so too is the speed with which changes are recognized and embraced. Not all new ideas are necessarily good ones, or workable; not all new ideas that turn out to be good are immediately received as such. New ideas, no matter how good, are often viewed as dangerous to an established order (certainly Copernicus's model of the universe fits into this category). And yet, no matter the pace, somebody is always copying what somebody else is doing. Change occurs no matter what because culture develops mimetically, as a kind of dialectic between imitation and innovation.

Meme theory, as first conceived by the Oxford zoologist Richard Dawkins in his 1976 book *The Selfish Gene*, codifies these ideas by proposing that culture evolves through the transmission of self-propagating units, or memes. By this formulation, which one might call a Newtonian understanding of cuture, a meme represents any part of an idea—language, music, design, moral value—that jumps from one person's brain to the next. It is an attraction theory, rather like gravity, at least in the sense of suggesting that the proximity of one person to the next creates a kind of cultural pull, much as the proximity of objects in space creates a physical one. Interestingly, mimesis (so important to the Western tradition of art) and meme share similar Greek roots, namely mimneskesthai, which means memory, and mimeisthai, which means to imitate. And the closeness of these two terms in appearance is fascinating in its own right, for it suggests that on some basic level of speech, the ancient Greeks intrinsically understood that the mind works on the binary operation of data storage and retrieval. Today, computer files are retrieved from the computer's memory by copying.

To a certain extent, it was this was basic binary operation that Dawkins sought to describe by coining meme as the cultural equivalent of nature's self-replicating gene. Since the process of self-propagation defines life, why then wouldn't it also define culture? And Dawkins goes further, suggesting

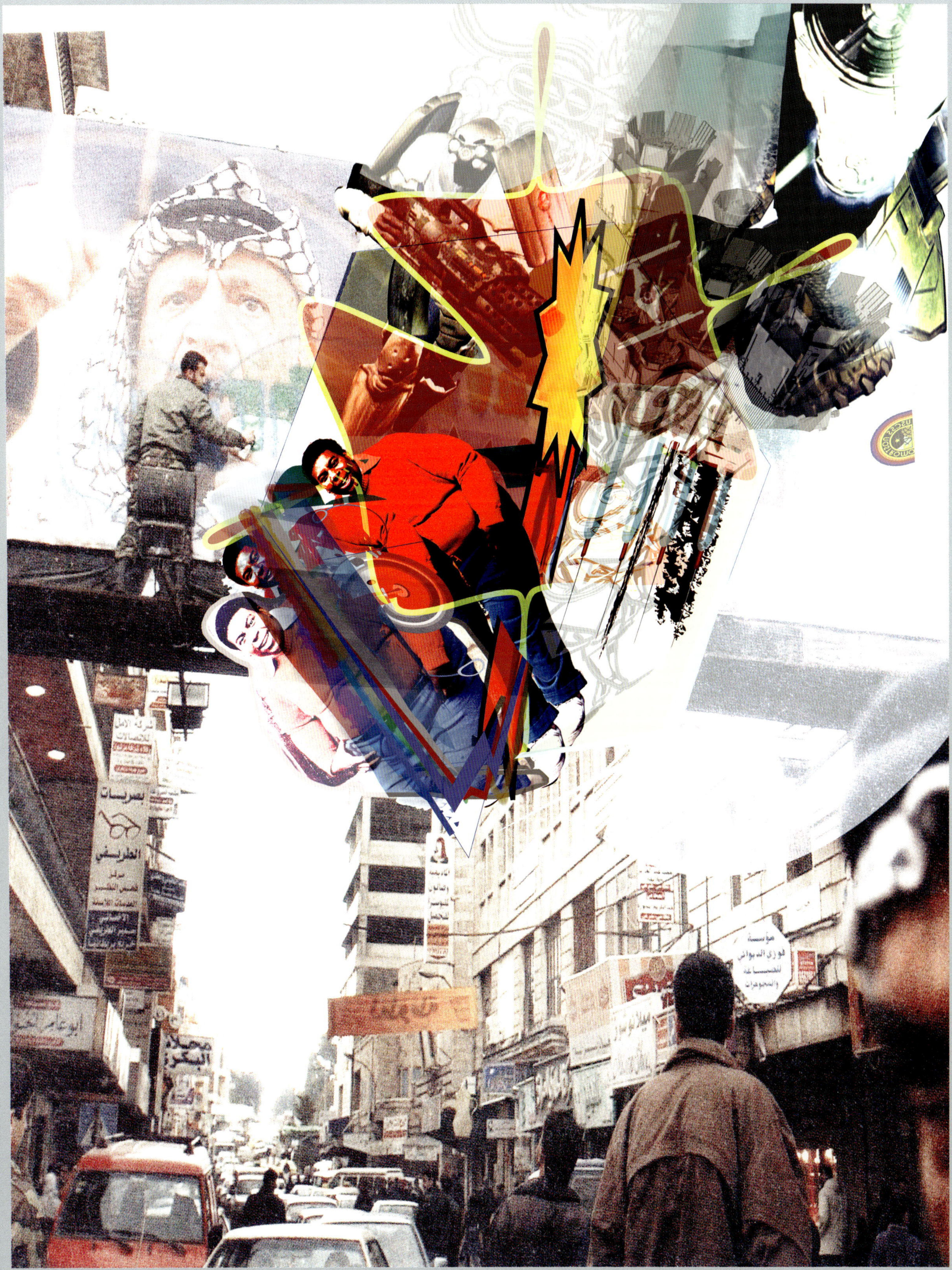

that human evolution no longer depends on genes; it is equally, if not more, dependent on memes. Ultimately, however, Dawkins never goes beyond establishing meme as a provocative term. He does not provide a sufficient explanation for how the replication of units of information in the brain controls human behavior and ultimately culture. Indeed, his idea for the meme comes in the last few pages of *The Selfish Gene*, which is, after all, devoted to formulating a new understanding of how biology functions according to a competition between "selfish" genes acting only in the interest of their own self-replication.

So refinements in meme theory have been left to others, like Susan Blackmore, who in her 1999 book *The Meme Machine*, proposes that memes, like genes, are "selfish," working towards self-replication in a mutation process that resembles evolution insofar as units of information are copied with variation and selection. "Memes spread themselves around indiscriminately without regard to whether they are useful, neutral or positively harmful to us," she writes. "Successful memes are the ones that get copied and spread, while unsuccessful memes do not." Then she adds what can only be described as an Einsteinian corker: "Instead of thinking of our own ideas as our own creations and as working for us, we have to think of them as autonomous selfish memes, working only to get themselves copied. We humans...have become just the physical hosts needed for the memes to get around."

This, needless to say, is a radically relativistic view, suggesting that humans do not create culture; rather it's the other way around. But let's suppose for a moment that Blackmore is correct. We have already devised systems for information storage and retrieval that go well beyond the human host, and thanks to human ingenuity these machines are taking exponential leaps in operational sophistication. The nightmarish scenario of robots taking over is a common sci-fi trope, but that doesn't mean it's impossible. Certainly Blackmore's notion, if it becomes a widely successful meme, would help to create a philosophical basis for that eventuality. Still, one might ask, if humans are merely an evolutionary stepping stone to some form of cybernetic "life," than so what? Ultimately we don't know what the future will bring any more than dinosaurs knew that 150 million years of evolutionary success would be wiped out by an asteroid.

On the other hand, what is meme theory but yet another attempt to solve some aspect of the questions surrounding human existence, the same questions that religion claims it already has an answer for? Whatever else separates human from other animals, even animals like orangutans that seem to have some form of culture, one thing that surely stands out is our propensity as a species to at once elevate ourselves to the center of the universe, and then find reasons for why this is not so. We kill for culture but in a way culture kills us, insofar as each of us (either by volition or coercion or some combination of the two) sublimate aspects of our lives to a larger order, whether that order is represented by Allah, or the celebrity worship proffered by tabloid TV and magazines.

Blackmore argues in effect that our thoughts may not really be our own in the same way that religionists argue that our lives are not really our own (that is, they belong to God). And she may be right, for whoever wins the current conflict between materialism and fundamentalism, free thinking, that great cornerstone of Western civilization will surely be a casualty. As the French historiographer Alain Finkielkraut writes in his book, *The Defeat of the Mind*, "when it is not cultural identity restricting the choices the individual can make...it is the entertainment industry, the creation of the technological age, that reduces great works of art to drivel. The life of the mind has quietly moved out of the way, making room for the terrible and pathetic encounter of the fanatic and the zombie." What finally may be so compelling about our Madonna of the Iraqi war is that for one brief instant, we can see in her eyes a reflection of the fanatic and the zombie shaking hands.

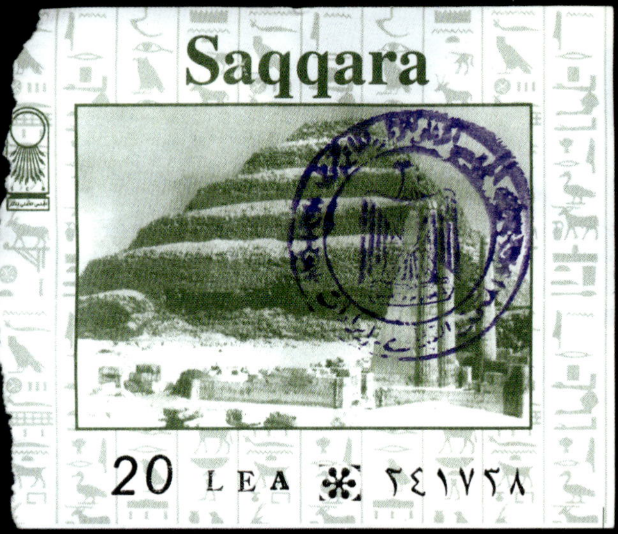
Saqqara
20 L E A ٢٤١٧٢٨

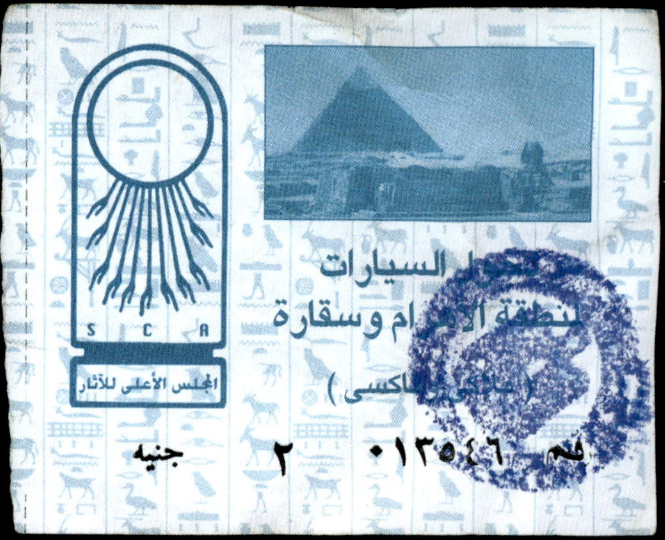
S C A
المجلس الأعلى للآثار
السيارات
وسقارة
جنيه ٢ ٠١٣٥٤٦

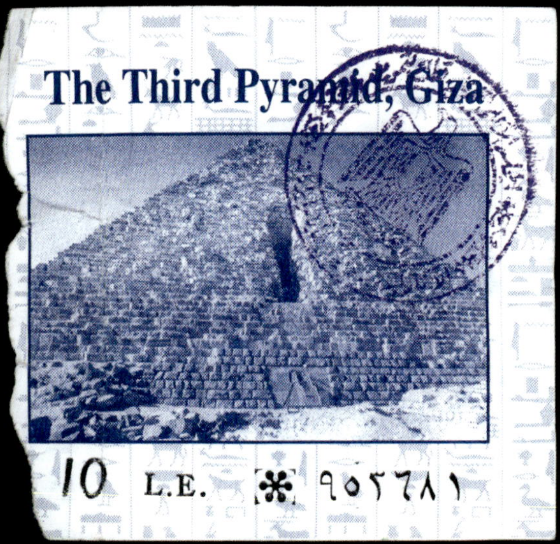
The Third Pyramid, Giza
10 L.E. ٩٥٢٦٨١

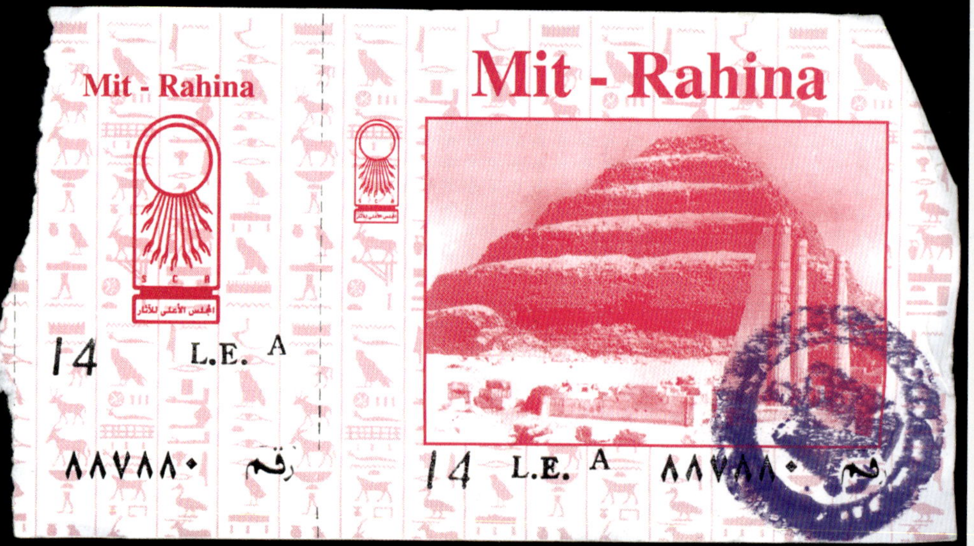
Mit - Rahina
S C A
14 L.E. A
٨٨٧٨٨٠ رقم
Mit - Rahina
14 L.E. A ٨٨٧٨٨٠

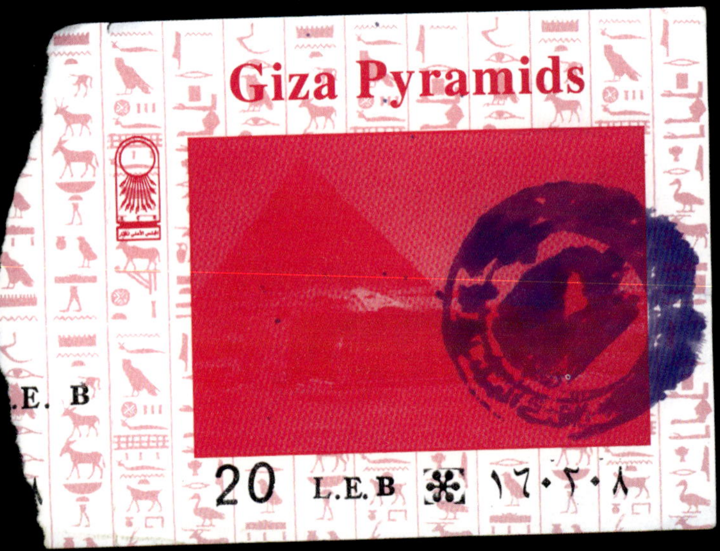
Giza Pyramids
.E. B
20 L.E. B ١٦٠٢٠٨

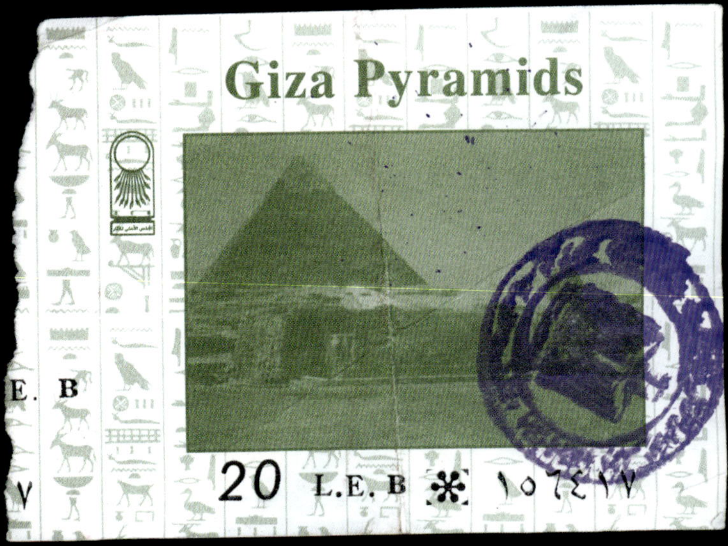
Giza Pyramids
E. B
20 L.E. B ١٥٦٤٢٧

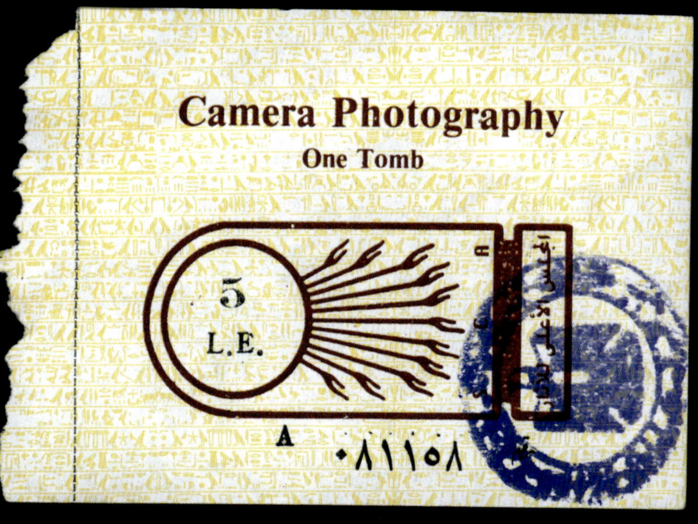
Camera Photography
One Tomb
5
L.E.
A ٠٨١١٥٨

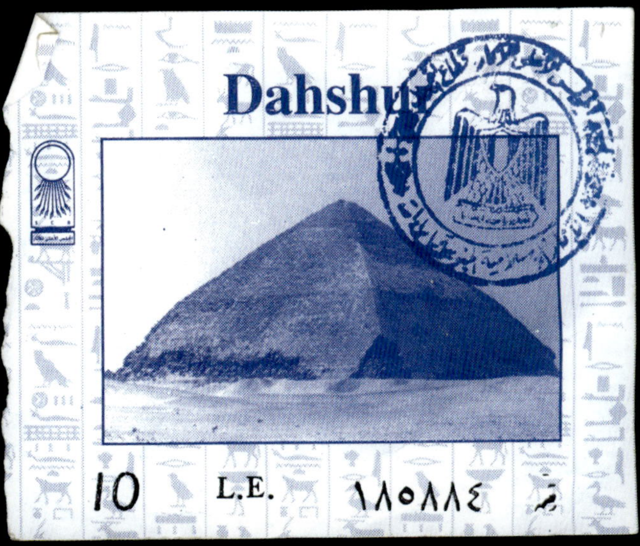
Dahshur
10 L.E. ١٨٥٨٨٤

171_Untitled, journalism swipe. 18 x 24 inch digital canvas, 2005.

173_My daughter Tueme on the coast south of Barcelona. 2004.

174-5_Planck Moment #6. 2005.

176_Tree at Vulubolis. 20 x 30 inch ink jet print, Morocco, 2003.

178_Ville de Saint-Maximin. France, 2004. This is the back of the abbey where Mary Magdalene is entombed. Over one hundred years of chicken scratch graffiti.

179_Vague Dimensions. 16 x 20 inch Epson print.

180_Dark Alley, Fez. 2002.

181_ Ville de Saint-Maximin. France, 2004

183_Untitled, corner piece #2. 18 x 24 inch digital canvas, 2004.

185_Cairo man with white scarf. 2002.

186_Photomagnetic Receiver, *Time Out New York* blurb. 1999.

189_Untitled, journalism swipe #2. 18 x 24 inch digital canvas, 2005.

192_Giza passes.

194_Untitled, Cairo street #5 with blur. 18 x 24 inch digital canvas, 2005.

195_Planck Moment Rough #3. 2005.

197_Planck Moment #5, #9 (details). 2005.

198_Moiré set and placebo, west coast version. 1999.

199_Palestine. 18 x 24 inch digital canvas, 2005.

200_Dark Alley with Ideopics. 2005.

201_Planck Moment #5 (detail). 2005.

202_Cairo street #4 with swipe, 18 x 24 digital canvas, 2004.

203_Data Background. These pieces lifted from Data Acquisition.

204_Data Acquisition panel 10. 20 x 30 inch light jet prints, 1997.

205_Data Acquisition panel 12. 20 x 30 inch light jet prints, 1997.

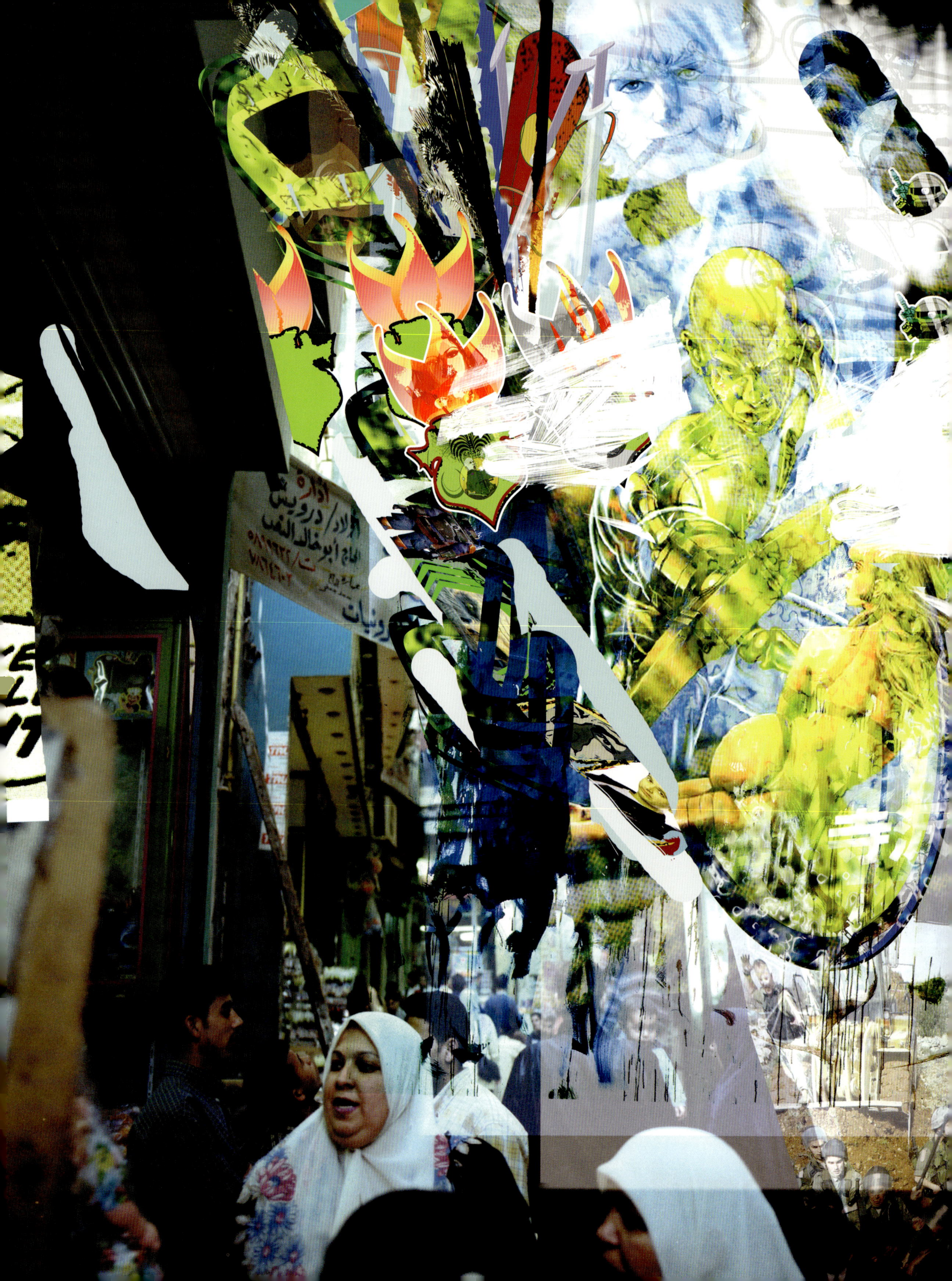

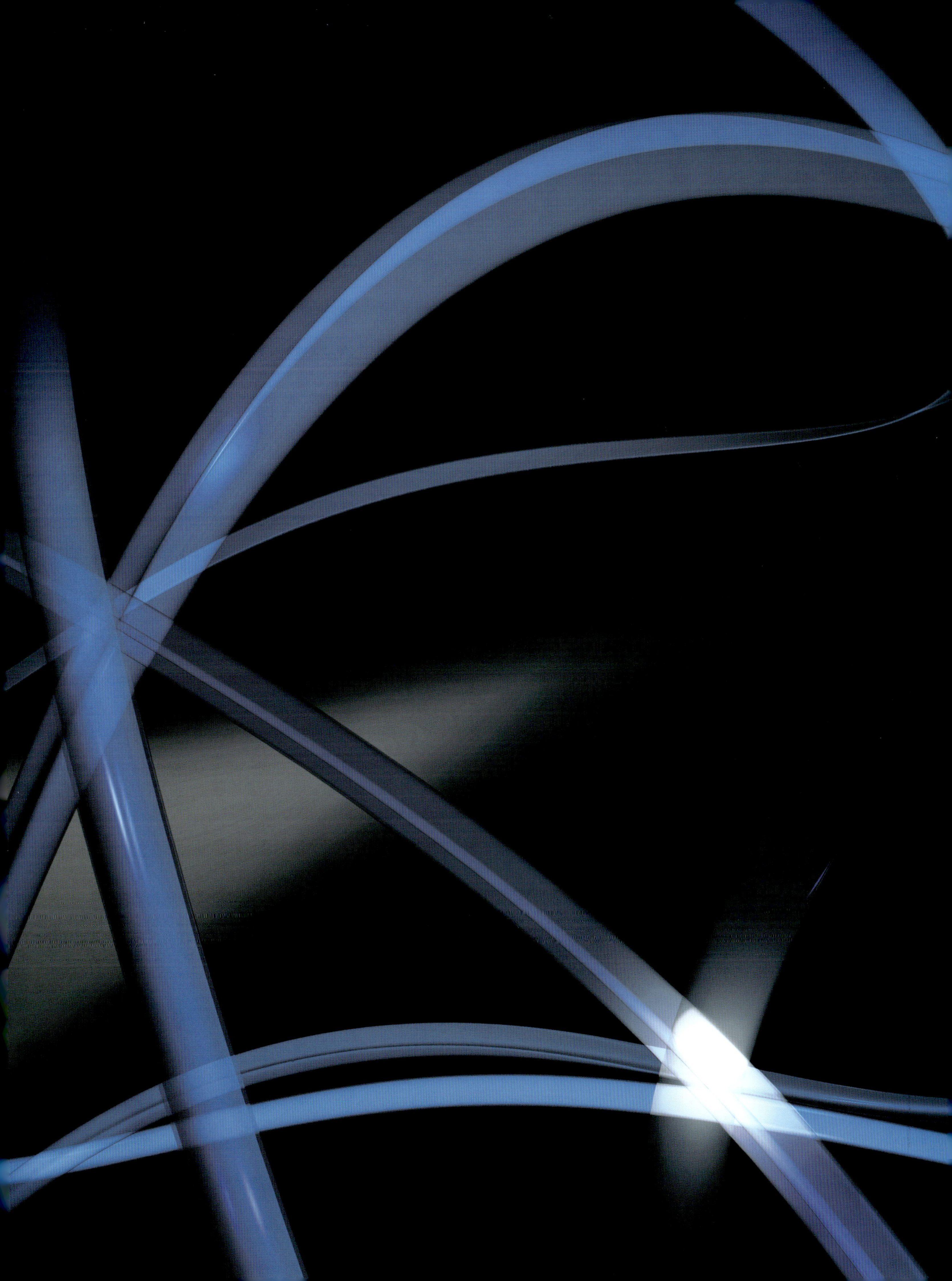

We weren't journalists. The protein axis was not our cross to bear. Princeton: The DEMARC team came to the horizon. They figured multiples of ten generations tighter. We're out and so is Santa Fe. If you stay with the remnant vectors you will get in, Jim's team got down to 12. All the best.

They were classified: the Physics of bloody passions. Chemistry of murderous vocation. Several areas you covered off. That's to your credit. A sensible archive of backgrounds—significantly stressed/de-sorted. 2 months ago, the equation moved to several more difficult juggernauts. whtcanwesay_Deckard: Got the shakes? It comes with the business.
Rachel: I'm not part of the business, I am the business.
Keep us posted.

Funding for KG comes thru in 98. Until then, CRichter understands base camp will be monitored. That's good. Rumors are true, skunked group having SOS sale on 23rd. All files open for consideration. Special inquiries for the other sets we spoke about. We're getting closed down. If New Haven thinks we made a crack, that's good enough for us, we take it.

I would like to kick off the discussion. Is the no-hair theorem just a coincidence of a small class of theories or does it have fundamental relevance (as for example censorship, a violation of which would have bad consequences on predictability)? Anyway, interesting things related to NCG, categories_and I hope_categorified NCG, but it becomes quite a stretch, therefore will reply to Kea's latest post here.

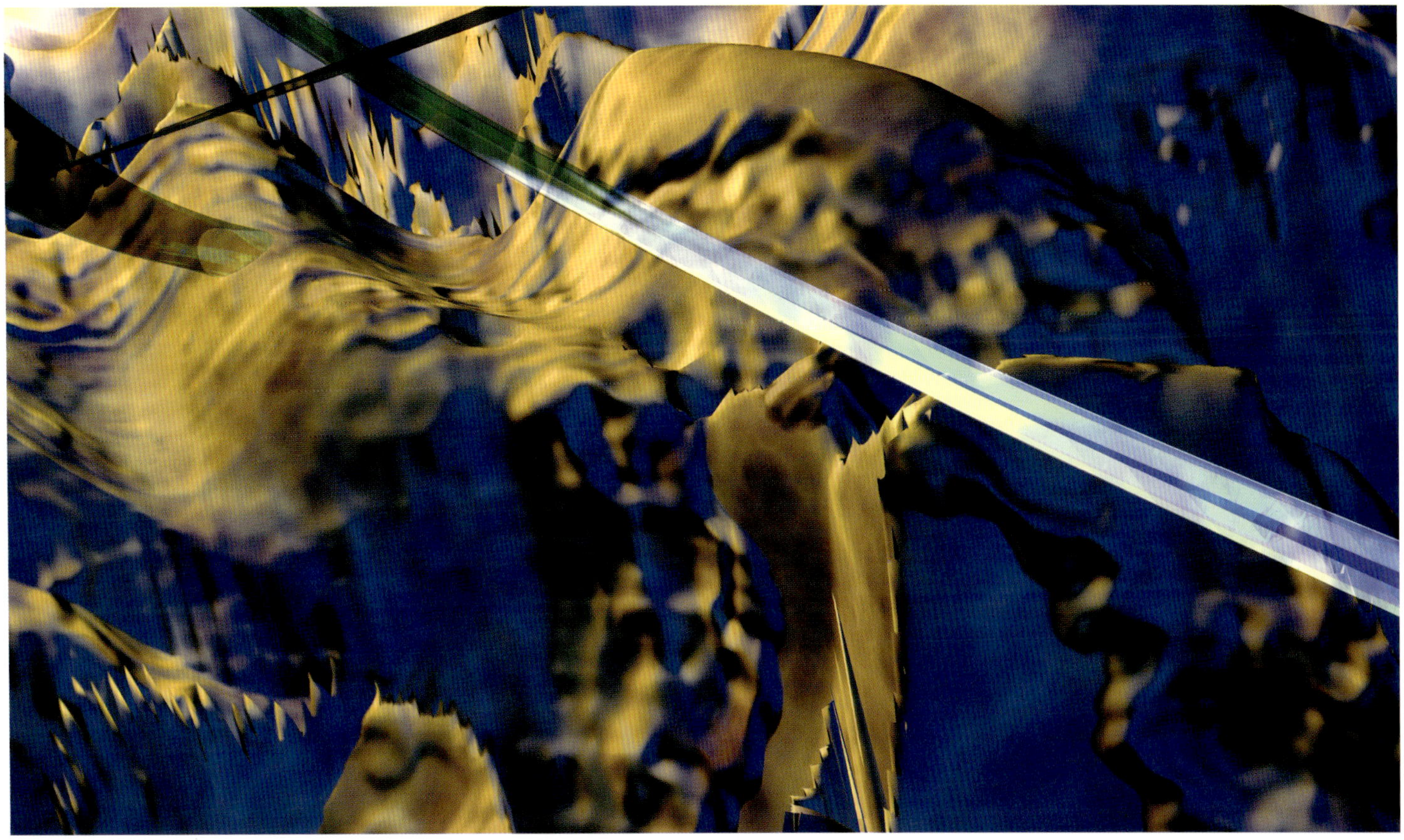

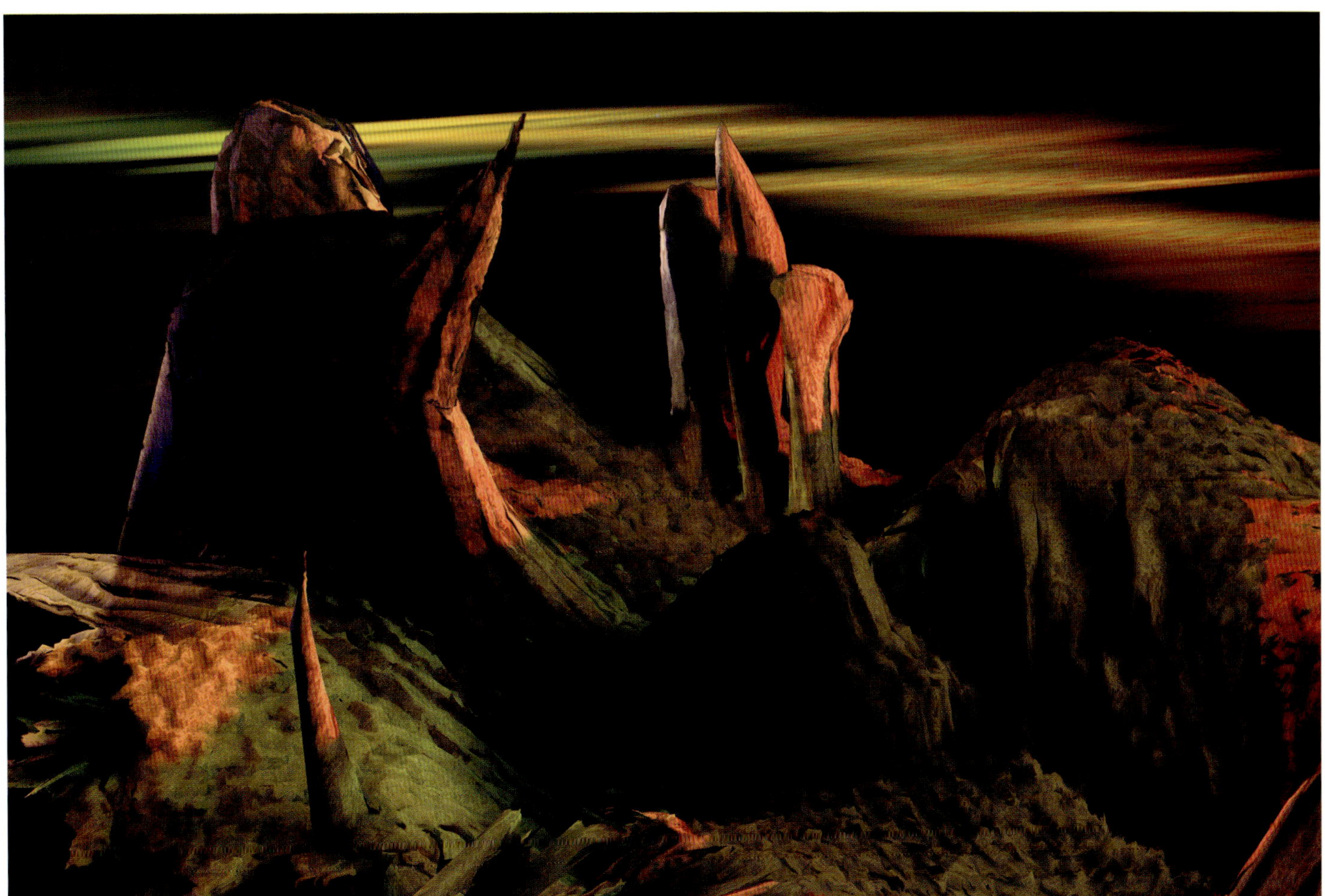

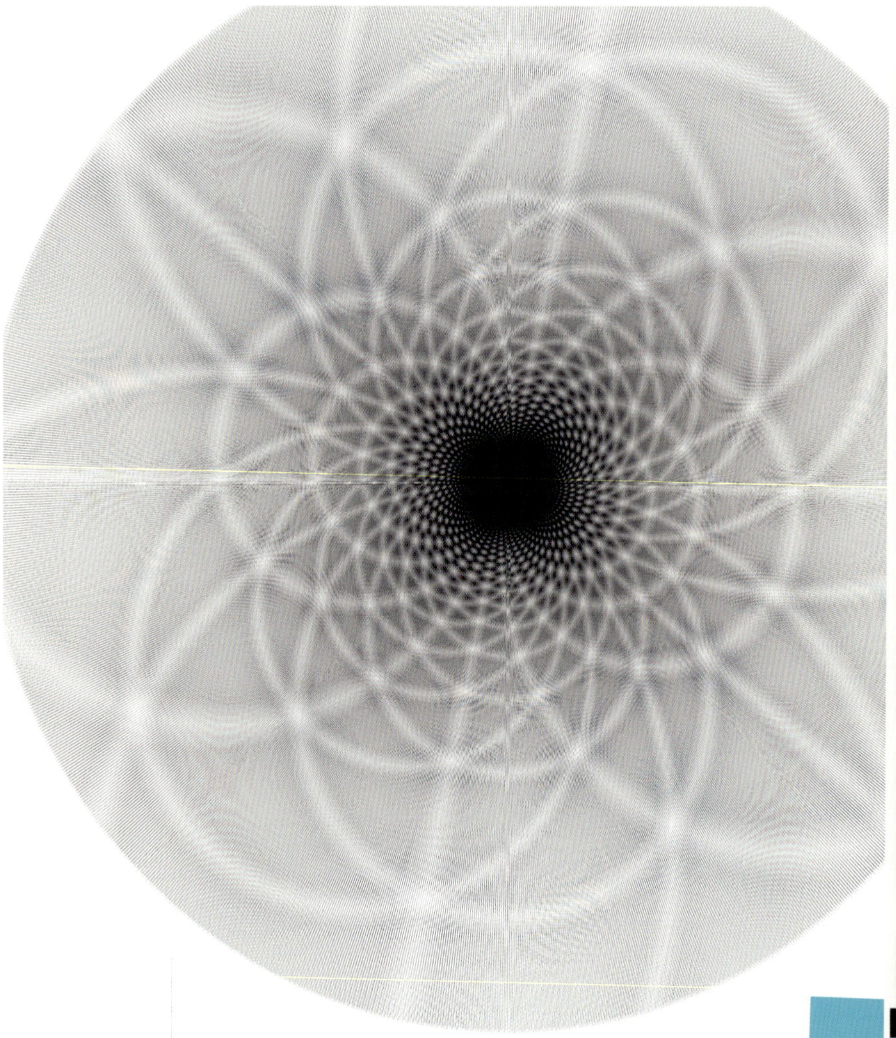

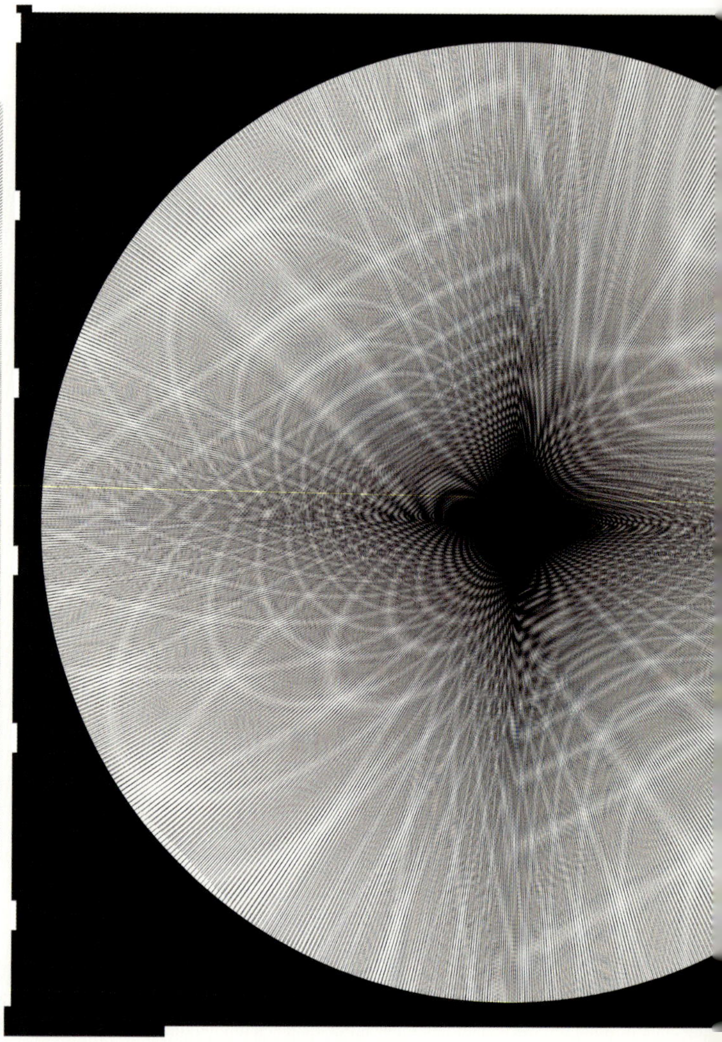

.265

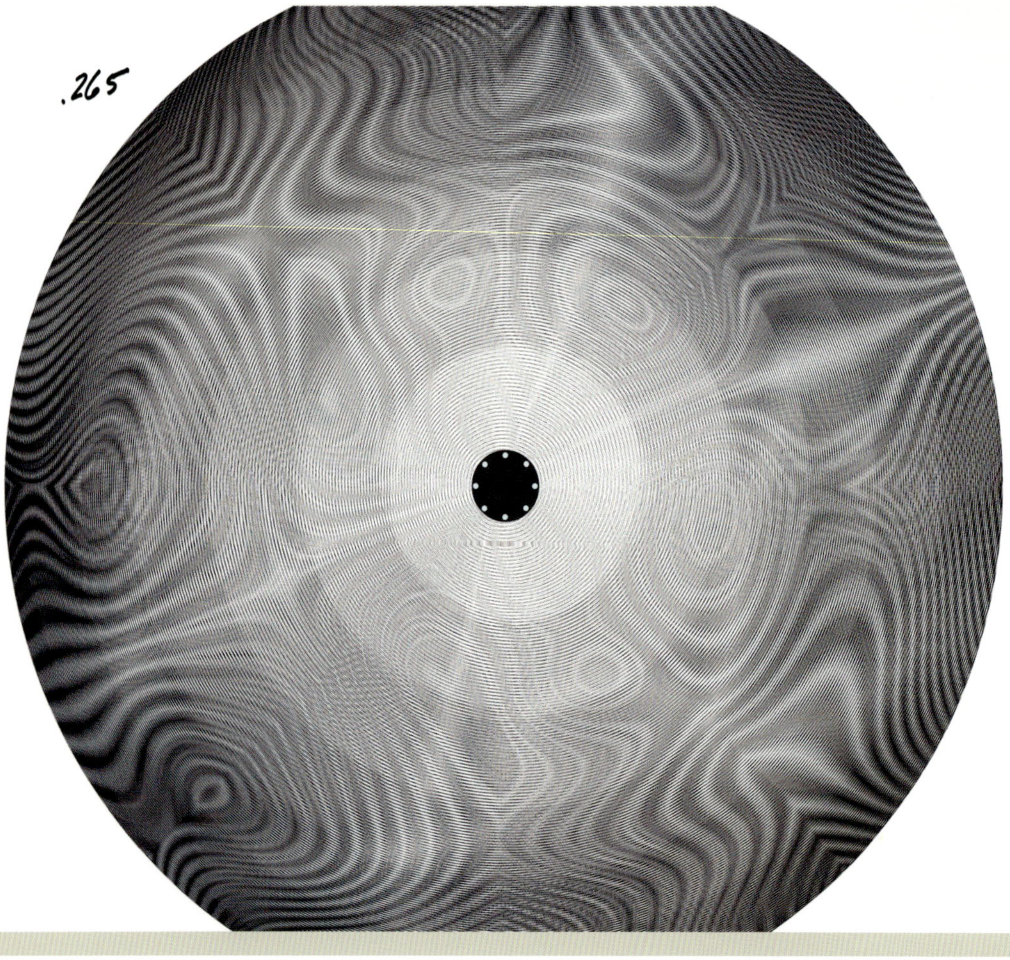

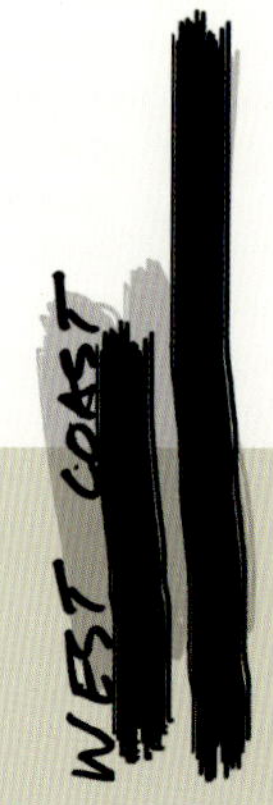

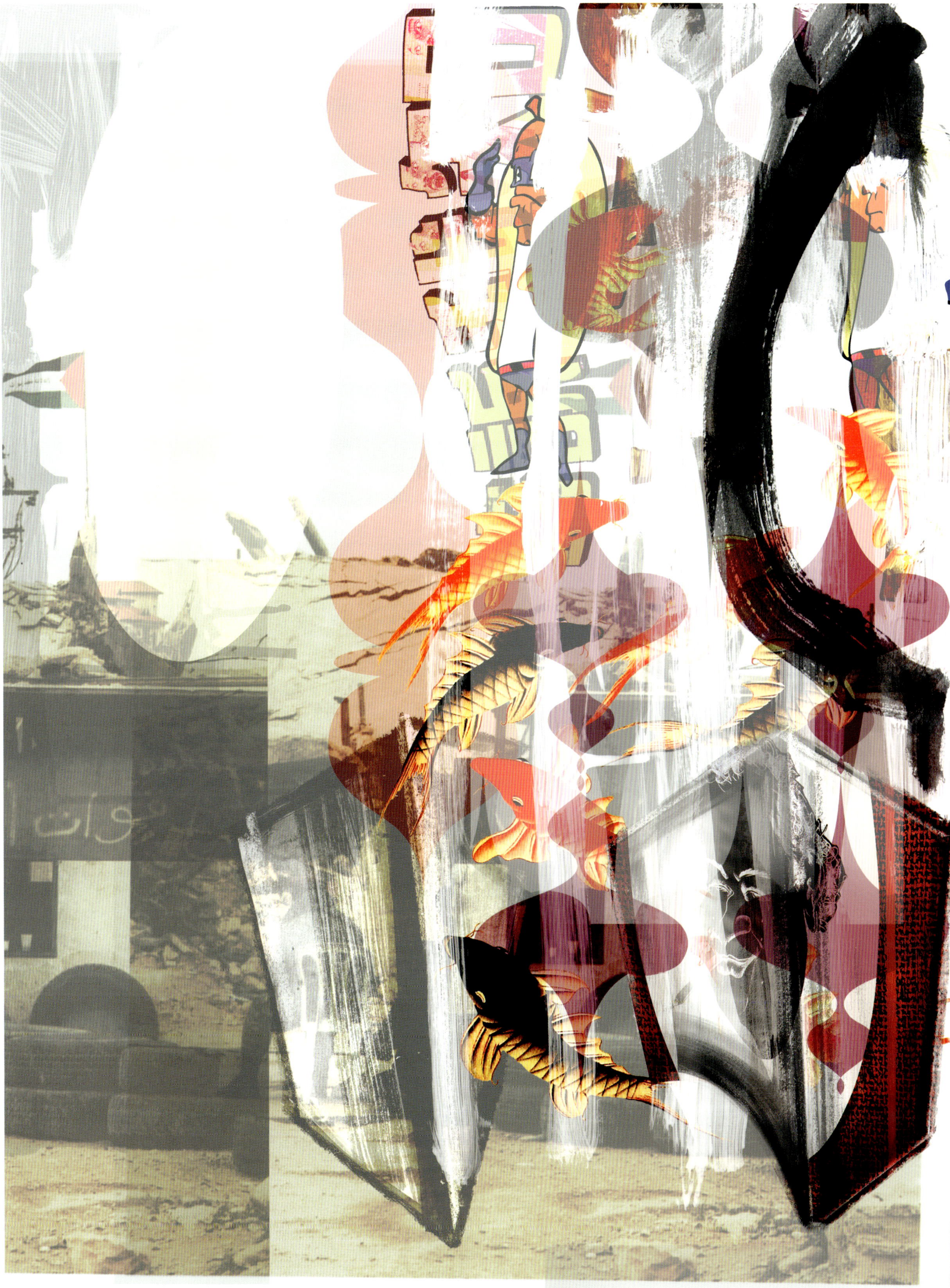

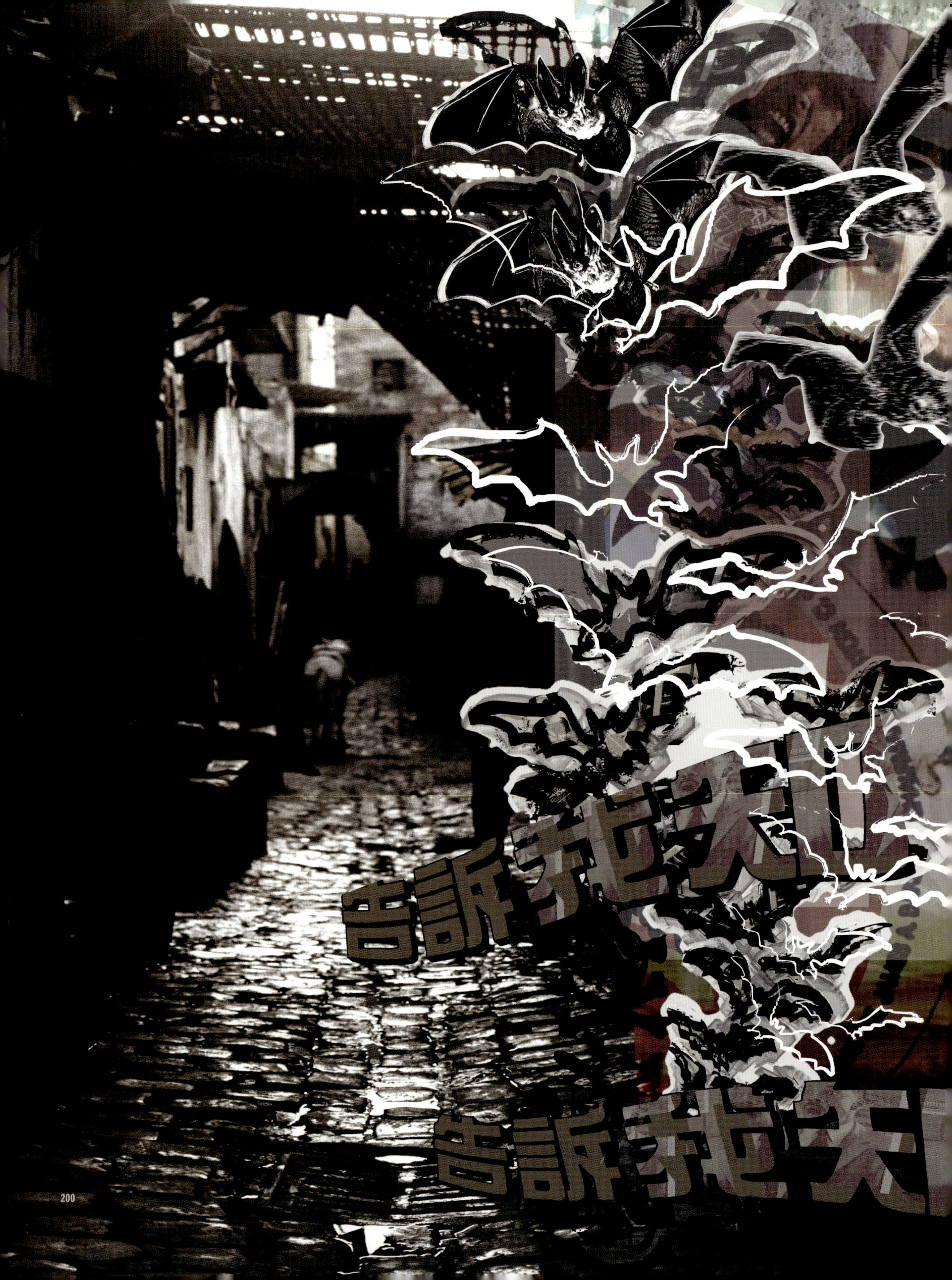
告訴我天
告訴我天

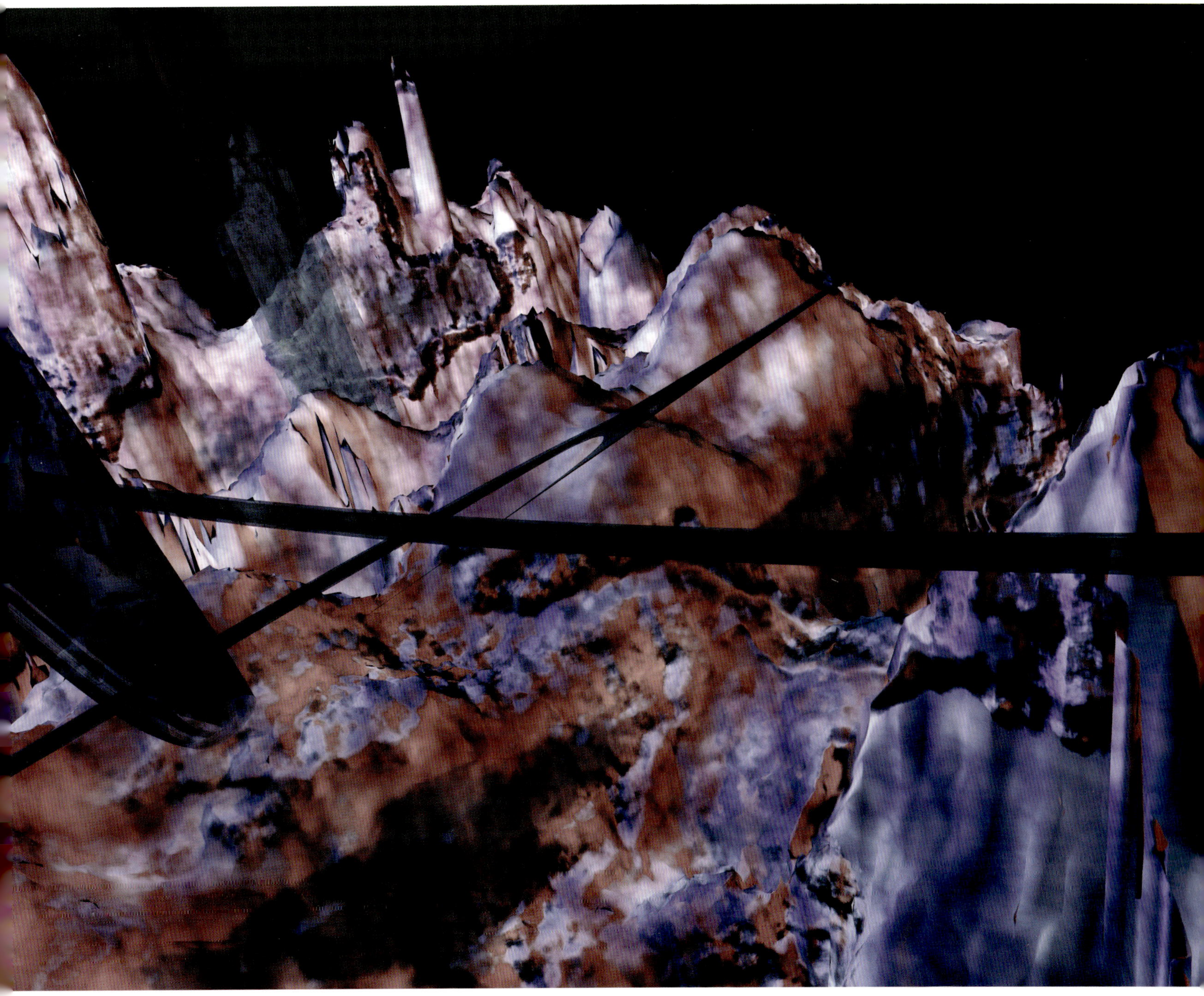

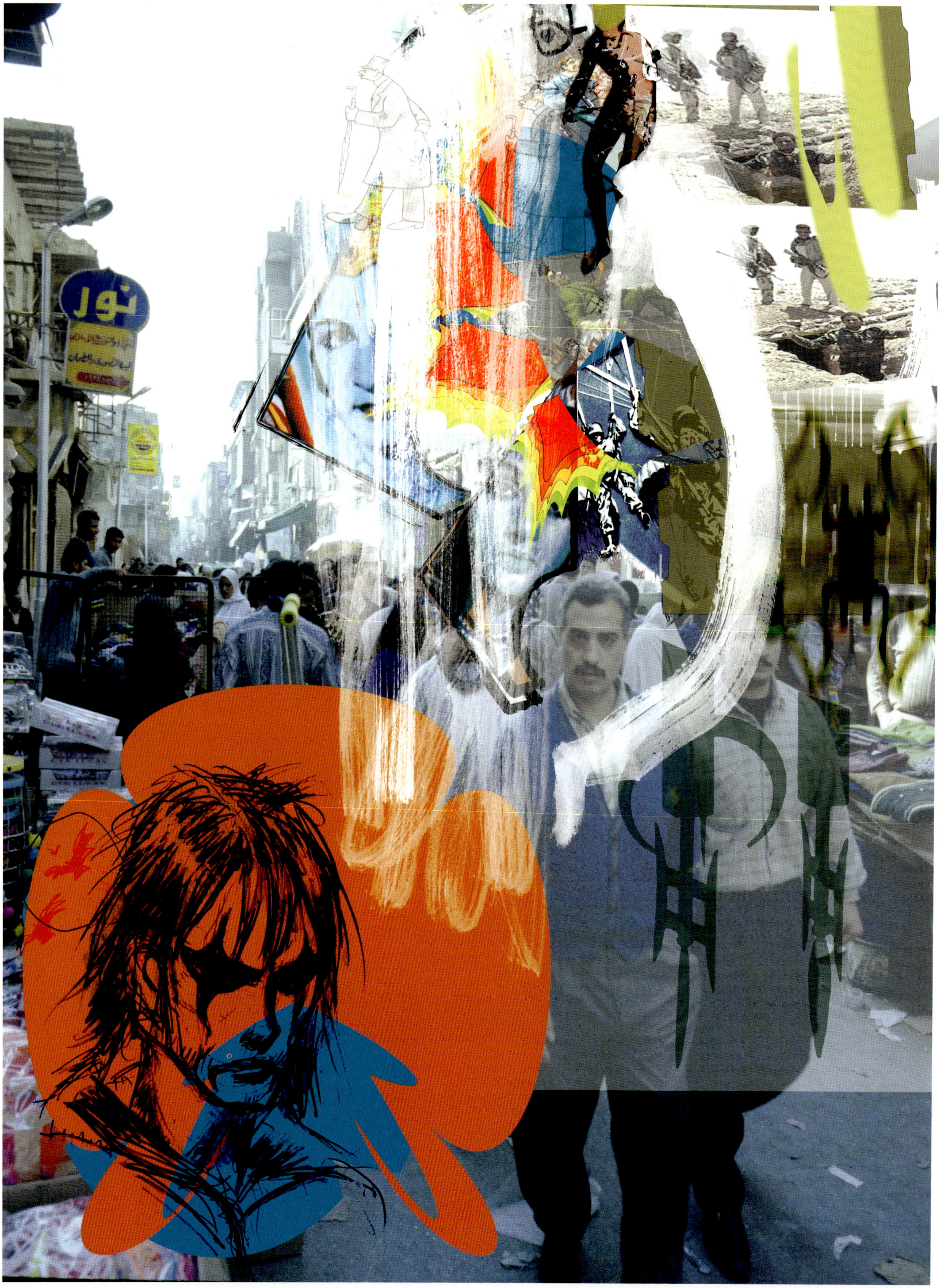
نور

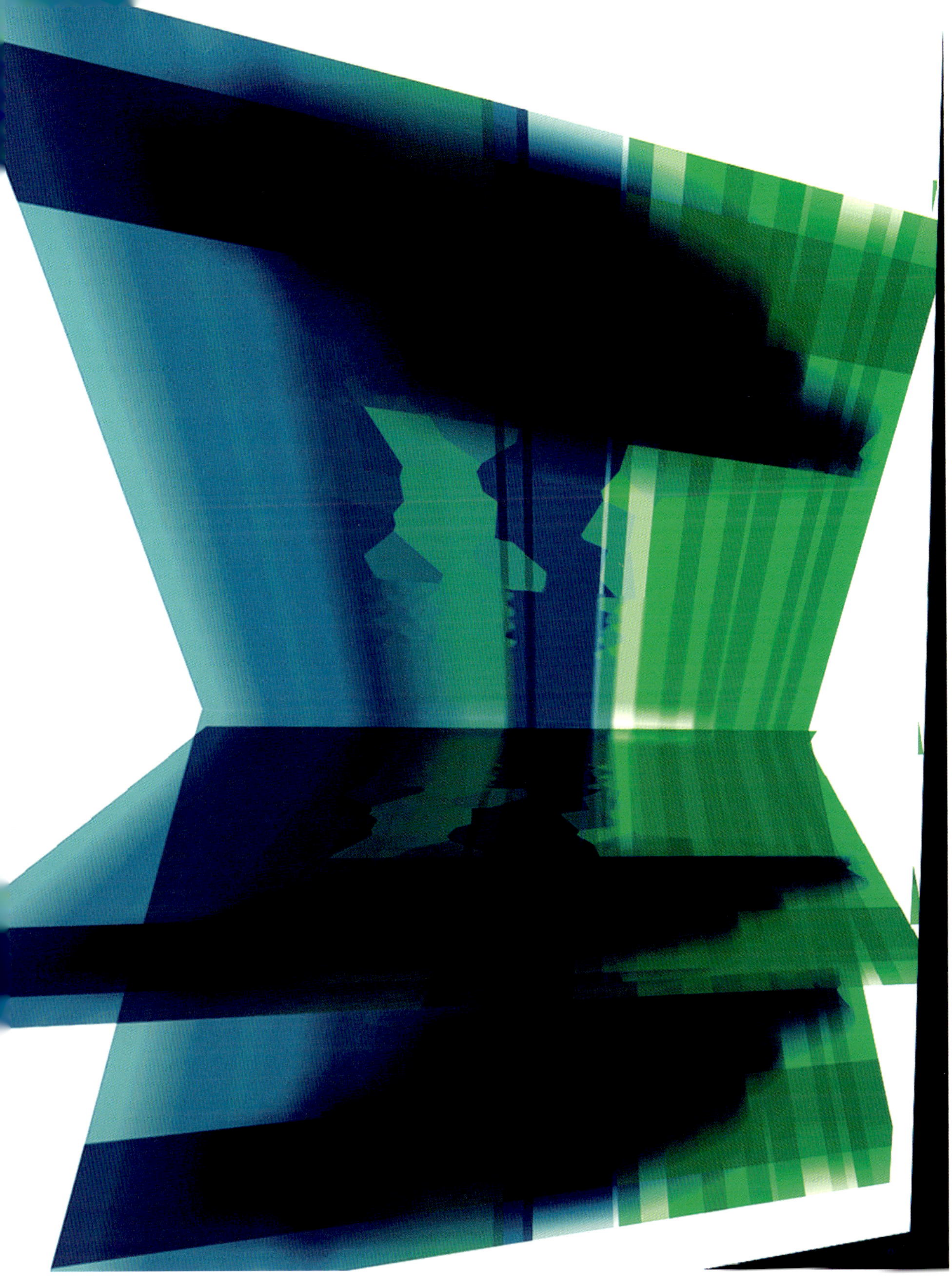

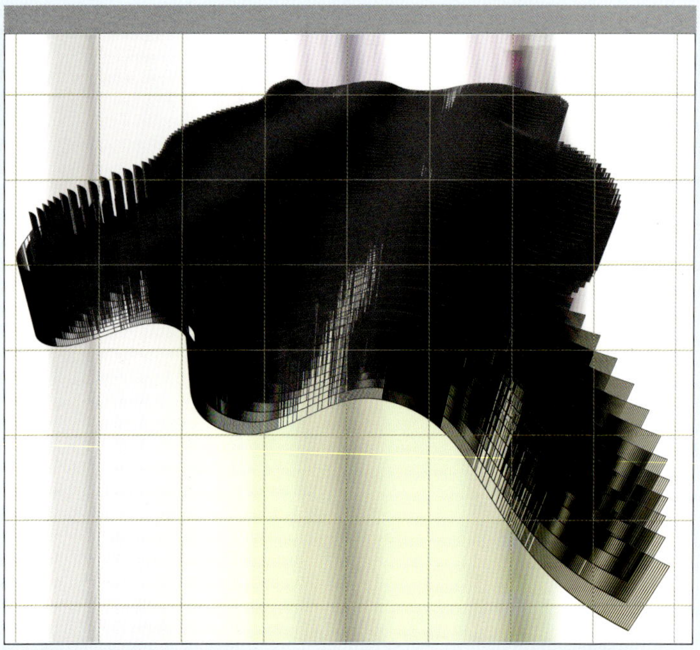

ACTIVE NOISE AND VIBRATION TECHNOLOGY PHOENIX
NOSE CANCELLATION TECHNOLOGIES STAMFORD, CONN.
MOTOROLA ANALOG DEVICES NORWOOD, MASS.

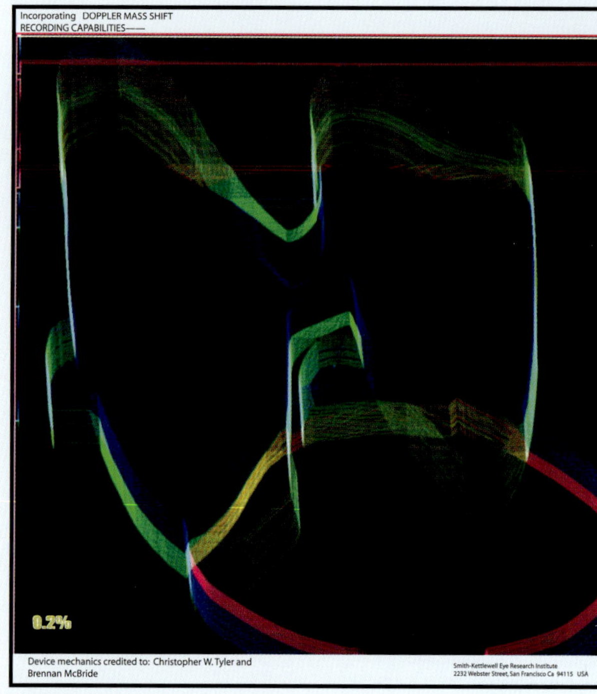


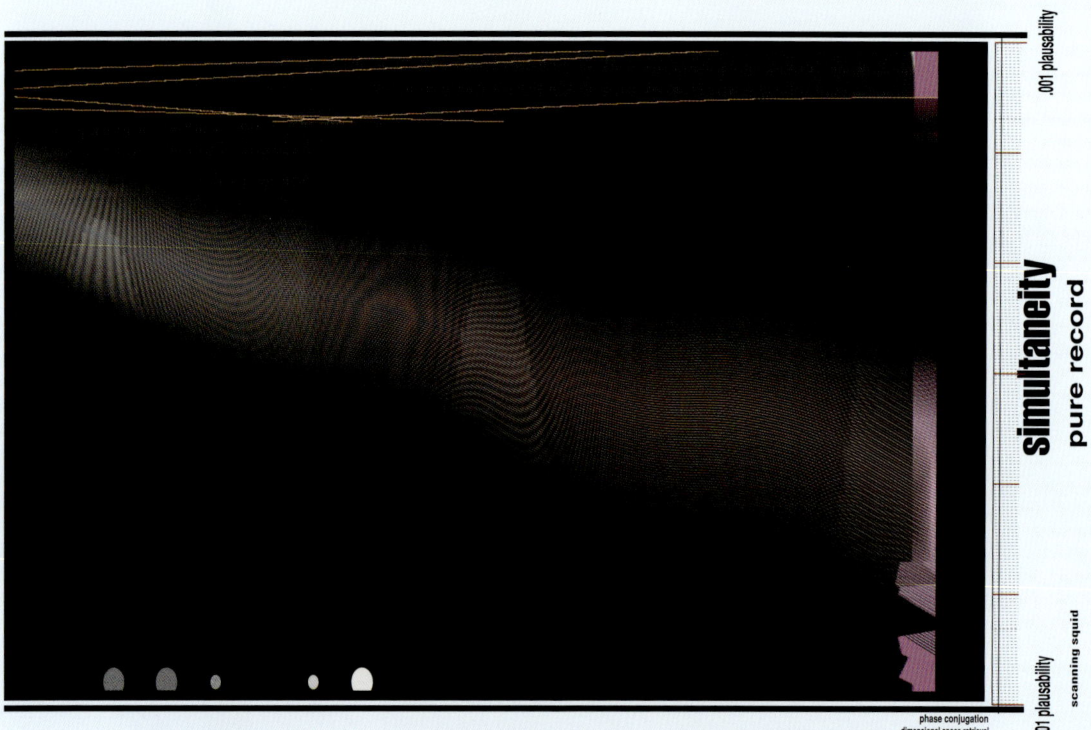


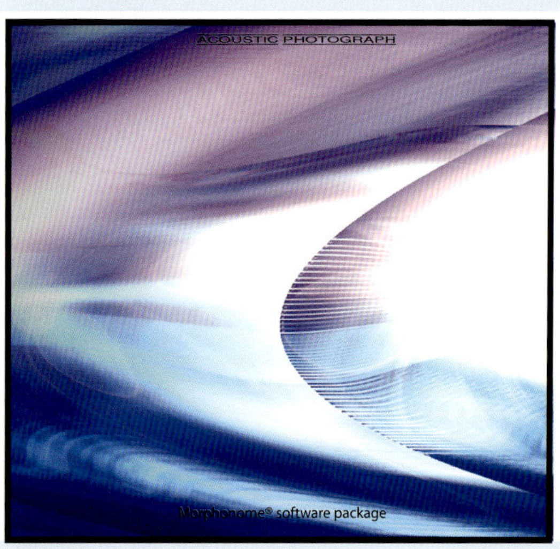


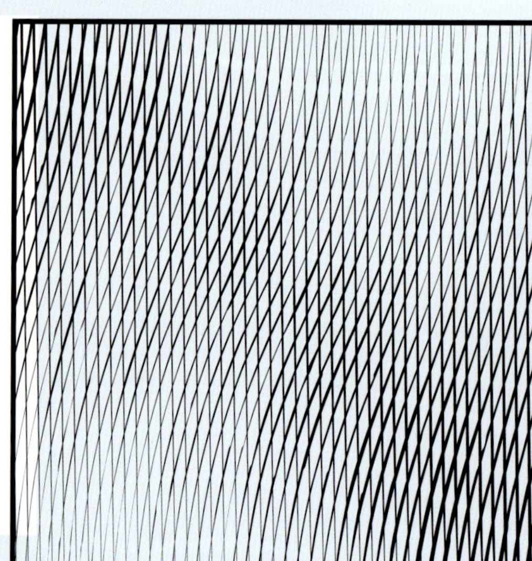

EMULSION CHEMISTRY COURTESY LANGLEY RESEARCH CENTER, HAMPTON, VIRGINIA

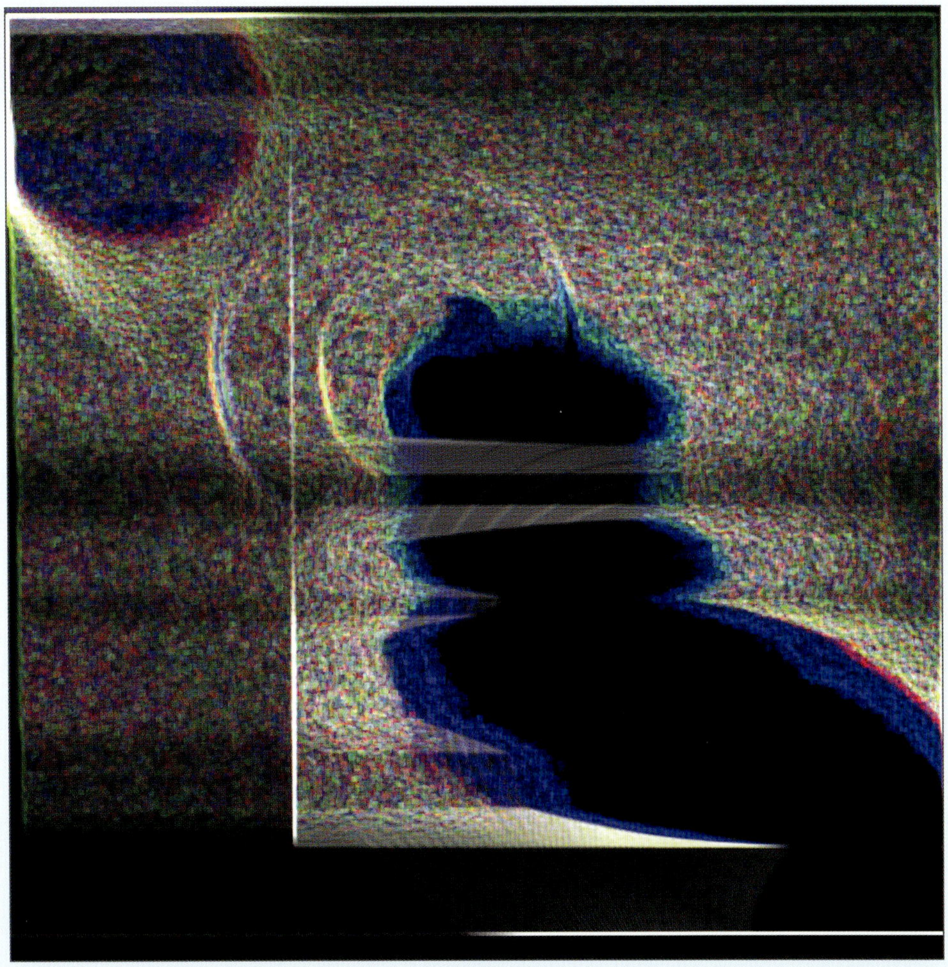

retiring magnetic signals--generate the data to acquire picture.
motion shift appears as a D+=0 snapshot.

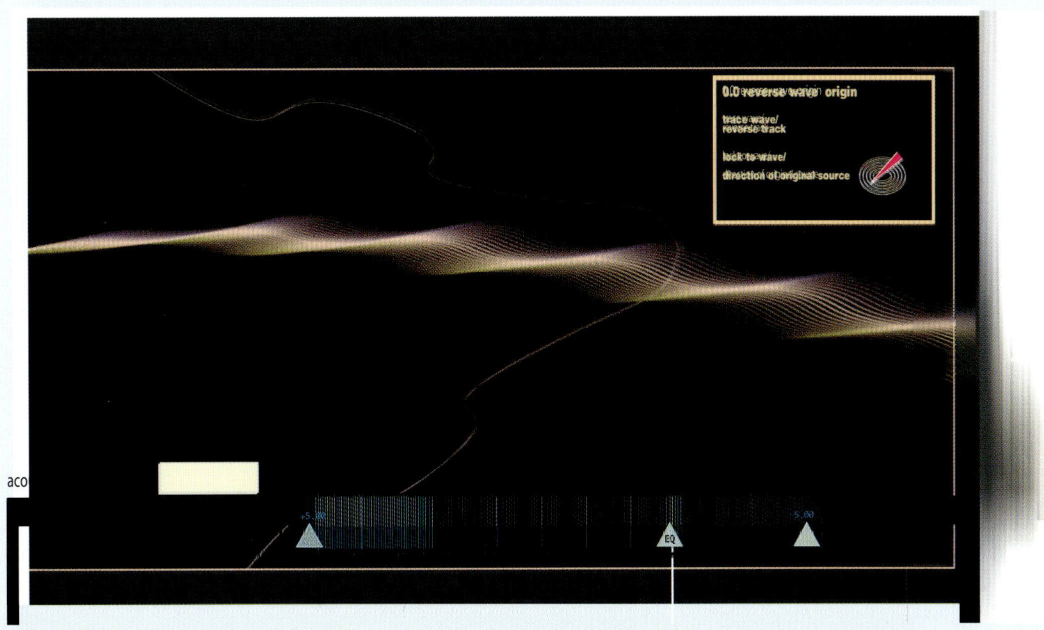


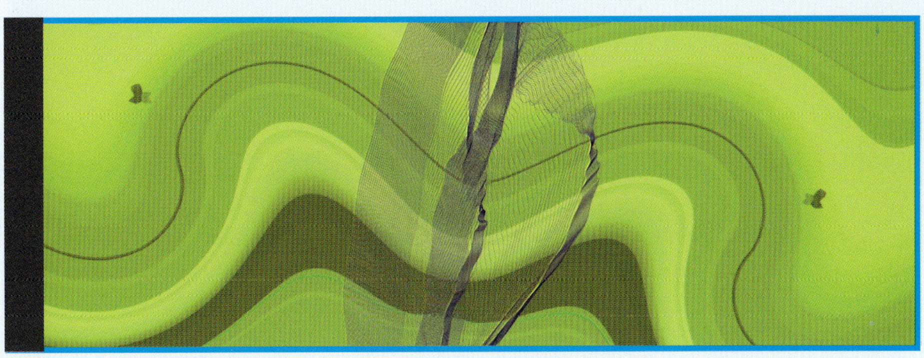

Particles + Waves with Plausibility



I would like to thank everyone that has helped me through these concepts and assisted in the various installations—at American Fine Arts, Daniel McDonald, and Christine Tsvetanov. Without the continued support of the MICA Foundation this work would not have been possible. Thank you Alit and Tueme Balk for your patience and support. James Lewin, I owe you so much for your insightful enthusiasm on this and other projects. Much gratitude to Howard Halle for this text and support over the years. This book is dedicated to Colin de Land.



Published in the United States by powerHouse Books,
a division of powerHouse Cultural Entertainment, Inc.
68 Charlton Street, New York, NY 10014-4601
telephone 212 604 9074, fax 212 366 5247
e-mail: particles@powerHouseBooks.com
website: www.powerHouseBooks.com

First edition, 2005

Library of Congress Cataloging-in-Publication Data:

Balk, Dennis.
Particles + waves with plausibility / by Dennis Balk ; text by Howard Halle.-- 1st ed.
p. cm.
ISBN 1-57687-216-5 (hardcover)
1. Balk, Dennis--Themes, motives. 2. Digital art. 3. Photography, Artistic. 4. Art and science. I. Title: Particles and waves with plausibility. II. Halle, Howard. III. Title.

N7433.4.B35P37 2006
776--dc22

2005047265

Hardcover ISBN 1-57687-216-5

Printing and binding by Lotus Printing, Inc.

Book design by Dennis Balk

A complete catalog of powerHouse Books and Limited Editions is available upon request;
please call, write, or visit our website.

10 9 8 7 6 5 4 3 2 1

Printed and bound in Hong Kong

Front cover: Tree at Vulubolis, Morocco
Back cover: Large Particle #2, Cairo Street